PDE
Dynamics

Mathematical Modeling and Computation

The SIAM series on Mathematical Modeling and Computation draws attention to the wide range of important problems in the physical and life sciences and engineering that are addressed by mathematical modeling and computation; promotes the interdisciplinary culture required to meet these large-scale challenges; and encourages the education of the next generation of applied and computational mathematicians, physical and life scientists, and engineers.

The books cover analytical and computational techniques, describe significant mathematical developments, and introduce modern scientific and engineering applications. The series will publish lecture notes and texts for advanced undergraduate- or graduate-level courses in physical applied mathematics, biomathematics, and mathematical modeling, and volumes of interest to a wide segment of the community of applied mathematicians, computational scientists, and engineers.

Appropriate subject areas for future books in the series include fluids, dynamical systems and chaos, mathematical biology, neuroscience, mathematical physiology, epidemiology, morphogenesis, biomedical engineering, reaction-diffusion in chemistry, nonlinear science, interfacial problems, solidification, combustion, transport theory, solid mechanics, nonlinear vibrations, electromagnetic theory, nonlinear optics, wave propagation, coherent structures, scattering theory, earth science, solid-state physics, and plasma physics.

PDE Dynamics
An Introduction

Christian Kuehn

Technical University of Munich
Munich, Germany

Society for Industrial and Applied Mathematics
Philadelphia

Publications Director	Kivmars H. Bowling
Acquisitions Editor	Elizabeth Greenspan
Developmental Editor	Gina Rinelli Harris
Managing Editor	Kelly Thomas
Production Editor	Ann Manning Allen
Copy Editor	Matthew Bernard
Production Manager	Donna Witzleben
Production Coordinator	Cally A. Shrader
Compositor	Cheryl Hufnagle
Graphic Designer	Doug Smock

Library of Congress Cataloging-in-Publication Data
Names: Kuehn, Christian, author.
Title: PDE dynamics : an introduction / Christian Kuehn (Technical University of Munich, Munich, Germany).
Other titles: Partial differential equation dynamics
Description: Philadelphia : Society for Industrial and Applied Mathematics, [2019] | Series: Mathematical modeling and computation ; 23 | Includes bibliographical references and index.
Identifiers: LCCN 2018056824 (print) | LCCN 2018060895 (ebook) | ISBN 9781611975666 | ISBN 9781611975659 (print)
Subjects: LCSH: Differential equations, Partial. | Differential equations.
Classification: LCC QA377 (ebook) | LCC QA377 .K838 2019 (print) | DDC 515/.353–dc23
LC record available at *https://lccn.loc.gov/2018056824*

 is a registered trademark.

Contents

Preface

This book is intended to foster the interchange of ideas between partial differential equations (PDEs) and dynamical systems. It is designed for beginning graduate students and researchers in the area to obtain an overview of the myriad possibilities to apply dynamical systems techniques to PDEs as well as to highlight the impact of PDE methods on dynamical systems. In recent years, the two fields have seemingly drifted further apart in teaching than may seem strictly necessary. A first sequence of introductory PDE courses tends to focus on the existence and regularity aspects while dynamical systems courses tend to focus on applications to ordinary differential equations (ODEs) or discrete-time iterated maps. Although it would be excellent if students would take all possible courses, this is rarely practical. Hence, one often encounters PDE practitioners well versed in functional-analytic inequality-based arguments or those more comfortable with a geometry-based dynamics approach. The combination is relatively rare. This book aims to change this situation by highlighting how the different approaches are intertwined. A natural alternative title for the book could have been

A Concise Introduction to the Interactions between Dynamical Systems and Partial Differential Equations

but I found the title too lengthy in the end; nevertheless, it still describes the intentions quite well. Since many prefaces of mathematics books tend to be slightly dull, I decided to try to summarize principles set out for this book in a different format:

- **Applications:** Although in such a short book no real-world cutting-edge applications can be treated in depth, the generic techniques are designed to be very applicable for broad classes of PDEs. My strategy was to cover PDEs, which have turned out to be incredibly applicable in the past. Of course, the additional assumption is that it pays off to study *classics* to be prepared for the future.

- **Background:** As background knowledge a first course in PDEs is helpful, e.g., one based upon the introductory parts of [186, 472]. Some basic knowledge of ODEs [24, 270] is naturally desirable as well. Further additional background is often elementary, and one may look up the required results along the way.

- **Choice:** The references show that there is certainly *a lot* of choice for an introduction to PDE dynamics. I have tried to omit topics which place *too much strain on more specialized background*.

- **Design:** The focus of the book is to introduce key concepts and important analysis strategies at the interface between dynamical systems and PDE. The idea is to condense, contrast, and combine different viewpoints to maximize the preparation for research, but still keep the size of the book relatively short.

- **Examples:** All results are illustrated by concrete PDE examples. This approach automatically introduces a key suite of benchmark PDEs, which have guided the development of the general theory.

- **Figures:** Figures are kept deliberately simple and noncomputational in the main part of the text; see the appendices for basic computational tools. The approach allows one to directly utilize the book for blackboard/classroom use.

- **Genre:** A very broad selection of topics is presented to allow beginners an initial overview of the potential directions. However, the book is obviously *not* a monograph.

- **Highlights:** Since the book does not even attempt to be exhaustive in its area, several highlight results have been selected. Certainly, this selection is highly debatable, but it does exemplify the rapid growth of the field over the last couple of decades.

- **Ideas:** It is often necessary to learn a toolbox of "standard" tricks, proof strategies, or computations in a new area. This book tries to introduce as much as possible from this standard toolbox. However, the high density of ideas necessitates that we demonstrate each new technique often only once, or at most twice. So one has to adapt the lecturing/reading pace accordingly.

- **Jokes:** Albeit (hopefully!) being "fun-to-read" from a scientific perspective, jokes have been omitted as they almost surely do not transmit well in mathematics writing; and the last sentence provides a relevant example for the keen probabilist.

- **Kickoff:** As an author, I would be particularly pleased if this book would form not the end of a journey for you. Rather the opposite outcome would be desirable. The concept is to push you towards deeper questions at the very vibrant interface of dynamics and PDEs.

- **Lectures:** Each chapter corresponds approximately to a lecture of ninety minutes. Some chapters come as packages of 2–4 connected units; the overall structure is outlined in the section "Course Design" following the Preface. The style allows lecturers to relatively freely combine topics into a course that fits their needs and interfaces well with other courses taught within the curriculum.

- **Multipurpose:** Each chapter could also be the basis of a seminar talk. Furthermore, the book is suitable for self-study to get an overview of the area. For beginning graduate or advanced undergraduate students, it is also an excellent step to actually see all the basic mathematics from the first few semesters *in action*.

- **Notation:** The notation has been kept simple, in the hope of avoiding confusion with too many symbols and variable names. In particular, one objective was to make to the notation relatively easy to use on the blackboard, and to have a fully consistent scheme for the main global objects, yet potentially reuse variable names for auxiliary local objects.

- **Omissions:** There have been many cuts. I shall just mention four cases to illustrate the problem: (1) Focusing more on fluid dynamics was tempting as it is a topic deeply connected to the development of PDEs and dynamical systems.

(2) Integral/nonlocal PDEs would have been an interesting field to explore. (3) PDEs arising in geometric analysis, e.g., in the context of Ricci flow or general relativity, were also an intriguing option. (4) There are many links to stochastic problems, which I have—from a personal perspective extremely reluctantly—decided to omit.

- **Proofs:** As many proofs as possible are included, particularly if there are key ideas and new strategies contained within a proof. However, in a topic as broad as PDE dynamics, and taking into account the introductory approach, we have to refer to specialized books for very long technical arguments. However, if a proof is omitted, then a precise reference is provided, where a proof can be found.

- **Questions:** For each chapter, three exercises are provided. There are many more potential exercises that could have been selected. The idea is that for a lecture course or self-study, it is usually very unclear which of the many exercises one should try. Here the solution is simple: just do all of them if a chapter is covered in a course or relevant for your research. Most of the exercises are not difficult and are designed to just keep you thinking about the material.

- **References:** At the end of each chapter, some references are mentioned to spark interest in further reading. The reference list is certainly incomplete. Since this is an introduction to PDE dynamics, I decided to have a bias in this book towards citing other, generally more monographic texts or review papers for certain subclasses of PDE problems.

- **Style:** The format is based around relatively short units/chapters. The principle upon which this assumption is based centers around the unavoidable fact that readers may want to only delve into a subset of all topics. Although this means introducing some notations again, I feel this tradeoff towards modularity is well worth it.

- **Termination:** Since I simply do not need the letters U–Z to convey what I wanted to say about the structure of this book, I shall simply take the liberty to terminate this part of the introduction rather unexpectedly.

In addition to the previous remarks, I have now the pleasure to thank Luca Arcidiacono, Tobias Böhle, Marcel Braukhoff, Paul Carter, Annalisa Iuorio, Tobias Jawecki, Chris Münch, Anne Pein, Stefan Portisch, Pedro Aceves Sanchez, Elisabeth Schiessler, Lara Trussardi, Hannes Uecker, and Julian Westermeier for alerting me to several mistakes and misprints in previous versions of this book. Parts of this book have been used for a course at Vienna University of Technology in fall/winter 2014/2015 and for a course at Technical University of Munich (TUM) in fall/winter 2016/2017 and I would like to thank all students for bearing with me while I tried out this relatively new format. In fact, providing a broad, yet concise, introductory account for the next generation of PDE dynamics researchers was a key motivation to write the book in the first place.

Regarding the figures in this book, I am indebted to Andreas Burkhart, who converted my hand-drawn scribbles into sustainable vector graphics and postscript images.

Although it is clear to most readers, despite the support of colleagues as well as the editorial staff at SIAM, let me emphasize that I am entirely responsible for all remaining errors and inaccuracies remaining in this version of the book. I am going to maintain a web page to provide improvements and bug fixes, so please report any errors or inaccuracies you find to me via email.

Course Design

This extended preface is (mostly) intended for anyone who plans to teach a course from this book or use it as a basis for a seminar. If you are unfamiliar with the material or uncertain about your preferred topics, you can directly proceed to Chapter 1 or just start browsing. It is easy to then come back to the remarks here later as needed.

Dependencies

To make course design and self-study simple, I have deliberately tried to achieve two goals:

(I) There is a simple *linear storyline*. In particular, one can just use the book page-by-page starting from page 1. There are direct motivation connections from one chapter to the next one. Therefore, a simple option to design a course is to lecture in the same way. Just start and see how far you get in the allocated time.

(II) Despite the existence of a linear storyline, the book is *fully modular*. This means that there are blocks which are technically and thematically sufficiently independent. Figure 1 provides an overview of the different thematic blocks, which are composed of several chapters. In Figure 1, dotted lines indicate weak dependencies, while only the two bold lines arising from semigroup theory are somewhat essential dependencies.

Course Plans

There are many ways to design a course based upon the book; see Figure 1 for an overview. Here I shall simply *suggest* a few possible options I found useful. In general, I managed to roughly explain one chapter within 1.5 lecture hours. Obviously, some chapters are slightly shorter, while others take a bit longer. The actual classroom pace depends upon the audience quite a bit. Let us assume that you have a certain number of actual lecture hours available; then the following options are possible either for a one-semester course (options (1a), (1b), (1c), (1/2)), or for a two-semester course (options (1/2), (2a), (2b), (2c)):

(1a) $\approx$ **15 hours**: Quickly cover parts of Chapters 1–3. Since time is limited, consider just teaching the first three basic blocks, i.e., Chapters 4–13. It seems reasonable to omit all, or parts, of Chapters 10–11 to also introduce semigroups from Chapter 14. I would probably refrain from trying to cover later chapters with so little time available, but it might be possible to do so if students already come in with more advanced knowledge.

(1b) $\approx$ **20 hours**: In this case, one can aim to cover everything from Chapters 1–14. Or one can follow plan (1a), and then in the end augment the course by

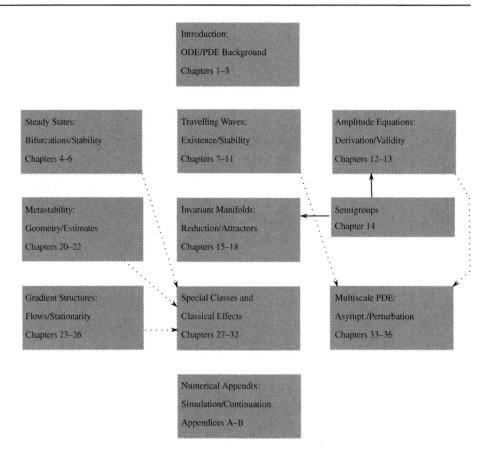

Figure 1. *Thematic/technical blocks are shown as gray boxes. Dotted lines = (very) weak dependency; bold lines = strong dependency.*

one of the following three-chapter blocks: Chapters 15–17, Chapters 20–22, or Chapters 23–25. If there is time left, two of the last three mentioned blocks have additional chapters; see Figure 1.

(1c) ≈ **25 hours**: With this length, one finally starts to have some flexibility. I have taught a course just based upon Chapters 1–17. Personally, I think that completely omitting geometric aspects from Chapters 15–21 would not be ideal if one has 25 hours available, so one should try to cover some ideas involving manifolds, e.g., Chapters 15–17. Then it might be best to just go topic-by-topic and pick a few selected aspects from Chapters 23–36.

(1/2) ≈ **30 hours**: With this amount of time, one option is to try to cover Chapters 1–19 completely. Then one can augment this with either the block of Chapters 20–22 on metastability, or with asymptotic gradient-type dynamics from Chapters 23–26. If one wants to focus on multiscale problems, it is also possible to cover Chapters 1–14, then switch to Chapters 33–36 and then finish with Chapters 20–21.

(2a) ≈ **40 hours**: Now it is possible to cover Chapters 1–26, potentially one has to go quite quickly through the introduction parts in Chapters 1–3 in this case. In principle, one can also omit gradient-type dynamics in Chapters 23–26 completely and replace it with more normal form, pattern-forming, and nonstationary dynamics from Chapters 27–32, or with multiscale PDE from Chapters 33–36.

(2b) ≈ **50 hours**: Almost everything can be covered. Probably one has to omit approximately five chapters out of the total 38 (also counting the numerical appendices). Individual chapters which can be omitted without much loss of continuity are 10, 11, 22, 26–32. Therefore, I would recommend to just lecture for about 20–25 hours following a plan (1b), (1c), or (1/2). Then one can reassess how to plan out the second half or second semester of the course.

(2c) ≈ **60 hours**: Everything can be covered relatively easily, most likely also including the appendices on numerical methods and some computer work.

Computer demonstrations could also be added into the previous course plans into slots, where it might not be feasible to start a new topic or go in depth on a thematic block, e.g., in the last lecture, the first lecture, or before holiday breaks. Furthermore, it might be helpful if you definitely want to include certain concepts and/or equations to take a look at the index to make sure you include the right chapters.

Chapter 1

A Whirlwind Introduction

The goal is to study the dynamics of prototypical models of **partial differential equations (PDEs)** for some unknown function u with

$$x \in \Omega \subseteq \mathbb{R}^N, \qquad t \in [0, T), \qquad u = u(x, t), \qquad u : \Omega \times [0, T) \to \mathbb{R},$$

where Ω is a domain, usually taken as bounded with sufficiently smooth **boundary** $\partial \Omega$ or as $\Omega = \mathbb{R}^N$. Furthermore, $T > 0$ is a final time up to which we study the dynamics; frequently, the one-dimensional spatial case $x \in \mathbb{R}$ and long-time dynamics with $T = +\infty$ will be considered as starting points. The **closure** and **interior** of Ω will be denoted by $\overline{\Omega}$ and Ω°. The following notations for derivatives will be used:

$$\Delta u := \sum_{n=1}^{N} \frac{\partial^2 u}{\partial x_n^2}, \quad \partial_t u := \frac{\partial u}{\partial t}, \quad \partial_{x_n}^k u := \frac{\partial^k u}{\partial x_n^k}, \quad \partial_{x_i x_j} u := \frac{\partial^2 u}{\partial x_i \partial x_j},$$

$$\nabla u := (\partial_{x_1} u, \dots, \partial_{x_N} u)^\top,$$

where $(\cdot)^\top$ denotes transpose (so the **gradient** ∇u is a column vector) and Δu is the **Laplacian** of u. We shall use $|\cdot|$ to denote absolute value and $\|\cdot\|$ for norms, where the **Euclidean norm** is always understood as the default one on any finite-dimensional normed space. A similar remark applies to **inner products**, denoted by $\langle \cdot, \cdot \rangle$, where the Euclidean inner product is the default choice. As standard function space notation we use $C^k(\Omega, \mathbb{R}) =: C^k$ for k-**times continuously differentiable functions** and $L^p(\Omega, \mathbb{R}) =: L^p$ for the Lebesgue space with norm

$$\|g\|_{L^p} = \left(\int_\Omega |g(x)|^p \, \mathrm{d}x \right)^{1/p}, \qquad p \in [1, \infty),$$

and L^∞ is endowed with a norm using the essential supremum over Ω $\|g\|_{L^\infty} := \mathrm{ess\ sup}_{x \in \Omega} |g(x)|$.

Example 1.1. The **Nagumo** or **Real-Ginzburg-Landau (RGL)** equation

$$\partial_t u = \Delta u + u(1 - u)(u - p), \tag{1.1}$$

where $p \in \mathbb{R}$ is a parameter, is a classical model in nonlinear PDE theory. The Nagumo equation initially arose as a simplification of the Hodgkin-Huxley model in neuroscience for electric impulse propagation in axons. The RGL name arose in the context of amplitude (or modulation) equations; see Chapter 12. Sometimes (1.1) is also referred to as the **Allen-Cahn equation**. We augment (1.1) by the **initial condition** $u(x, 0) = u_0(x)$ for a given sufficiently smooth function $u_0 : \Omega \to \mathbb{R}$. Furthermore, if $\Omega \subsetneq \mathbb{R}^N$ has a boundary then we consider suitable boundary conditions such as **Dirichlet boundary conditions**

$$u(x, t) = g(x) \quad \text{for } x \in \partial\Omega,$$

where $\partial\Omega$ denotes the boundary of Ω, or **Neumann boundary conditions**

$$(\vec{n} \cdot \nabla u)(x, t) = g(x) \quad \text{for } x \in \partial\Omega,$$

where $\vec{n}$ is the outer unit normal vector to $\partial\Omega$, and $g : \mathbb{R}^N \to \mathbb{R}$ is assumed to be sufficiently smooth. Usually, the precise boundary conditions for the Nagumo equation, as well as for all the other equations to be discussed, will not be our main focus in this book. We shall use **homogeneous** conditions $g(x) \equiv 0$ or even **periodic boundary conditions**

$$x \in \mathbb{T}^N := \mathbb{R}^N / \mathbb{Z}^N \ (= \text{the } N\text{-dimensional torus})$$

to simplify the problem and to focus on the basic *dynamical* features. ♦

In particular, we shall always implicitly assume from now on that "sufficiently nice" initial and boundary conditions are chosen for the problem at hand. Once the boundary and/or initial condition are needed for a particular calculation, we specify them explicitly.

Example 1.2. The **stationary** version of the Nagumo/RGL equation (1.1) is obtained by setting $\partial_t u = 0$ and hence is given by

$$0 = \Delta u + u(1 - u)(u - p), \qquad u = u(x). \tag{1.2}$$

Nonlinear PDEs of the form $0 = \Delta u + f(u, p)$, such as (1.2), arise frequently in applications, e.g., nonlinear elasticity, mathematical biology or theoretical physics. ♦

Although the cubic nonlinearity of the Nagumo equation and the Laplace operator Δ are natural choices to model **reaction** and **diffusion** respectively, there are many other choices that are natural to study.

Example 1.3. One option is to consider a quadratic nonlinearity instead, which yields the so-called **Fisher-Kolmogorov-Petrovskii-Piskounov (FKPP) equation**

$$\partial_t u = \Delta u + u(1 - u). \tag{1.3}$$

Instead of changing the reaction term, one frequently encounters other linear operators, not just Δ. A typical example is the **Swift-Hohenberg equation**

$$\partial_t u = -(1 + \Delta)^2 u + f(u), \tag{1.4}$$

where $f(u)$ is often taken as a cubic or cubic-quintic nonlinearity in u, and depends usually on a parameter $p \in \mathbb{R}$ as well. In applications, it is frequently natural to consider additional components. A classical example is the **FitzHugh-Nagumo equation**

$$\begin{aligned} \partial_t u &= \Delta u + u(1 - u)(u - p_1) - v + p_2, \\ \partial_t v &= p_3(u - p_4 v), \end{aligned} \tag{1.5}$$

where $p_j \in \mathbb{R}$, $j \in \{1, 2, 3, 4\}$ are parameters and $v = v(x, t)$. ♦

The field of nonlinear spatio-temporal evolution equations and their dynamical analysis is vast (to say the least). We are going to focus on key examples but one should always keep in mind that even the list of important examples is extremely long. Here we shall list a few nonlinear PDEs, where dynamical systems techniques have turned out tremendously helpful (just look at the differences as well as similarities in structure of the equations for now):

$$\textbf{Allen-Cahn} \quad \partial_t u = \Delta u - f(u),$$

$$\textbf{Burgers} \quad \partial_t u = -u\partial_x u,$$

$$\textbf{Cahn-Hilliard} \quad \partial_t u = \Delta(f(u) - p\Delta u),$$

$$\textbf{Gross-Pitaevskii} \quad \mathrm{i}\partial_t \psi = -\Delta\psi + V(x)\psi + p\psi|\psi|^2, \quad \psi \in \mathbb{C},$$

$$\textbf{Keller-Segel} \quad \begin{pmatrix} \partial_t u \\ \partial_t v \end{pmatrix} = \begin{pmatrix} \nabla \cdot (\nabla u - u\nabla v) \\ \Delta v - v + u \end{pmatrix},$$

$$\textbf{Korteweg-de Vries} \quad \partial_t u = -u\partial_x u + \partial_x^3 u,$$

$$\textbf{Kuramoto-Sivashinsky} \quad \partial_t u = -\partial_x^4 u - \partial_x^2 u - u\partial_x u,$$

$$\textbf{Nonlinear Schrödinger} \quad \mathrm{i}\partial_t \psi = -\Delta\psi + p\psi|\psi|^2, \quad \psi \in \mathbb{C},$$

$$\textbf{Nonlinear wave} \quad \partial_t^2 u = \Delta u + f(u),$$

$$\textbf{Porous medium} \quad \partial_t u = \Delta(u^p), \quad p \in (0, +\infty),$$

$$\textbf{Swift-Hohenberg} \quad \partial_t u = -(1 + \Delta)^2 u + f(u),$$

where $\mathrm{i} := \sqrt{-1} \in \mathbb{C}$ denotes the **imaginary unit**. The list of PDEs above could be continued with Boltzmann, Boussinesq, Camassa-Holm, Euler, Euler-Lagrange, Fokker-Planck, Gierer-Meinhardt, Gray-Scott, Hamilton-Jacobi, Klein-Gordon, Landau-Lifshitz-Gilbert, Navier-Stokes, Perona-Malik, and many more! The *main point* is that a dynamical systems viewpoint can be useful to understand all of these PDEs better, and vice versa PDEs motivate many technical developments in dynamics. This two-way *interplay* is the primary theme of this book.

Example 1.4. (Example 1.3 continued) As a starting point, think about removing the Laplacian term from (1.3), which yields the ODE

$$\partial_t u = u(1 - u), \qquad u = u(t), \tag{1.6}$$

which has a *one-dimensional* **phase space** $u(t) \in \mathbb{R}$ for each t. However, if we consider (1.3), then $u(\cdot, t)$ is expected to be in a function space for each t. For example, one could aim to show that $\int_\Omega |u(x,t)|^q \, \mathrm{d}x < +\infty$ for certain initial data $u(x,0)$ so that $u(\cdot, t) \in L^q(\Omega)$ for some $q \in [1, +\infty)$. Clearly, $L^q(\Omega)$ is an *infinite-dimensional* phase space. ◆

The observations in the last example show that we definitely have to develop suitable methods to be able to understand infinite-dimensional dynamics with the philosophy of finite-dimensional methods. The next two chapters review some background material which is helpful to bring the different approaches closer together.

Chapter 2

Some ODE Theory and Geometric Dynamics

We have to recall some important results for dynamics of ODEs of the form

$$u' := \frac{\mathrm{d}u}{\mathrm{d}t} = f(u), \qquad u(0) =: u_0 \in \mathbb{R}^d. \tag{2.1}$$

Theorem 2.1. *(local existence and uniqueness; see, e.g., [270, Ch. 7,17]) Consider the ODE (2.1) with $f \in C^1 = C^1(\mathbb{R}^d, \mathbb{R}^d)$ (i.e., f is a continuously differentiable map). Then there exist $t_0 > 0$ and $u : (-t_0, t_0) \to \mathbb{R}^d$ such that*

$$u'(t) = f(u(t)) \ \text{for } t \in (-t_0, t_0) \quad \text{and} \quad u(0) = u_0, \tag{2.2}$$

i.e., u solves (2.1) for some open time interval containing $t = 0$.

Theorem 2.1 is completely local in time, and says almost *nothing* about what the solutions to (2.1) actually do. Of course, it is critical to have such knowledge for any application for which the model was written down in the first place. The next classical theorem does only slightly better.

Theorem 2.2. *(continuous dependence; see, e.g., [270, Ch. 7,17]) Consider (2.1) with $f \in C^1$ and f with Lipschitz constant κ on an open bounded set $\mathcal{U} \subset \mathbb{R}^d$. Suppose u, v both solve (2.1) and remain in $\mathcal{U}$ for all $t \in [0, T]$; then*

$$\|u(t) - v(t)\| \leq \|u(0) - v(0)\| \mathrm{e}^{\kappa T}, \tag{2.3}$$

i.e., solutions may diverge at most exponentially from each other; see also Figure 2.1.

Remark: As we have done already in Theorem 2.2, we shall always reserve the notation e ≈ 2.71828 for **Euler's number**.

Theorems 2.1–2.2 *do not provide* enough information on solutions if one is interested in applications. Nonlinear ODEs are already very difficult and the dynamics can be tremendously complicated. Even one-dimensional examples are interesting.

Example 2.3. Consider the one-dimensional ODE

$$u' = f(u), \qquad u : \mathbb{R} \to \mathbb{R}.$$

If the equation is linear $f(u) = pu$ for a parameter $p \in \mathbb{R}$, then $u(t) = u(0)\mathrm{e}^{pt}$ and we can easily draw the phase portraits as in Figure 2.2.

4

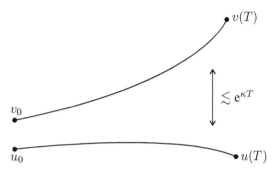

Figure 2.1. *Illustration of potential divergence of two nearby initial conditions for ODEs as discussed in Theorem 2.2. Note that the result is an upper bound, i.e., trajectories may exhibit any behavior between converging towards each other to diverging exponentially fast.*

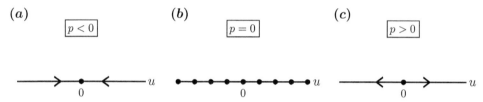

Figure 2.2. *Phase portraits for the linear ODE $u' = pu$. (a) Stable steady state at $u^* = 0$. (b) An entire line of steady states. (c) Unstable steady state at $u^* = 0$.*

For general nonlinear equations, closed-form solutions in classical elementary functions usually do not exist. Even if they do, it is frequently more insightful to argue more abstractly. For example, consider the ODE

$$u' = u(p - u) = f(u). \tag{2.4}$$

The **steady states** (or **equilibrium points**) u^* are obtained by setting $u' = 0$, i.e., solving $f(u) = 0$, so that $u_1^* = 0$ and $u_2^* = p$ as shown in Figure 2.3.

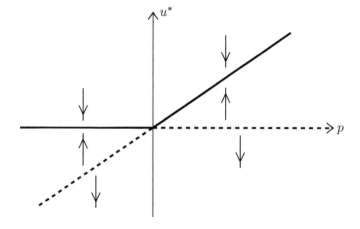

Figure 2.3. *Steady states for the ODE (2.4) plotted as a function of the parameter p. Solid lines are locally stable while dashed lines are unstable. For p fixed, we can view each vertical slice as a phase portrait for the ODE; cf. Figure 2.2.*

It is an extremely important idea to first study the dynamics *locally* near steady states. Consider

$$u = u_2^* + \varepsilon w = p + \varepsilon w$$

for some small $\varepsilon > 0$. Then we have

$$u' = (p + \varepsilon w)' = \varepsilon w'$$

as well as

$$u(p - u) = (p + \varepsilon w)(p - p - \varepsilon w) = -\varepsilon pw + \mathcal{O}(\varepsilon^2) \approx -\varepsilon pw.$$

So we can hope to study the linear system $w' = -pw$ locally near $w = 0$ to obtain stability results for $u_2^* = p$. Note that we can also derive this system via direct linearization as

$$w' = (\mathrm{D}_u f)(u_2^*)w = f'(p)w = (p - 2p)w = -pw. \tag{2.5}$$

From (2.5), the local flow near p is given as in Figure 2.3. The stability of $u_2^* = p$ changes as p passes through zero, i.e., u_2^* is **locally asymptotically unstable** for $p < 0$ and **locally asymptotically stable** for $p > 0$. Similarly, we can obtain results for $u_1^* = 0$. Then another very helpful view is to consider the (p, u) plane and draw a **bifurcation diagram** as shown in Figure 2.3. ♦

To define what we really mean by a bifurcation, we need a definition. By Theorem 2.1, the ODE (2.1) generates a **flow** φ on the time interval $\mathcal{T} := (-t_0, t_0)$ formally given by

$$\varphi : \mathbb{R}^d \times \mathcal{T} \to \mathbb{R}^d, \qquad \varphi(u_0, t) = u(t). \tag{2.6}$$

Hence, the ODE (2.1) generates a **dynamical system** $(\mathcal{X}, \mathcal{T}, \varphi)$, where $\mathcal{X} = \mathbb{R}^d$ is the **phase space**, $\mathcal{T}$ the time domain, and φ satisfies for $v \in \mathcal{X}$ the conditions

$$\varphi(v, 0) = v, \qquad \varphi(v, t + s) = \varphi(\varphi(v, s), t) \quad \forall t, s \in \mathcal{T}.$$

We can ask, when should two flows be considered as "equivalent"?

Definition 2.4. A dynamical system (A) is **topologically equivalent** to another one (B) if there is a homeomorphism of phase space mapping trajectories of (A) to trajectories of (B) preserving the direction of time.

In one situation, concluding topological equivalence is almost immediate. Suppose $h : \mathbb{R}^d \to \mathbb{R}^d$ is a diffeomorphism, $v := h(u)$; then

$$v' = (\mathrm{D}h)u' = (\mathrm{D}h)f(u) = (\mathrm{D}h)f(h^{-1}(v)) \tag{2.7}$$

and it is easy to show that $v' = g(v) := (\mathrm{D}h)f(h^{-1}(v))$ and $u' = f(u)$ yield topologically equivalent flows. Unfortunately, the converse is not true, and existence of a topological equivalence does not imply the existence of a diffeomorphic coordinate change h. Furthermore, there are subtle differences between continuous-time and discrete-time dynamical systems, which are not of interest here; see references at the end of this chapter.

Example 2.5. Consider as another example the **fold** (or **saddle-node**) **bifurcation**, which is exemplified by the one-dimensional system

$$u' = p + u^2 = f(u). \tag{2.8}$$

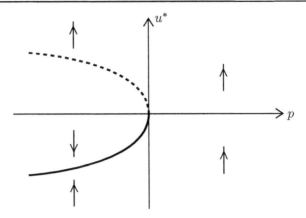

Figure 2.4. *Bifurcation diagram of the fold bifurcation normal form* (2.8).

Steady states are $u_1^* = \sqrt{-p}$, $u_2^* = -\sqrt{-p}$, which exist for $p < 0$, collide at $p = 0$, and disappear for $p > 0$. Local stability is easily checked from the linearization

$$w' = (D_u f)(u^*)w = 2u^*w = \pm 2\sqrt{-p}\, w, \tag{2.9}$$

so u_1^* is unstable and u_2^* is stable as shown in the bifurcation diagram in Figure 2.4. Clearly, the phase portraits for $p < 0$, $p = 0$, and $p > 0$ are not topologically equivalent. ♦

It should be noted that (2.8) is a **normal form** for a fold bifurcation, i.e., generic fold bifurcations in other coordinates can be qualitatively transformed to (2.8).

Definition 2.6. The appearance of a topologically nonequivalent phase portrait under parameter variation is called a **bifurcation**.

Studying bifurcations for PDEs will be one of the main themes to follow. Another core topic is to take a geometric viewpoint of phase space.

Example 2.7. Consider the two-dimensional linear ODE

$$\begin{pmatrix} u_1' \\ u_2' \end{pmatrix} = \underbrace{\begin{pmatrix} -1 & 0 \\ 0 & 1 \end{pmatrix}}_{=:A} \begin{pmatrix} u_1 \\ u_2 \end{pmatrix}, \tag{2.10}$$

which has solutions $u_1(t) = u_1(0)e^{-t}$, $u_2(t) = u_2(0)e^{t}$ and a **saddle** steady state at $u^* = (0,0)$; see Figure 2.5(a). The eigenspaces of A,

$$E^{\mathrm{s}}(u^*) := \{u = (u_1, u_2)^\top \in \mathbb{R}^2 : u_2 = 0\}, \tag{2.11}$$
$$E^{\mathrm{u}}(u^*) := \{u = (u_1, u_2)^\top \in \mathbb{R}^2 : u_1 = 0\}, \tag{2.12}$$

are **invariant** under the flow, i.e., trajectories cannot enter or leave them. Furthermore, the dynamics is directed away from u^* in $E^{\mathrm{u}}(u^*)$ and towards u^* in $E^{\mathrm{s}}(u^*)$; see Figure 2.5(a). ♦

More generally, one should not only look at linear spaces but (smooth) manifolds.

Definition 2.8. Let φ be a flow associated to an ODE (2.1) with a steady state u^*. Define the **stable** and **unstable manifolds** by

$$W^{\mathrm{s}}(u^*) := \{v \in \mathbb{R}^d : \varphi(v, t) \to u^*, \quad \text{as } t \to +\infty\},$$
$$W^{\mathrm{u}}(u^*) := \{v \in \mathbb{R}^d : \varphi(v, t) \to u^*, \quad \text{as } t \to -\infty\}.$$

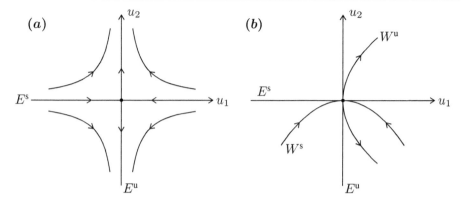

Figure 2.5. *(a) Phase portrait of the linear ODE (2.10) with a saddle steady state at the origin. The stable and unstable eigenspaces (2.11)–(2.12) are shown as well. (b) Perturbed situation sketching the stable and unstable manifolds as discussed in Theorem 2.11. These manifolds are tangent to the linear eigenspaces.*

Definition 2.9. A steady state u^* of the ODE (2.1) is called **hyperbolic** if $(D_u f)(u^*) \in \mathbb{R}^{d \times d}$ has eigenvalues λ_i with $\mathrm{Re}(\lambda_i) \neq 0$ for all $i \in \{1, 2, \ldots, d\}$.

The next two classical theorems show that local hyperbolicity implies persistence of qualitative dynamics and nice perturbation geometry.

Theorem 2.10. *(**Hartman-Grobman Theorem**; see, e.g., [529, Thm. 9.9]) Suppose the ODE (2.1) has a hyperbolic steady state u^*; then its flow is locally near u^* topologically equivalent to the flow of the linearized system $w' = D_u f(u^*)w$.*

Not only does the qualitative dynamics perturb very nicely, when the linearized and the fully nonlinear systems are compared, but so does the geometry.

Theorem 2.11. *(**Stable-Unstable Manifold Theorem**; see, e.g., [320, Sec. 6.2]) Suppose the ODE (2.1) has a hyperbolic steady state u^* and $(D_u f)(u^*)$ has k real-part negative and $d - k$ real-part positive eigenvalues with corresponding eigenspaces $E^{\mathrm{s}}(u^*)$ and $E^{\mathrm{u}}(u^*)$ for the linearized system. Then there exists a neighborhood $\mathcal{U}$ of u^* with local stable and unstable manifolds $W^{\mathrm{s}}_{\mathrm{loc}}(u^*)$ and $W^{\mathrm{u}}_{\mathrm{loc}}(u^*)$:*

$$W^{\mathrm{s}}_{\mathrm{loc}}(u^*) = \{v \in \mathcal{U} : \varphi(v, t) \to u^* \text{ as } t \to \infty \text{ and } \varphi(v, t) \in \mathcal{U} \; \forall t \geq 0\},$$
$$W^{\mathrm{u}}_{\mathrm{loc}}(u^*) = \{v \in \mathcal{U} : \varphi(v, t) \to u^* \text{ as } t \to -\infty \text{ and } \varphi(v, t) \in \mathcal{U} \; \forall t \leq 0\}.$$

Furthermore, $W^{\mathrm{s}}_{\mathrm{loc}}(u^*)$ and $W^{\mathrm{u}}_{\mathrm{loc}}(u^*)$ are tangent to $E^{\mathrm{s}}(u^*)$ and $E^{\mathrm{u}}(u^*)$ at u^* and are as smooth as f.

Remark: For more background on smooth **manifolds**, their **tangent spaces**, and related basic differential geometry topics see [365].

Note that Theorem 2.11 does relate algebra, geometry, analysis, and dynamics near hyperbolic steady states; see also Figure 2.5(b). Of course, one would want like to apply, and if necessary generalize, geometric ODE techniques to PDEs. Hence, one may ask, which classes of PDEs are the most immediate targets for this generalization? In the next chapter, we recall some basic PDE theory providing us good hints about which classes we may want to look at first.

Exercise 2.12. Use separation of variables to show that the ODE $u' = u^2$ for $u \in \mathbb{R}$ has solutions becoming unbounded in finite time, i.e., $t_0 \neq \infty$ in Theorem 2.1. $\Diamond$

Exercise 2.13. Consider $u' = Au$ for some matrix $A \in \mathbb{R}^{2 \times 2}$ and classify the stability of $u^* = 0$ based upon the trace and determinant of A. $\Diamond$

Exercise 2.14. Derive the bifurcation diagram for the **pitchfork bifurcation** normal form $u' = u(p - u^2)$ with $u \in \mathbb{R}$ and $p \in \mathbb{R}$. $\Diamond$

Background and Further Reading: There are many excellent books on dynamical systems and ODEs. Very readable introductions suited for self-study are [270, 515]. Bifurcation theory for ODEs is well documented in [247, 351]; see also [255]. Classical sources for ODE theory are [24, 253]. More advanced texts covering the interplay between ODEs and dynamical systems are [25, 236, 106, 373, 401, 455, 510, 549, 563], while discrete-time iterated maps are more focused upon in [15, 478]. More abstract theoretical introductions of dynamical systems can be found in [71, 320]. Deriving all the results and techniques presented in this chapter is almost an entire lecture course by itself, so we have just stated the basic results in this book as motivating starting points.

Chapter 3

Some PDE Theory and Functional Analysis

PDE theory naturally starts at the linear level with the classical **transport, Laplace, heat**, and **wave** equations:

$$\partial_t u + p \cdot \nabla u = 0, \quad \Delta u = 0, \quad \partial_t u = \Delta u, \quad \partial_t^2 u = \Delta u, \tag{3.1}$$

where $p \in \mathbb{R}^N$ is a vector of given parameters. For these PDEs many different techniques are available to find explicit solutions; see Exercise 3.10. To prove existence and regularity of broader classes of PDEs, additional methods are required. For example, consider the linear **elliptic PDE** on a bounded domain Ω given by

$$\begin{cases} Lu := \nabla \cdot (a(x)\nabla u) + b(x) \cdot \nabla u + c(x)u = f(x) & \text{in } \Omega, \\ \qquad\qquad\qquad\qquad\qquad\qquad\quad\ u = 0 & \text{on } \partial\Omega, \end{cases} \tag{3.2}$$

where $a(x) \in \mathbb{R}^{N \times N}$ is a uniformly positive definite matrix, $b(x) \in \mathbb{R}^N$, $f(x) \in \mathbb{R}$, $c(x) \in \mathbb{R}$, Ω is a domain with smooth boundary, and we assume that all mappings are sufficiently smooth in x. Despite the strong smoothness assumption, one obstacle with (3.2) is, how do we know *a priori* that $u = u(x)$ is actually in $C^2(\Omega, \mathbb{R}) =: C^2$ to make the Laplacian well defined classically? We multiply the PDE in (3.2) by a **test function** $v \in C_c^\infty(\Omega, \mathbb{R}) =: C_c^\infty$ (smooth functions with compact support) with $v = 0$ on $\partial\Omega$ and integrate. This yields

$$\int_\Omega fv \, \mathrm{d}x = \int_\Omega -\nabla v \cdot a(x)\nabla u + vb(x) \cdot \nabla u + c(x)uv \, \mathrm{d}x, \tag{3.3}$$

where integration by parts has been used. Hence, one should look at spaces where the formula (3.3) is well defined.

Definition 3.1. Let $u, \tilde{u}$ be locally integrable, i.e., $u, \tilde{u} \in L_{\text{loc}}^1(\Omega)$; then $\tilde{u}$ (also denoted by $\partial_{x_j} u$) is the **weak (partial) derivative** of u with respect to x_j, provided $\int_\Omega u(\partial_{x_j} v) \, \mathrm{d}x = -\int_\Omega \tilde{u}v \, \mathrm{d}x$ for all test functions v.

The definition naturally generalizes to multi-indices $\alpha = (\alpha_1, \dots, \alpha_N)^\top \in (\mathbb{N}_0)^N$ and confusion rarely arises with the overloaded notation $\partial_{x_j} u$ used for the weak derivative and the strong/classical derivative.

Definition 3.2. The **Sobolev space** $W^{k,p}(\Omega)$ consists of all $u \in L_{\text{loc}}^1(\Omega)$ such that, for each multi-index α with $\sum_j \alpha_j := |\alpha| \leq k$, we have $\mathrm{D}^\alpha u \in L^p(\Omega)$. We endow

$W^{k,p}(\Omega)$ with the norm

$$\|u\|_{W^{k,p}(\Omega)} := \begin{cases} \left(\sum_{|\alpha|\leq k} \int_\Omega |\mathrm{D}^\alpha u|^p \, \mathrm{d}x\right)^{1/p} & \text{for } p \in [1,+\infty), \\ \sum_{|\alpha|\leq k} \mathrm{ess \ sup}_\Omega |\mathrm{D}^\alpha u| & \text{for } p = +\infty. \end{cases} \qquad (3.4)$$

We let $W_0^{k,p}(\Omega)$ be the closure of $C_c^\infty(\Omega)$ in $W^{k,p}(\Omega)$ and denote $H^k(\Omega) := W^{k,2}(\Omega)$.

The strategy is to view (3.3) as the **weak formulation** of the original PDE, solve the equation in a suitable Sobolev space, and then prove regularity.

Definition 3.3. Define the function space $H_0^1(\Omega)$ as the completion with respect to $\|\cdot\|_{H^1(\Omega)}$ of $C_c^\infty(\Omega)$.

Intuitively, we would like to think of H_0^1 as just H^1 including zero boundary conditions [186, Sec. 5.5].

Definition 3.4. $u \in H_0^1(\Omega)$ is a **weak solution** of (3.2) if (3.3) holds for all $v \in H_0^1(\Omega)$.

Although it is quite a bit of work, the final result is that, for elliptic PDEs with smooth coefficients on a smooth bounded domain, a unique weak solution exists [186, Sec. 6.2]; in fact, several of the assumptions can be substantially weakened as well. The idea to prove regularity is based upon a bootstrap argument.

Theorem 3.5. (*higher-interior regularity, [186, Thm. 2, Sec. 6.3]*) Suppose $a_{ij}, b_j, c \in C^{m+1}(\Omega)$ for all i,j and $f \in H^m(\Omega)$. Assume u is a weak solution of (3.2); then $u \in H_{\mathrm{loc}}^{m+2}(\Omega)$.

Plainly, Theorem 3.5 shows that we *gain* two weak derivatives upon solving the elliptic PDE in comparison to the input data. Obviously, not all PDEs have this nice effect of smoothing; consider, e.g., the transport equation given above. How does one pass to strong classical differentiability?

Theorem 3.6. (*Sobolev inequality, [186, Thm. 6, Sec. 5.6]*) Consider any $u \in W^{k,p}(\Omega)$ and suppose q satisfies $1/q = 1/p - k/N$. If $k < N/p$ then $u \in L^q(\Omega)$ and

$$\|u\|_{L^q(\Omega)} \leq C\|u\|_{W^{k,p}(\Omega)}. \qquad (3.5)$$

for some constant $C = C(k, p, N, \Omega)$. Furthermore, if $k > n/p$ then $u \in C^r(\overline{\Omega})$ for $r = k - \mathrm{floor}(n/p) - 1$, where $\mathrm{floor}(n/p)$ is the biggest integer smaller than n/p.

So the second part of Theorem 3.6 tells us that having enough weak derivatives eventually yields smoothness.

Theorem 3.7. (*elliptic regularity*) If the input data a, b, c, f, Ω for (3.2) are smooth, then a weak solution u satisfies $u \in C^\infty(\Omega)$.

Proof. Combine higher-order interior weak regularity with the Sobolev inequality. $\square$

Results similar to Theorem 3.7 hold also for involving the boundary, so that $u \in C(\overline{\Omega})$. For a **parabolic PDE**,

$$\begin{cases} \partial_t u = \nabla \cdot (a(x)\nabla u) + b(x) \cdot \nabla u + c(x)u + f(x) & \text{in } \Omega, \\ \qquad\qquad\qquad\qquad\qquad\qquad\qquad\qquad\quad u = 0 & \text{on } \partial\Omega, \end{cases} \qquad (3.6)$$

sufficiently smooth input also yields smooth solutions, which we then refer to as **parabolic regularity**. Similar theory also exists in the case of linear **hyperbolic PDEs** such as

$$\begin{cases} \partial_t^2 u = \nabla \cdot (a(x)\nabla u) + b(x) \cdot \nabla u + c(x)u + f(x) & \text{in } \Omega \times (0, T], \\ \qquad\qquad\qquad\qquad\qquad\qquad\qquad\qquad\quad u = 0 & \text{on } \partial\Omega \times [0, T], \\ \qquad\qquad\qquad\qquad u = g_1, \quad \partial_t u = g_2 & \text{on } \Omega \times \{t = 0\}, \end{cases} \tag{3.7}$$

i.e., smooth input implies smooth solutions. Unfortunately, for PDEs beyond involving elliptic operators L and/or nonlinear problems, much less is known and one has to check carefully which regularity applies. Here, we shall usually refer to these results when we need them. Once regularity is established, the most classical tool for further analysis of elliptic (and parabolic) PDEs is the following:

Theorem 3.8. *(**strong maximum principle**, [186, Thm. 4, Sec. 6.4]) Suppose $c \leq 0$, Ω is connected, and $u \in C^2 \cap C(\overline{\Omega})$.*

- *If u is a **subsolution** of (3.2), i.e., $-Lu \leq 0$ in Ω, and u attains a nonnegative maximum over $\overline{\Omega}$ in Ω°, then u is constant within Ω.*

- *If u is a **supersolution** of (3.2), i.e., $-Lu \geq 0$ in Ω, and u attains a nonpositive minimum over $\overline{\Omega}$ in Ω°, then u is constant within Ω.*

In particular, we have bounds on elliptic and parabolic problems based upon sub- and super-solutions and boundary data. It is an important strategy in several dynamical systems problems for PDEs to construct suitable sub- and super-solutions. Another strategy to obtain a priori estimates which we briefly state are general Poincaré-type inequalities such as the following.

Theorem 3.9. *(**Poincaré inequality**, [186, Thm. 3, Sec. 5.6]) Suppose Ω is a bounded domain and $u \in W_0^{1,p}(\Omega)$ for some $p \in [1, N)$. Then for some constant $C = C(p, N, \Omega)$ we have*

$$\|u\|_{L^p(\Omega)} \leq C\|\nabla u\|_{L^p(\Omega)}. \tag{3.8}$$

Essentially (3.8) helps in certain arguments since it may be easier to control the weak gradient (or some higher derivative) of a PDE a priori than estimating the solution u (or a lower derivative) directly.

In summary, classical PDE theory hints at the fact that it might be a natural step to start with nonlinear elliptic problems such as

$$0 = \Delta u + f(u, p), \tag{3.9}$$

where $p \in \mathbb{R}$ is a parameter changing the nonlinearity; note that this naturally extends the problem class from Theorem 3.7 and the nonlinearity $f(u, p)$ links to the ODE Examples 2.3 and 2.5. Hence, we take (3.9) as motivating example for our first main theme: the analysis of steady states under parameter variation.

Exercise 3.10. Solve at least two of the classical linear PDEs (3.1). You may select, but carefully specify, a domain, boundary conditions, and/or initial data. ◊

Exercise 3.11. Prove the strong maximum principle in the case $N = 1$ and $c = 0$. How far can you generalize the arguments of your proof? ◊

Exercise 3.12. Consider the **logarithmic Sobolev inequality** for a function $u \in H^1(\mathbb{R}^N)$ given by

$$\frac{1}{\pi} \int_{\mathbb{R}^N} \|\nabla u(x)\|^2 \, dx \geq \int_{\mathbb{R}^N} |u(x)|^2 \ln \left(\frac{|u(x)|^2}{\|u\|^2_{L^2(\mathbb{R}^N)}} \right) \, dx + N \|u\|^2_{L^2(\mathbb{R}^N)}. \quad (3.10)$$

(a) Find a function $u \in H^1(\mathbb{R}^N)$, not involving any exponentials, which satisfies (3.10) with strict inequality. (b) Determine the constant k so that $u(x) = \exp(k\|x\|^2)$ satisfies (3.10) with equality. $\Diamond$

Background and Further Reading: The material in this chapter is quite classical and can be found in [186]; see also [472] for a general introduction, and consider [232] for elliptic problems and [211] for parabolic equations. Many helpful analysis basics for PDEs can be found in [375], and a broad exposition to maximum principles is [466]. There are many other textbooks providing introductions to PDE techniques with functional-analytic flavor; we just refer to a few of these [70, 297, 308, 514, 523, 524, 525]. For a landmark contribution to linear PDE, we refer to the series of monographs [277, 275, 276, 278].

Chapter 4

Implicit Functions and Lyapunov-Schmidt

The first step to understand steady states of PDEs under parameter variation is to lift certain finite-dimensional analysis theorems to the infinite-dimensional setting. Let X, Y, Z be real Banach spaces. We want to study the problem

$$F(u,v) = 0, \qquad F : X \times Y \to Z, \quad (u,v) \in X \times Y. \tag{4.1}$$

It helps to think of the simple case $Y = \mathbb{R}$ and to interpret v as a parameter. One may write, e.g., the stationary Nagumo equation (1.2) in the form (4.1)

$$F(u,p) = \Delta u + u(1-u)(u-p) = 0 \tag{4.2}$$

with $p = v \in \mathbb{R}$ and for suitable Banach spaces X, Z; also compare this to our motivating problem (3.9).

Definition 4.1. $F : X \times Y \to Z$ is **Fréchet differentiable** in X at (u_0, v_0), if there exists a bounded linear operator $D_u F(u_0, v_0) \in \mathcal{L}(X, Z)$ such that

$$\lim_{h \to 0} \frac{\|F(u_0 + h, v_0) - F(u_0, v_0) - D_u F(u_0, v_0)h\|_Z}{\|h\|_X} = 0.$$

Of course, a similar definition applies to differentiation in v, and we just write $F \in C^1$ if both Fréchet derivatives exist and are continuous. Sometimes we shall write $(D_u F)(u_0, v_0)$ to indicate with correct brackets where the derivative is evaluated. Mostly, the shorter notation $D_u F(u_0, v_0)$, or even just $D_u F$, will be used.

Remark: It will be helpful to think of derivatives as linear operators. Fix v_0 and observe that $D_u F(u_0, v_0) \in \mathcal{L}(X, Z)$. Furthermore, $D_{uu} F(u_0, v_0) \in \mathcal{L}(X, \mathcal{L}(X, Z))$, which is a space we can identify with bounded bilinear maps $\mathcal{L}_2(X, Z)$ so that we may write $D_{uu} F(u_0, v_0)[\cdot, \cdot]$ as taking two arguments in X. Obviously this generalizes to higher derivatives and multilinear maps [145, p. 46].

Theorem 4.2. *(Implicit Function Theorem [145, Ch. 4, Sec. 15]) Suppose $F \in C^1(X \times Y, Z)$, (u_0, v_0) satisfies (4.1), and $(D_u F)(u_0, v_0)$ is bijective. Then there exists a neighborhood $\mathcal{U} \times \mathcal{V}$ of (u_0, v_0) and a continuous map $f : \mathcal{V} \to \mathcal{U}$ such that $f(v_0) = u_0$ and*

$$F(f(v), v) = 0 \quad \text{for all } v \in \mathcal{V}. \tag{4.3}$$

Moreover, all solutions in $\mathcal{U} \times \mathcal{V}$ are of the form (4.3); see also Figure 4.1.

14

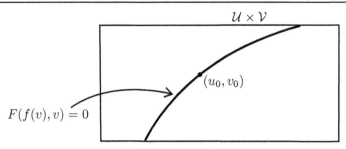

Figure 4.1. *Illustration of the Implicit Function Theorem 4.2.*

Theorem 4.2 is a natural generalization of the Implicit Function Theorem on $\mathbb{R}^d$ [481, p. 224]. All the assumptions on F and its derivatives only have to hold locally. Furthermore, if F has more regularity, e.g., $F \in C^k$, then we also get $f \in C^k$.

Example 4.3. Consider the **transcritical bifurcation** (2.4) again:

$$F(u,p) = u(p - u), \qquad u \in \mathbb{R} = X, \ p \in \mathbb{R} = Y, \ Z = \mathbb{R}. \qquad (4.4)$$

Observe that $(u, p) = (0, p)$ always solves (4.4). Now we calculate

$$(\mathrm{D}_u F)(0,p) = p - 2 \cdot 0 = p \qquad \text{and} \qquad (\mathrm{D}_p F)(0,p) = 0.$$

So the Implicit Function Theorem applies using $\mathrm{D}_u F$ as long as $p \neq 0$ and fails at the bifurcation value $p = 0$. In particular, for $p \neq 0$ the last part of the Implicit Function Theorem guarantees local *uniqueness* of a *branch* of solutions while two solution branches cross at $p = 0$; see Figure 2.3. ♦

Definition 4.4. Consider $F : \mathcal{U} \subset X \to Z$ with F Fréchet differentiable and let $u_0 \in \mathcal{U} \subset X$. F is a nonlinear **Fredholm operator** if the following conditions hold:

(D1) $\dim(\mathcal{N}[(\mathrm{D}_u F)(u_0)]) < \infty$, where $\mathcal{N}[\cdot]$ denotes the **nullspace** (or **kernel**),

(D2) $\mathrm{codim}(\mathcal{R}[(\mathrm{D}_u F)(u_0)]) < \infty$, where $\mathcal{R}[\cdot]$ denotes the **range**,

(D3) the range is closed, i.e., $\mathcal{R}[(\mathrm{D}_u F)(u_0)]$ is closed in Z,

where $\mathrm{codim}(S) := \dim(Z/S)$ for a subspace S in Z. Then define the **Fredholm index** as

$$\text{Fredholm index} := \dim(\mathcal{N}[(\mathrm{D}_u F)(u_0)]) - \mathrm{codim}(\mathcal{R}[(\mathrm{D}_u F)(u_0)]).$$

It can be shown that the Fredholm index is independent of $u_0 \in \mathcal{U}$; essentially Fredholm operators have small nullspace and miss just a small part of the range; see Figure 4.2.

The next goal is to reduce the general infinite-dimensional problem

$$F(u,v) = 0, \qquad F : \mathcal{U} \times \mathcal{V} \subset X \times Y \to Z \qquad (4.5)$$

to a more tractable finite-dimensional problem. We usually assume $\mathcal{V} \subset \mathbb{R}$ (or $\mathbb{R}^n$ for some n) and it remains to reduce the u component. This will be achieved using the **Lyapunov-Schmidt method**. Assume that

$$F(u_0, v_0) = 0, \quad F \in C^1, \quad F(\cdot, v_0) : X \to Z \text{ is a Fredholm operator.}$$

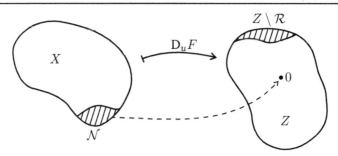

Figure 4.2. *Illustration of a Fredholm operator F as introduced in Definition 4.4. Essentially, we can think of these operators as having a linearization $\mathrm{D}_u F$ "acting nicely" on large parts of the domain and range.*

Then it is relatively easy to show that

$$X = \mathcal{N}[(\mathrm{D}_u f)(u_0, v_0)] \oplus X_0 =: \mathcal{N} \oplus X_0,$$
$$Z = \mathcal{R}[(\mathrm{D}_u f)(u_0, v_0)] \oplus Z_0 =: \mathcal{R} \oplus Z_0,$$

where $\mathcal{N}$ and Z_0 are finite-dimensional by Definition 4.4(D1)–(D2). Then define projections

$$\begin{aligned} P : X &\to \mathcal{N} & &\text{along } X_0, \\ Q : Z &\to Z_0 & &\text{along } \mathcal{R}. \end{aligned}$$

Lemma 4.5. *P, Q are continuous.*

Proof. We just apply the **closed graph theorem** (recall that $T : X \to Z$ is continuous if and only if the graph $\{(x, z) : Tx = z\}$ is closed; see [482, p. 51] or [127, p. 91]). □

Theorem 4.6. *(Lyapunov-Schmidt Reduction) Under the assumptions in this chapter, there exists a neighborhood $\mathcal{U} \times \mathcal{V}$ of (u_0, v_0) such that $F(u, v) = 0$ is equivalent in $\mathcal{U} \times \mathcal{V}$ to the finite-dimensional problem*

$$\Phi(\tilde{u}, v) = 0, \qquad (\tilde{u}, v) \in \tilde{\mathcal{U}}_1 \times \mathcal{V}_1 \subset \mathcal{N} \times \mathcal{V}, \tag{4.6}$$

and Φ is continuous with $\Phi(\tilde{u}_1, v_1) = 0$ for $(\tilde{u}_1, v_1) \in \tilde{\mathcal{U}}_1 \times \mathcal{V}_1$.

Remark: Note that $\tilde{u} \in \mathcal{N}$, and $v_1 \in \mathbb{R}^n$ (for some fixed $n \in \mathbb{N}_0$) by assumption, really means that (4.6) is finite-dimensional. A more precise expression for the **bifurcation function** Φ will be provided below.

Proof. (of Theorem 4.6) The equation $F(u, v) = 0$ is equivalent to the system

$$\begin{aligned} QF(Pu + (\mathrm{Id} - P)u, v) &= 0, \\ (\mathrm{Id} - Q)F(Pu + (\mathrm{Id} - P)u, v) &= 0. \end{aligned} \tag{4.7}$$

Define

$$\tilde{u} := Pu, \qquad w := (\mathrm{Id} - P)u, \qquad G(\tilde{u}, w, v) := (\mathrm{Id} - Q)F(\tilde{u} + w, v). \tag{4.8}$$

We work locally near (u_0, v_0) so it is natural to also define

$$\tilde{u}_0 := Pu_0, \qquad w_0 := (\mathrm{Id} - P)u_0 \tag{4.9}$$

and observe that $G(\tilde{u}_0, w_0, v_0) = 0$ by (4.7). The key observation of the proof is that

$$(\mathrm{D}_w G)(\tilde{u}_0, w_0, v_0) = (\mathrm{Id} - Q)(\mathrm{D}_u F)(u_0, v_0) : X_0 \to \mathcal{R} \quad \text{is bijective.}$$

In particular, $\mathrm{D}_w G$ acts nicely on the infinite-dimensional parts of X and Z; here the finiteness assumptions of the Fredholm property are absolutely crucial. Now we may just apply the Implicit Function Theorem 4.2 and obtain

$$\psi : \tilde{\mathcal{U}}_1 \times \mathcal{V}_1 \to X_0, \quad \psi(\tilde{u}, v) = w, \quad \psi(\tilde{u}_0, v_0) = w_0. \tag{4.10}$$

Inserting $\psi(\tilde{u}, v) = w$ into (4.7) yields

$$0 = QF(\tilde{u} + w, v) = QF(\tilde{u} + \psi(\tilde{u}, v), v), \tag{4.11}$$
$$0 = (\mathrm{Id} - Q)F(\tilde{u} + w, v) = (\mathrm{Id} - Q)F(\tilde{u} + \psi(\tilde{u}, v), v), \tag{4.12}$$

where (4.12) holds by construction and (4.11) is a finite-dimensional problem,

$$\Phi(\tilde{u}, v) := QF(\tilde{u} + \psi(\tilde{u}, v), v) = 0. \tag{4.13}$$

Continuity of Φ also follows from the Implicit Function Theorem. $\square$

Corollary 4.7. *Consider the same setup as in Theorem 4.6 and $F \in C^1$. Then $\psi \in C^1$, $\Phi \in C^1$, and*

$$\mathrm{D}_{\tilde{u}}\psi(\tilde{u}_0, v_0) = 0 \in \mathcal{L}(\mathcal{N}, X_0), \tag{4.14}$$
$$\mathrm{D}_{\tilde{u}}\Phi(\tilde{u}_0, v_0) = 0 \in \mathcal{L}(\mathcal{N}, Z_0). \tag{4.15}$$

Proof. Regularity is clear and just follows from the Implicit Function Theorem. The idea is to just differentiate (4.12) with respect to $\tilde{u}$ so

$$(\mathrm{Id} - Q)(\mathrm{D}_u F)(\tilde{u} + \psi(\tilde{u}, v), v)[\mathrm{D}_{\tilde{u}}\psi + \mathrm{Id}_{\mathcal{N}}] = 0.$$

Evaluating at $(\tilde{u}_0, v_0)$, and using $(\mathrm{D}_u F)\mathrm{Id}_{\mathcal{N}} = 0$, yields

$$(\mathrm{Id} - Q)(\mathrm{D}_u F)(\tilde{u}_0 + w_0, v_0)[\mathrm{D}_{\tilde{u}}\psi(\tilde{u}_0, v_0)] = 0.$$

Since $\mathrm{D}_{\tilde{u}}\psi(\tilde{u}_0, v_0)$ maps into X_0, which is complementary to $\mathcal{N}$, the last expression will only vanish if $\mathrm{D}_{\tilde{u}}\psi(\tilde{u}_0, v_0) = 0$, so (4.14) holds. Differentiating Φ is similar and yields (4.15). $\square$

Exercise 4.8. Carry out the last step in the proof of Corollary 4.7. $\Diamond$

Exercise 4.9. Let $F(u, p) := -\Delta u + f(u, p)$, where $f(u, p) = u(p - u)$ for all $p \in \mathbb{R}$. Consider $-\Delta$ as an operator for $x \in [0, \pi] =: \mathcal{I}$ on $H_0^1(\mathcal{I}) \cap H^2(\mathcal{I})$, i.e., with Dirichlet boundary conditions

$$-\Delta u = -\partial_x^2 u, \quad u(0) = 0 = u(\pi). \tag{4.16}$$

Prove that the Fréchet derivative $L(u_0, p_0) := (\mathrm{D}_u F)(u_0, p_0)$ is $L(u_0, p_0) = -\Delta + p_0 - 2u_0$, when we view $-\Delta$ as a mapping from $H_0^1([0, \pi]) \cap H^2([0, \pi])$ to $L^2([0, \pi])$. $\Diamond$

Exercise 4.10. Calculate the **eigenvalues** λ_j and **eigenfunctions** $e_j = e_j(x)$ of $L(0, p_0)$ from Exercise 4.9; i.e., when do we have $L(0, p_0)e_j = \lambda_j e_j$? $\Diamond$

Background and Further Reading: The material in this chapter is based upon [333], which is a very detailed, but not necessarily easy to digest, account of bifurcation theory for wide classes PDEs. Other classic accounts for the topic are [113, 289]; see also [502] for the many uses of Lyapunov-Schmidt methods. Since (nonlinear) functional analysis is helpful background, we mention a few sources in this area [53, 127, 145, 363, 482, 572, 573] as well as a very detailed series of monographs [570, 571, 574, 575].

Chapter 5

Crandall-Rabinowitz and Local Bifurcations

The next step is to prove a result which we can actually apply to find local bifurcations of certain classes of PDEs. Consider

$$F(u, p) = 0, \qquad F : X \times \mathbb{R} \to Z, \qquad F(0, p) \equiv 0, \tag{5.1}$$

where the last condition means that we always have a **trivial** (or **homogeneous**) **solution branch**. Furthermore, we assume throughout this chapter that

(A1) $\dim(\mathcal{N}[\mathrm{D}_u F(0, p_0)]) = 1 = \mathrm{codim}(\mathcal{R}[\mathrm{D}_u F(0, p_0)])$ so $F(\cdot, p_0)$ is a nonlinear Fredholm operator of index zero at p_0,

(A2) $F \in C^3$ in an open neighborhood of the trivial branch.

We view $\mathrm{D}_{up} F$ as an element of $\mathcal{L}(X, Z)$ as discussed in the remark following Definition 4.1. We may shift the parameter, if necessary, and assume without loss of generality that the interesting point, where the Implicit Function Theorem fails, is at $p_0 = 0$. The next theorem is a fundamental result in the field:

Theorem 5.1. *(**Crandall-Rabinowitz Theorem**) Consider* (5.1) *and assume that (A1)–(A2) hold as well as*

$$\mathcal{N}[\mathrm{D}_u F(0, 0)] = \mathrm{span}[e_0], \quad (\mathrm{D}_{up} F)(0, 0)e_0 \notin \mathcal{R}[\mathrm{D}_u F(0, 0)] \tag{5.2}$$

*for $e_0 \in X$ and $\|e_0\|_X = 1$. Then there is a **nontrivial branch** of solutions described by a C^1 curve through $(u, p) = (0, 0)$,*

$$\{(u(s), p(s)) \; : \; s \in (-s_0, s_0), \; (u(0), p(0)) = (0, 0)\}, \tag{5.3}$$

which satisfies $F(u(s), p(s)) = 0$ locally, and all solutions in a neighborhood of $(0, 0)$ are either the trivial solution or on the nontrivial curve (5.3).

Remark: An illustration of the **bifurcation** point at $(u, p) = (0, 0)$ is shown in Figure 5.1, together with a few different situations we expect to appear from the finite-dimensional cases discussed in Chapter 2. However, note that for PDEs many works avoid developing an extension of topological equivalence as in Definition 2.6 to define a **bifurcation** point. Instead, one commonly requires failure of the Implicit Function Theorem in combination with the appearance of a nontrivial curve of solutions.

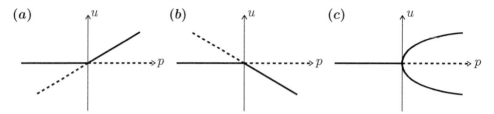

Figure 5.1. *Three basic examples for bifurcation points at the origin with a simple eigenvalue crossing. (a)–(b) Transcritical bifurcations. (c) Supercritical pitchfork bifurcation.*

Proof. (of Theorem 5.1) The idea is to apply the Implicit Function Theorem to get the nontrivial curve, as we have done before in similar situations. Using the Lyapunov-Schmidt Reduction from Theorem 4.6, the bifurcation function Φ satisfies

$$\Phi(\tilde{u}, p) = 0, \qquad \Phi : \tilde{\mathcal{U}}_1 \times \mathcal{V}_1 \to Z_0, \quad \dim(Z_0) = 1.$$

Using Theorem 4.6 and Corollary 4.7, as well as the associated notations, we also have

$$\psi(0, p) = 0, \qquad D_p\psi(0, p) = 0, \qquad \Phi(0, p) = 0, \tag{5.4}$$

in a suitable neighborhood of $p = 0$. Using the last observation about Φ gives us

$$\Phi(\tilde{u}, p) = \int_0^1 \frac{\mathrm{d}}{\mathrm{d}t} \Phi(t\tilde{u}, p) \, \mathrm{d}t \quad \Rightarrow \quad \Phi(\tilde{u}, p) = \int_0^1 D_{\tilde{u}}\Phi(t\tilde{u}, p)\tilde{u} \, \mathrm{d}t. \tag{5.5}$$

Next, let $\tilde{u} = se_0$, $s \in (-s_0, s_0)$, and observe that we get nontrivial solutions to $\Phi(\tilde{u}, p) = 0$ if we can solve

$$\tilde{\Phi}(s, p) := \int_0^1 D_{\tilde{u}}\Phi(tse_0, p)e_0 \, \mathrm{d}t = 0 \tag{5.6}$$

for $s \neq 0$. To solve the last equation, we want to use the Implicit Function Theorem, and this requires computing a derivative, so we carry out the following main computation of the proof. We need

$$D_p((D_{\tilde{u}}\Phi)(\tilde{u}, p)e_0) \overset{(4.13)}{=} D_p(Q(D_uF)(\tilde{u} + \psi(\tilde{u}, p), p)(e_0 + (D_{\tilde{u}}\psi)(\tilde{u}, p)e_0))$$
$$= \underbrace{Q(D_{uu}F)(\tilde{u} + \psi(\tilde{u}, p), p)[e_0 + (D_{\tilde{u}}\psi)(\tilde{u}, p)e_0, D_p\psi(\tilde{u}, p)]}_{=:(\mathrm{I})}$$
$$+ \underbrace{Q(D_{up}F)(\tilde{u} + \psi(\tilde{u}, p), p)(e_0 + (D_{\tilde{u}}\psi)(\tilde{u}, p)e_0)}_{=:(\mathrm{II})}$$
$$+ \underbrace{Q(D_uF)(\tilde{u} + \psi(\tilde{u}, p), p)((D_{p\tilde{u}}\psi)(\tilde{u}, p)e_0))}_{=:(\mathrm{III})}$$

and evaluate the last expression at $(\tilde{u}, p) = (0, 0)$. Then $(\mathrm{I}) = 0$ since $(D_p\psi)(0, 0) = 0$ by (5.4). Furthermore, $(\mathrm{III}) = 0$ since Q projects along $\mathcal{R}$ to the complement of the range of the linearized problem. So we are left with (II). Now we can compute

$$(D_p\tilde{\Phi})(0, 0) = \int_0^1 (D_pD_{\tilde{u}}\Phi)(0, 0)e_0 \, \mathrm{d}t$$
$$= Q(D_{up}F)(0, 0)e_0 \neq 0 \in Z_0, \tag{5.7}$$

where the last conclusion about the nonequality with zero follows from the assumption (5.2). Finally, applying the Implicit Function Theorem 4.2 yields a curve $\varphi : (-s_0, s_0) \to \mathcal{V}_1$, such that $\varphi(0) = 0$ and $\tilde{\Phi}(s, \varphi(s)) = 0$ near $s = 0$. This implies

$$\Phi(se_0, \varphi(s)) = s\tilde{\Phi}(s, \varphi(s)) = 0$$

and the curve given by

$$(u(s), p(s)) = (se_0 + \psi(se_0, \varphi(s)), \varphi(s)), \qquad s \in (-s_0, s_0),$$

has all the required properties we stated in the theorem. $\qquad\qquad\qquad\square$

Frequently, the situation in the Crandall-Rabinowitz Theorem is also described as **bifurcation from a simple eigenvalue**, which makes sense as $\dim \mathcal{N} = 1$ by assumption, so the interesting eigenspace indeed has a simple eigenvalue.

Corollary 5.2. *The tangent vector to the nontrivial solution curve at $(u, p) = (0, 0)$ is given by $(e_0, \dot{p}(0))$.*

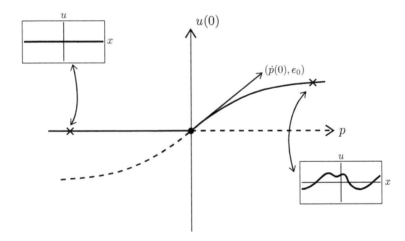

Figure 5.2. *Sketch of the Crandall-Rabinowitz Theorem 5.1; several choices, i.e., not only $u(0)$, would be possible to provide a coarse-grained visualization of u on the vertical axis; see, e.g., Figure B.3 for a computational bifurcation diagram. Here we find a bifurcation from a homogeneous solution (cross and inset on the left) at the origin to a nontrivial branch containing nonhomogeneous solutions (cross and inset on the right). The tangent vector at the bifurcation point illustrates the setting of Corollary 5.2.*

The proof of the corollary will be left as an exercise; see also Figure 5.2. The next step is to determine the shape of the nontrivial solution curve more precisely, which requires us to evaluate the derivative $\dot{p}(0) = \frac{d}{ds}p(s)\big|_{s=0}$. It is helpful to get another representation of the projection $Q : Z \to Z_0$ to do this calculation.

Lemma 5.3. *Suppose $Z_0 = \operatorname{span}[g_0]$, $g_0 \in Z$, $\|g_0\|_Z = 1$; then*

$$Qz = \langle z, e_0' \rangle g_0 \qquad \text{for some } e_0' \in Z' \text{ and for all } z \in Z, \tag{5.8}$$

*where $\langle \cdot, \cdot \rangle$ denotes the **duality pairing** between Z and the **dual space** Z'; furthermore, we have $\langle g_0, e_0' \rangle = 1$.*

Proof. We apply the **Hahn-Banach Theorem** [482, p. 56–61] (or more precisely an immediate consequence of the Hahn-Banach Theorem) to obtain a vector $e_0' \in Z'$ such that

$$\langle g_0, e_0' \rangle = 1, \qquad \langle z, e_0' \rangle = 0 \quad \forall z \in \mathcal{R}[D_u F(0,0)].$$

The result now follows immediately. $\square$

The following result is an elegant formula that can, however, be difficult to evaluate in some situations.

Theorem 5.4. *(first-order bifurcation curve approximation) The derivative $\dot{p}(0)$ of the nontrivial solution curve from Theorem 5.1 is given by*

$$\dot{p}(0) = -\frac{1}{2} \frac{\langle (D_{uu}F)(0,0)[e_0, e_0], e_0' \rangle}{\langle (D_{up}F)(0,0)e_0, e_0' \rangle}. \tag{5.9}$$

Proof. Recall that $\tilde{\Phi}(s, p(s)) = 0$, locally near $s = 0$, and differentiate

$$\frac{\mathrm{d}}{\mathrm{d}s} \tilde{\Phi}(s, p(s)) \Big|_{s=0} = (D_s \tilde{\Phi})(0,0) + (D_p \tilde{\Phi})(0,0)\dot{p}(0) = 0. \tag{5.10}$$

We proved in (5.7) that $(D_p \tilde{\Phi})(0,0) \neq 0$ and computed an expression for it in terms of F. So it remains to calculate

$$(D_s \tilde{\Phi})(0,0) = \int_0^1 (D_{\tilde{u}\tilde{u}}\Phi)(0,0)[e_0, te_0] \, \mathrm{d}t = \frac{1}{2}Q(D_{uu}F)(0,0)[e_0, e_0], \tag{5.11}$$

where we have used that $(D_{\tilde{u}}\Phi) = Q(D_u F)\mathrm{Id}_{\mathcal{N}}$ and bilinearity of the second derivative operator; here $\mathrm{Id}_{\mathcal{N}}$ denotes the **identity map** restricted to $\mathcal{N}$. Using (5.10), (5.11), and Lemma 5.3 yields the result. $\square$

If $\dot{p}(0) \neq 0$ then we expect the structure of a **transcritical bifurcation** similar to Example 2.3. So let us make a sanity check.

Example 5.5. Consider the ODE (2.4) again:

$$u' = u(p - u), \qquad F(u,p) := u(p - u). \tag{5.12}$$

So we have $X = \mathbb{R}$, $Y = \mathbb{R}$, and $Z = \mathbb{R}$. The Fréchet derivatives just become usual partial derivatives and the dual pairing just becomes the inner product (i.e., just multiplication in $\mathbb{R}$ here). Furthermore, the point of interest clearly is $(u, p) = (0, 0)$, where the Implicit Function Theorem fails as

$$(D_u F)(0,0) = (\partial_u F)(0,0) = 0, \quad (D_p F)(0,0) = (\partial_p F)(0,0) = 0.$$

So $\mathrm{span}[e_0] = \mathcal{N}[D_u F(0,0)]$ for $e_0 = 1$ and

$$(D_{up}F)(0,0)e_0 = (\partial_{up}F)(0,0)e_0 = 1 \cdot 1 = 1 \notin \mathcal{R}[D_u F(0,0)] = \{0\}.$$

So the Crandall-Rabinowitz Theorem, Theorem 5.1, applies. In addition, we can easily use the formula (5.9):

$$\dot{p}(0) = -\frac{1}{2} \frac{\langle (D_{uu}F)(0,0)[e_0, e_0], e_0' \rangle}{\langle (D_{up}F)(0,0)e_0, e_0' \rangle} = -\frac{1}{2} \frac{(-2) \cdot 1}{1 \cdot 1} = 1.$$

Hence, we recover the result in Figure 2.3. ♦

Example 5.5 demonstrates that we may expect formula (5.9) to be quite widely applicable, and easy to calculate, if X is a Hilbert space. Sometimes the explicit calculations can even be done in quite general Banach spaces, but these examples are beyond what we can present here.

Theorem 5.6. *(second-order bifurcation curve approximation [333, Sec. I.6]) The second derivative $\ddot{p}(0)$ of the nontrivial solution curve from Theorem 5.1 is given by*

$$\ddot{p}(0) = -\frac{1}{3}\frac{\langle (\mathrm{D}_{uuu}F)(0,0)[e_0, e_0, e_0], e_0' \rangle}{\langle (\mathrm{D}_{up}F)(0,0)e_0, e_0' \rangle}. \tag{5.13}$$

In the Crandall-Rabinowitz Theorem case, when $\dot{p}(0) = 0$ and $\ddot{p}(0) \neq 0$ we have a **pitchfork bifurcation**. The pitchfork is **subcritical** if $\ddot{p}(0) < 0$ and **supercritical** if $\ddot{p}(0) > 0$; see Figure 5.3.

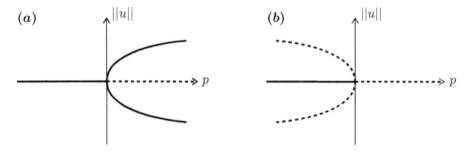

Figure 5.3. *Illustration of the pitchfork bifurcation; see also Theorem 5.1 and Theorem 5.6. (a) Supercritical pitchfork. (b) Subcritical pitchfork.*

One last step we need to verify is that many of the operators we are dealing with are indeed Fredholm. For example, consider again the problem

$$\partial_t u = \Delta u + f(u, p) = F(u, p), \tag{5.14}$$

where we could, e.g., take $f(u, p) = u(1 - u)(u - p)$ or $f(u, p) = u(p - u)$. Then the ansatz $u = u^* + \varepsilon w$ for a steady state u^*, say for $p = 0$, easily leads to the linearized problem

$$\partial_t w = \Delta w + (\mathrm{D}_u f)(u^*, 0)w = (\mathrm{D}_u F)(u^*, 0)w. \tag{5.15}$$

So a typical class of operators we should deal with are of the form "Laplacian + lower order terms." Recall the definition of elliptic operators from Chapter 3 and consider the differential operator

$$Lu := \sum_{i,j=1}^{N} a^{ij}(x)\partial_{x_i}\partial_{x_j}u + \sum_{j=1}^{N} b^j(x)\partial_{x_j}u + c(x)u$$

for $x \in \Omega \subset \mathbb{R}^N$ with sufficiently regular (say, smooth) coefficient functions, and assume that it is **uniformly elliptic**, i.e., there exists a constant $K > 0$ such that

$$\sum_{i,j=1}^{N} a^{ij}(x)\xi_i\xi_j \geq K\|\xi\|^2 \qquad \forall x \in \Omega, \xi \in \mathbb{R}^N.$$

The following is quite a remarkable and extremely useful fact.

Theorem 5.7. (*ellipticity and Fredholm operators*) *Suppose* Ω *is bounded with smooth boundary. Let the* $\mathcal{D}(L) := H^2(\Omega) \cap H_0^1(\Omega)$ *be the* **domain** *of L and consider it as an operator* $L : \mathcal{D}(L) \to L^2(\Omega)$*; then L is a Fredholm operator of index zero.*

Remark: Of course, by elliptic regularity [186, Sec. 6.3], we already know that the eigenfunctions we are going to get are not only in some Sobolev space but are actually classical smooth solutions.

Proof. (Sketch; see also [333]) By standard elliptic PDE theory, one may see that the operator $L - c\,\mathrm{Id} : \mathcal{D}(L) \to H^0(\Omega)$ is bounded and bijective for a suitable constant $c \in \mathbb{R}$ (indeed, just look at solving $Lu - cu = g$). Then one may see that the operator

$$(L - c\,\mathrm{Id})^{-1} =: K_c, \qquad K_c : L^2(\Omega) \to L^2(\Omega)$$

is **compact** (the image of a bounded sequence under K_c has a convergent subsequence); indeed, the compactness of the operator follows from the compact embedding of $H_0^1(\Omega)$ into $L^2(\Omega)$ (this embedding is special case of the **Rellich-Kondrachov Theorem** [186, p. 272]; see also the proof of Corollary 15.6 for a more detailed argument). Next, for $g \in L^2(\Omega)$ we can compute

$$Lu = g \quad \Leftrightarrow \quad (\mathrm{Id} + cK_c)u = K_c g. \tag{5.16}$$

One alternative characterization of Fredholm operators is that they are precisely those which are invertible up to a compact operator (which accounts for the kernel and range properties we used above to define Fredholm operators). Hence, one sees that $\mathrm{Id} + cK_c$ is Fredholm. Furthermore, (5.16) then implies

$$\dim(\mathcal{N}[L]) = \dim(\mathcal{N}[\mathrm{Id} + cK_c]) = n < \infty$$

as well as

$$g \in \mathcal{R}[L] \quad \Leftrightarrow \quad K_c g \in \mathcal{R}[\mathrm{Id} + cK_c].$$

One now just has to show that the range also has codimension n to get that the index is zero. From the decomposition

$$g = (\mathrm{Id} + cK_c)g - cK_c g$$

one finds

$$L^2(\Omega) = \mathcal{R}[\mathrm{Id} + cK_c] + \mathcal{R}[K_c] \quad \Rightarrow \quad L^2(\Omega) = \mathcal{R}[\mathrm{Id} + cK_c] + K_c(Z_0)$$

for some n-dimensional space $Z_0 \subset L^2(\Omega)$ with $\mathcal{R}[L] \cap Z_0 = \{0\}$. Then it is relatively easy to conclude with a few further steps (check it!) that $\mathrm{codim}(\mathcal{R}[L]) = n$. $\square$

It should be noted that, for more general PDEs, there is a crucial tradeoff between the choice of function spaces and the ability to show that the relevant operators are Fredholm; in fact, the last proof shows that, even for relatively simple elliptic PDEs, one already has to work quite hard to verify all properties.

Exercise 5.8. Prove Corollary 5.2. $\Diamond$

Exercise 5.9. Write down a one-dimensional ODE $u' = f(u, p_1, p_2)$ which has a pitchfork bifurcation upon varying p_1 through 0 and the pitchfork changes from sub- to super-critical if p_2 is varied through 0. $\Diamond$

Exercise 5.10. Let $x \in (0, \pi) =: \Omega$ and consider the problem

$$0 = \partial_x^2 u + f(u, p), \qquad (5.17)$$

where ∂_x^2 is understood as an operator with the domain $H^2(\Omega) \cap H_0^1(\Omega)$, i.e., consider homogeneous Dirichlet boundary conditions. Now try to apply the results in this chapter to the two cases (a) $f(u, p) = u(p - u)$ and (b) $f(u, p) = u(p - u^2)$. As a more advanced question (optional), what do you expect to happen if we perturb f by $f(u, p) + \delta \tilde{f}(u)$ with $\tilde{f}(0) \neq 0$? This last case is sometimes called **imperfection** and might be easier to understand if you consider the ODE case first. $\Diamond$

Background and Further Reading: The material in this chapter is based upon [333] and the last exercise/question is motivated by [289]. The original work of Crandall and Rabinowitz is [132, 133]; see also [467, 491], and for the well-known **Chafee-Infante** test problem see [101, 263]. Another approach to local bifurcations for PDEs is to use center manifold theory and normal forms [258, 263]. We shall encounter the philosophy of the invariant manifold approach in a slightly different context in Chapters 16–17 and Chapter 32. The idea to iteratively simplify a PDE towards a simpler normal form will be discussed in the Hamiltonian context in Chapter 27; furthermore, normal form ideas are also key to amplitude equations as considered in Chapters 12–13. For bifurcation problems, it is also very useful to explore associated numerical algorithms, as calculating local bifurcation points can become tedious and/or impossible quite quickly; see Appendix B for an introduction to numerical bifurcation analysis of PDEs.

Chapter 6

Stability and Spectral Theory

Having analyzed the local *existence* of nontrivial solution branches at bifurcations, the next question we have to address is *stability* for the evolution problem

$$\partial_t u = F(u, p), \qquad u : [0, +\infty) \to X, \quad u = u(t) \in X, \quad p \in \mathbb{R}. \tag{6.1}$$

Recall that the **spectrum** $\sigma(A)$ of a linear operator A consists of all elements $\lambda \in \mathbb{C}$ such that $(A - \lambda \operatorname{Id})$ is not invertible or the inverse $(A - \lambda \operatorname{Id})^{-1}$ is not a bounded operator; be careful, as this may crucially depend on the choice of spaces for A and the associated domain for A involving the boundary conditions. Implicitly we understand that if A is not a closed operator, we always look at its closure.

Definition 6.1. A solution branch $(u^*(p), p)$ for $F(u, p) = 0$ is called (**linearly**) **stable** at p^* if $\sigma(D_u F(u^*, p^*))$ is properly contained in the left half of the complex plane.

We continue with the notation from Chapter 5 and assume in addition that the Banach space X is continuously embedded into the Banach space Z. Consider the case of a simple eigenvalue $\lambda(s)$ with $\lambda(0) = 0$, which occurs in the Crandall-Rabinowitz Theorem 5.1. In particular, simple eigenvalue means that $\mathcal{N}[D_u F(0,0)] = \operatorname{span}[e_0]$ implies $e_0 \notin \mathcal{R}[D_u F(0,0)]$. This implies we have a decomposition

$$Z = \mathcal{N}[D_u F(0,0)] \oplus \mathcal{R}[D_u F(0,0)]$$

and an associated induced decomposition

$$X = \mathcal{N}[D_u F(0,0)] \oplus (X \cap \mathcal{R}[D_u F(0,0)]).$$

Note that this identifies the projection Q used in the Lyapunov-Schmidt method to prove Crandall-Rabinowitz as $Q : Z \to \mathcal{N}[D_u F(0,0)]$.

We can determine local stability of the trivial branch from this eigenvalue if we assume that $\sigma(D_u F(0, p(s))) - \{\lambda(s)\}$ is properly contained in the left-half complex plane for $s \approx 0$; see Figure 6.1(a). Parametrize λ by p such that $\lambda = \lambda(p)$, $\lambda(0) = 0$, and consider the **eigenvalue perturbation**

$$(D_u F)(0, p)(e_0 + w(p)) = \lambda(p)(e_0 + w(p)), \qquad w(0) = 0. \tag{6.2}$$

Differentiating (6.2) with respect to p and evaluating at $p = 0$ yields

$$(D_{up} F)(0,0) e_0 + (D_u F)(0,0) \frac{dw}{dp}(0) = \frac{d\lambda}{dp}(0) \, e_0. \tag{6.3}$$

26

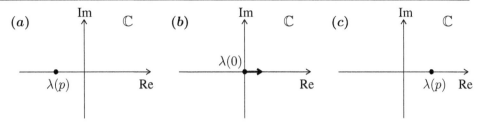

Figure 6.1. *Sketch of the crossing of a single eigenvalue through the imaginary axis upon varying the parameter p. It is important that in part (b) we can control that the eigenvalue really crosses with nonzero speed through the axis at the bifurcation parameter value p = 0.*

Recall that the dual pairing with $e_0' \in Z'$ satisfies the property that $\langle z, e_0' \rangle = 0$ for all $z \in \mathcal{R} = \mathcal{R}[(D_u F)(0,0)]$. Due to the identification of Z_0 with $\mathcal{N}[D_u F(0,0)]$ as discussed above, we can just apply the pairing $\langle \cdot, e_0' \rangle$ to (6.3) to obtain

$$\frac{d\lambda}{dp}(0) = \langle (D_{up}F)(0,0)e_0, e_0' \rangle.$$

In particular, the second condition in (5.2) in the Crandall-Rabinowitz Theorem has a reformulation

$$(D_{up}F)(0,0)e_0 \notin \mathcal{R}[D_u F(0,0)] \quad \Leftrightarrow \quad \frac{d\lambda}{dp}(0) \neq 0.$$

We may now *interpret* the nonvanishing condition of the derivative of $\lambda(p)$ at $p = 0$ as an eigenvalue **crossing with nonzero speed** upon variation of p or as a **transversality condition**; see Figure 6.1(b).

So far, we have implicitly assumed that we know the spectrum of the linear operator

$$(D_u F)(0,0) \in \mathcal{L}(X, Z). \tag{6.4}$$

However, for practical problems this may be far from trivial, so we consider a few standard cases where explicit calculations turn out to be possible. Before doing so, we briefly recall a classical finite-dimensional setting.

Example 6.2. Let $A \in \mathbb{R}^{d \times d}$ be a symmetric matrix and consider the linear ODE

$$u' = Au, \quad u = u(t) \in \mathbb{R}^d, \qquad \Rightarrow u(t) = e^{tA}u(0), \tag{6.5}$$

where we have just used the **matrix exponential**

$$e^{tA} := \sum_{k=0}^{\infty} \frac{(tA)^k}{k!}.$$

Without loss of generality (upon applying a coordinate change) we may assume that A is in diagonal form with eigenvalues λ_j, $j \in \{1, 2, \ldots, d\}$. If $\lambda_j < 0$ for all j then $u = 0 := (0, 0, \ldots, 0) \in \mathbb{R}^d$ is a (globally asymptotically) stable steady state. Suppose we may order the eigenvalues

$$\cdots \leq \lambda_2 < \lambda_1 < 0$$

and hence we see that $u(t) \approx K e^{\lambda_1 t} e_1$ as $t \to +\infty$, where $A e_1 = \lambda_1 e_1$. The behavior is dominated by the weakest attracting direction in the long-time limit. The rate of collapse onto this direction is given by the **spectral gap** $\lambda_2 - \lambda_1$. ◆

Now we can proceed to tackle more complicated yet still very classical examples.

Example 6.3. Consider the eigenvalue problem for the Laplacian on an interval

$$\frac{d^2}{dx^2} u = \lambda u \tag{6.6}$$

for $x \in [0, \rho]$. Denote the eigenfunctions by e_j and the associated eigenvalues by λ_j. We find

$$
\begin{aligned}
u(0) = 0 = u(\rho) &\Rightarrow \lambda_j = -(j\pi/\rho)^2, \ e_j(x) = \sin(j\pi x/\rho), \ j \geq 1, \\
\tfrac{du}{dx}(0) = 0 = \tfrac{du}{dx}(\rho) &\Rightarrow \lambda_j = -(j\pi/\rho)^2, \ e_j(x) = \cos(j\pi x/\rho), \ j \geq 0.
\end{aligned}
$$

Hence, the eigenvalues and eigenfunctions clearly depend crucially on the chosen boundary conditions. ◆

Example 6.4. Consider the Laplacian on a rectangle $\Omega = [0, \rho_1] \times [0, \rho_2]$ and the eigenvalue problem

$$\Delta u = \lambda u, \qquad u(x) = 0 \text{ for } x \in \partial\Omega. \tag{6.7}$$

Then the eigenvalues and eigenfunctions are easily checked to be

$$e_{jk}(x) = \sin(j\pi x_1/\rho_1)\sin(j\pi x_2/\rho_2), \quad \lambda_{jk} = -(j\pi/\rho_1)^2 - (k\pi/\rho_2)^2$$

for $j, k \geq 1$. ◆

However, beyond simple rectangular (or more generally hypercube) domains, giving explicit formulas for the eigenfunctions and eigenvalues of the Laplacian is usually not possible. Nevertheless, there are very remarkable abstract results.

Theorem 6.5. *(**spectrum of elliptic operators**; see [186, Sec. 6.5]) Consider the eigenvalue problem*

$$Lu = \lambda u \text{ in } \Omega, \qquad u = 0 \text{ on } \partial\Omega, \tag{6.8}$$

where L is uniformly elliptic as defined in (3.2); then L has a countable set of eigenvalues λ_j with $\lambda_j \to -\infty$ as $j \to \infty$.

Remark: One should be careful with sign conventions, as one also frequently finds in the literature results for the eigenvalue problem $-Lu = \lambda u$.

If one considers L on the domain $\mathcal{D}(L) = H_0^1(\Omega) \cap H^2(\Omega)$ with $L : \mathcal{D}(L) \to H^0(\Omega)$, it is not difficult to see that there are also associated bases of orthonormal eigenfunctions for the relevant Hilbert spaces associated to the eigenvalues.

Example 6.6. Let us return for $x \in \Omega$ to the problem

$$\partial_t u = \Delta u + f(u, p) = F(u, p), \qquad u = 0 \text{ on } \partial\Omega \tag{6.9}$$

from equation (5.14), where we assume that the trivial branch $(u, p) = (0, p)$ exists. The eigenvalue problem of the linearized PDE (see also (5.15)) for the trivial branch is

$$Lu = (\Delta + (D_u f)(0, p))u = \lambda u \quad \text{or} \quad \Delta u = (\lambda - (D_u f)(0, p))u, \tag{6.10}$$

where L is an elliptic operator. So we only have to deal with discrete eigenvalues in the spectrum by Theorem 6.5. On some domains, we can even calculate stability explicitly!

For example, if $\Omega = [0, \rho]$ then we see from (6.10) that the eigenvalues of the Laplacian $\Delta u = \tilde{\lambda} u$ are just shifted:

$$\tilde{\lambda}_j = \lambda_j - (\mathrm{D}_u f)(0, p)$$
$$\Rightarrow \lambda_j = \tilde{\lambda}_j + (\mathrm{D}_u f)(0, p)$$
$$\Rightarrow \lambda_j = -(j\pi/\rho)^2 + (\mathrm{D}_u f)(0, p)$$

for $j \geq 1$. In particular, we see that the **critical eigenvalue**, which is going to pass through the imaginary axis first, is λ_1. The condition for instability is $\lambda_1 > 0$ or

$$(\pi/\rho)^2 < (\mathrm{D}_u f)(0, p)$$

so increasing the domain size ρ will eventually destabilize the steady state, which is then referred to as a **long-wave instability**. Of course, we may also consider the case when ρ is fixed; then varying the parameter p could eventually destabilize the system. ◆

The next example shows that we do not only have to focus on elliptic PDEs and can treat other classes with very similar bifurcation-theoretic methods.

Example 6.7. Consider the **thin-film equation** (with constant surface tension) on a periodic domain:

$$\partial_t u = -\partial_x^4 u - \partial_x(f(u)\partial_x u) =: F(u), \quad x \in [0, 2\pi]/(0 \sim 2\pi) = \mathbb{T}^1, \qquad (6.11)$$

where $u = u(x, t)$ models the height of a thin fluid film on a substrate and f takes into account the substrate fluid interactions. Clearly any constant $u \equiv u^* \in \mathbb{R}^+$ is a steady state for (6.11). The linearization at u^* is

$$\partial_t w = (\mathrm{D}_u F)(u^*)w = [[\mathrm{D}(-\partial_x^4 - f(\cdot)\partial_x^2 - f'(\cdot)(\partial_x)^2)](u^*)]w,$$
$$= -\partial_x^4 w - f(u^*)\partial_x^2 w + \underbrace{f'(u^*)w \cdot \partial_x^2 u^* + \cdots}_{=0},$$

where we have used that evaluating operators such as ∂_x and ∂_x^2 on constants u^* yields zero (it is a good exercise to check the last calculation by setting $u := u^* + \varepsilon w$). In the context in which we work here, it makes sense to only study perturbations w, which have mean zero. Indeed, from the thin-film equation we have

$$\frac{\mathrm{d}}{\mathrm{d}t} \int_0^{2\pi} u(x, t) \, \mathrm{d}x = -\int_0^{2\pi} \partial_x^4 u + \partial_x(f(u)\partial_x u) \, \mathrm{d}x$$
$$= \left(-\partial_x^3 u - f(u)\partial_x u\right)\big|_0^{2\pi} = 0$$

by periodicity, so we have **mass conservation**, so meaningful perturbations should also have mass conservation. Substituting a **Fourier mode**

$$w_k = \exp(\sigma t)\exp(\mathrm{i}kx), \quad k \neq 0,$$

into the linearized problem yields

$$\sigma = -k^2(k^2 - f(u^*)). \qquad (6.12)$$

In fact, relations of the type (6.12) will be encountered several times; see, e.g., Chapters 8 and 12. Using (6.12), and assuming that f depends upon parameters, e.g., due to a parametrically changing substrate-fluid interaction, we can determine the stability of steady states. The sign of f is crucial: (a) if $f > 0$ then the second-order term is destabilizing, which is no surprise as it represents a "backward-heat-equation" term, and if (b) $f < 0$ then $\sigma < 0$, so the "forward-heat-equation" term stabilizes. ◆

In summary, we have now dealt with the most elementary cases for steady (or stationary) states of PDEs under parameter variation, including *local* existence and stability using bifurcation theorems and spectral theory.

Exercise 6.8. Construct the general solution of the second-order ODE

$$a\frac{\mathrm{d}^2 u}{\mathrm{d}x^2} + b\frac{\mathrm{d}u}{\mathrm{d}x} + cu = 0, \qquad u = u(x),\ x \in \Omega \subset \mathbb{R},\ a \neq 0,$$

using the roots of the equation $ar^2 + br + c = 0$. Use this result to derive the solutions of (6.6) for homogeneous Dirichlet, homogeneous Neumann, as well as homogeneous Robin

$$-\frac{\mathrm{d}u}{\mathrm{d}x}(0) + u(0) = 0 = -\frac{\mathrm{d}u}{\mathrm{d}x}(\rho) + u(\rho),$$

boundary conditions. ◊

Exercise 6.9. Consider (6.9) on $\Omega = (0, \pi)$ and look at the two cases (a) $f(u, p) = u(1 - u)(u - p)$ and (b) $f(u, p) = u(p - u)$. Investigate the local stability of the trivial branch of steady states for both cases under parameter variation of p. ◊

Exercise 6.10. Explicitly solve the **heat equation**

$$\partial_t u = \partial_x^2 u, \qquad x \in (0, \pi),$$

with Dirichlet boundary conditions and given smooth initial condition $u_0(x)$, using **separation of variables**, which is the ansatz $u(x, t) = u_1(x)u_2(t)$. In particular, compute how the eigenvalues of the Laplacian discussed above enter in the series solution obtained by separation of variables. ◊

Background and Further Reading: The material for supplementing Crandall-Rabinowitz is extracted from [333], while the spectral theory and thin-film example follows [357]. A good source for classical spectral theory results in applied mathematics is [142]. We also highlight one of the most fundamental monographs on linear operators and spectral theory [319]. Other excellent sources for spectral theory are [166, 167, 172, 501]. In fact, it is important to note that even the one-dimensional spatial case, which usually reduces to boundary value problems for ODEs, is already highly nontrivial [559]. A similar remark applies to the fundamental infinite-dimensional setting involving a self-adjoint operator on a Hilbert space [59].

Chapter 7

Existence of Travelling Waves

Having investigated stationary states, the next natural dynamical step is to see which classes of nonstationary solutions we can understand. Consider the one-dimensional reaction-diffusion PDE

$$\partial_t u = \partial_x^2 u + f(u), \qquad u = u(x,t) \in \mathbb{R}, \ x \in \mathbb{R}, \tag{7.1}$$

which we studied in Chapters 5 and 6 as an example several times. So which types of nonstationary solutions should we look for? Consider the **travelling wave ansatz**

$$u(x,t) = u(x - st) =: u(\xi), \qquad \text{i.e.,} \quad \xi := x - st, \tag{7.2}$$

where $s \in \mathbb{R}$ is the **wave speed** to be determined. Essentially (7.2) postulates that we only want to look for solutions u which depend upon a **moving frame** variable $\xi = x - st$. This approach is one instance of **spatial dynamics**, i.e., using a spatial variable as time. A wave profile moves "to the left" if $s < 0$, it moves "to the right" if $s > 0$, and it is a **standing wave** if $s = 0$; see Figure 7.1. Plugging (7.2) into (7.1) yields

$$-s\frac{\mathrm{d}u}{\mathrm{d}\xi} = \frac{\mathrm{d}^2 u}{\mathrm{d}\xi^2} + f(u), \tag{7.3}$$

where we just used the chain rule. It is clear that steady states $u(x,t) \equiv u^* \in \mathbb{R}$ of (7.1) are also steady states (or equilibria) for (7.3). It is those solutions where $u(\xi)$ is not constant which we are interested in here; see Figure 7.1. One may naively hope that (7.3) has explicit solution formulas. In generic situations, this is not the case. However, there are examples other than (7.1) where one is lucky to find nice closed-form solutions.

Example 7.1. Probably the most famous example where explicit formulas for waves exist is the **Korteweg-deVries (KdV) equation**

$$\partial_t u = -u\partial_x u - \partial_x^3 u. \tag{7.4}$$

Substituting (7.2) into (7.4) and rearranging terms yields

$$-su' + uu' + u''' = 0, \qquad \frac{\mathrm{d}}{\mathrm{d}\xi} = '. \tag{7.5}$$

Since $uu' = \frac{1}{2}(u^2)'$, the last equation can be integrated once

$$-su + \frac{1}{2}u^2 + u'' = c_1, \tag{7.6}$$

31

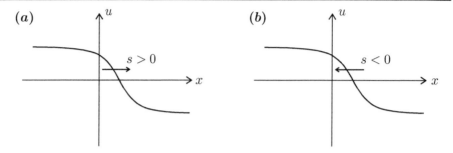

Figure 7.1. *Sketch of going into a travelling wave frame $\xi = x - st$, where a solution $u(x, t) = u(x - st)$ moves in a certain direction according to the sign of the wave speed s.*

where c_1 is a constant of integration. If we are only interested in waves which are doubly asymptotic as $\xi \to \pm\infty$ to zero (see also Figure 7.2(b)), with vanishing derivatives, we have the conditions

$$\lim_{\xi \to \pm\infty} u(\xi) = 0, \quad \lim_{\xi \to \pm\infty} u'(\xi) = 0, \quad \lim_{\xi \to \pm\infty} u''(\xi) = 0, \quad \cdots \qquad (7.7)$$

so the integration constant must be $c_1 = 0$ in this case. Multiplying (7.6) by u' gives

$$-suu' + \frac{1}{2}u^2u' + u''u' = -s\frac{1}{2}(u^2)' + \frac{1}{6}(u^3)' + \frac{1}{2}((u')^2)' = 0, \qquad (7.8)$$

which can be integrated. The integration constant is again zero by (7.7) so

$$-s\frac{1}{2}u^2 + \frac{1}{6}u^3 + \frac{1}{2}(u')^2 = 0 \quad \Leftrightarrow \quad 3(u')^2 = (3s - u)u^2. \qquad (7.9)$$

Under the further assumptions $u(\xi) \in (0, 3s)$ and taking the positive square root of the last expression, we get

$$\frac{\sqrt{3}}{u\sqrt{3s - u}}u' = 1. \qquad (7.10)$$

This equation is not quite easy enough to be integrated by hand, but upon the substitutions $v^2 = 3s - u$, $u' = -2vv'$ one finds

$$\frac{2\sqrt{3}}{3s - v^2}v' = -1 \quad \Rightarrow \quad \ln\left(\frac{\sqrt{3s} + v}{\sqrt{3s} - v}\right) = -\sqrt{s}\xi + c_2, \qquad (7.11)$$

where integration and the method of partial fractions have been used in the last step (see Exercise 7.6(a)). After a few further calculations, we find up to a shift of the profile of the wave that

$$u(\xi) = 3s\,\text{sech}^2\left[\frac{\sqrt{s}}{2}\xi\right] \quad \Rightarrow \quad u(x, t) = 3s\,\text{sech}^2\left[\frac{\sqrt{s}}{2}(x - st)\right], \qquad (7.12)$$

which is also called a **solitary wave** or **soliton**. The calculation showed that there are several special features of the KdV equation:

- it is **integrable** (formally, it can be viewed as an infinite-dimensional Hamiltonian dynamical system with a lot of conserved quantities; see also Chapter 27), which is echoed by the fact that we were able to determine some of the integrals of the moving frame ODE exactly;

- there is a wave for every wave speed s;

- higher solitary waves move faster due to the relation between the speed s and the amplitude $3s$.

In general, we cannot expect such a special calculation to hold, but it is a great starting point for **perturbation arguments**, i.e., one is interested in small perturbations of the KdV equation, which may not be integrable, but some integrable features persist under perturbation; see also Chapters 27 and 35. ◆

The travelling wave ansatz is interesting not only for equations which are first-order in the time derivative, as demonstrated by the next example.

Example 7.2. Consider the **sine-Gordon equation**

$$\partial_t^2 u = \partial_x^2 u - \sin u. \tag{7.13}$$

A similar procedure as for KdV works. Using the travelling wave ansatz and the assumption (7.7), a calculation (see Exercise 7.6) shows that

$$u(x,t) = 4 \arctan\left[\exp\left(-\frac{x - st}{\sqrt{1 - s^2}}\right)\right] \tag{7.14}$$

is a family of travelling wave solutions. ◆

We continue to make very important general observations about the ODEs obtained in the travelling wave frame and the full PDE system. To illustrate this in a more concrete case, consider again (7.3), which we can rewrite as a first-order system

$$\begin{aligned} u_1' &= u_2, \\ u_2' &= -su_2 - f(u_1). \end{aligned} \tag{7.15}$$

So which solutions of (7.15) correspond to nice bounded travelling wave profiles $u(x - st)$? Basically, bounded solutions connecting between the steady states $(u_1, u_2) = (a, 0)$ and $(u_1, u_2) = (b, 0)$ of (7.15) correspond to travelling wave solutions such that

$$\lim_{\xi \to -\infty} u(\xi) = a, \qquad \lim_{\xi \to +\infty} u(\xi) = b$$

as shown in Figure 7.2, with first derivatives tending to zero at infinity:

$$\lim_{\xi \to \pm\infty} u_1'(\xi) = 0, \qquad \lim_{\xi \to \pm\infty} u_2'(\xi) = 0.$$

We also call a and b the **end states** of the wave; see Figure 7.2. The next definition is valid for general ODEs and will be helpful for classifying travelling waves.

Definition 7.3. Consider the ODE $\frac{du}{d\xi} = f(u)$, $u \in \mathbb{R}^d$, with steady states u^* and $\tilde{u}^*$.

(D1) A solution $u(\xi)$ is called a **periodic orbit** (or periodic trajectory) of minimal **period** $\xi_T > 0$ if

$$u(\xi) = u(\xi + \xi_T) \tag{7.16}$$

and there is no smaller ξ_T such that (7.16) holds.

(D2) A solution $u(\xi)$ is called a **heteroclinic orbit** (or heteroclinic trajectory, or just heteroclinic) between u^* and $\tilde{u}^*$ if

$$\lim_{\xi \to -\infty} u(\xi) = u^*, \qquad \lim_{\xi \to +\infty} u(\xi) = \tilde{u}^*. \tag{7.17}$$

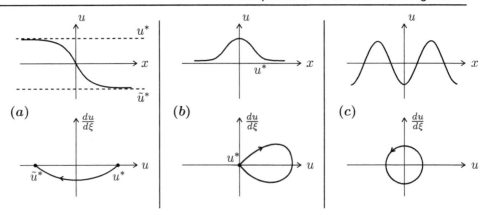

Figure 7.2. *Illustration of the correspondence between waves of one-component PDEs (upper row) to the spatial dynamics picture of the ODE arising as a travelling wave frame problem (lower row) with coordinate $\xi = x - st$ for a second-order one-component PDE. (a) Travelling front and heteroclinic orbit. (b) Travelling pulse and homoclinic orbit. (c) Travelling wave train and periodic orbit.*

(D3) A solution $u(\xi)$ is called a **homoclinic orbit** (or homoclinic trajectory, or just homoclinic) to u^* if

$$\lim_{\xi \to -\infty} u(\xi) = u^* = \lim_{\xi \to +\infty} u(\xi). \qquad (7.18)$$

Remark: Alternatively, one could also have expressed the previous definition using the definitions of α- and ω-**limit sets**, for the flow $\varphi(u_0, t)$ associated to $u' = f(u)$,

$$\begin{aligned} \alpha(\mathcal{U}) &:= \{u \in \mathbb{R}^d : \exists t_j, \, t_j \to -\infty \text{ s.t. } \varphi(u_0, t_j) \to u \text{ for some } u_0 \in \mathcal{U}\}, \\ \omega(\mathcal{U}) &:= \{u \in \mathbb{R}^d : \exists t_j, \, t_j \to +\infty \text{ s.t. } \varphi(u_0, t_j) \to u \text{ for some } u_0 \in \mathcal{U}\}. \end{aligned} \qquad (7.19)$$

For example, any point on a heteroclinic orbit from u^* to $\tilde{u}^*$ has as the α-limit set u^* and as the ω-limit set $\tilde{u}^*$; see also Figure 7.2(a).

Using Definition 7.3, the following observations/definitions are clear:

(H1) A heteroclinic orbit of the travelling wave ODEs corresponds to a **travelling front** solution of the associated PDE; see Figure 7.2(a).

(H2) A homoclinic orbit of the travelling wave ODEs corresponds to a **travelling pulse** solution of the associated PDE; see Figure 7.2(b).

(H3) A periodic orbit of the travelling wave ODEs corresponds to a **travelling wave train** solution of the associated PDE; see Figure 7.2(c).

Example 7.4. Suppose we study the Nagumo equation, i.e., (7.1) with $f(u) = u(1 - u)(u - p)$, say with $p \in (0, 1)$. Then (7.15) implies that the travelling wave ODE is

$$\begin{aligned} u_1' &= u_2, \\ u_2' &= -su_2 - u_1(1 - u_1)(u_1 - p), \end{aligned} \qquad (7.20)$$

where the only steady states occur for $u^{\mathrm{l}} = (0, 0)$, $u^{\mathrm{m}} = (p, 0)$, and $u^{\mathrm{r}} = (1, 0)$; see Figure 7.3.

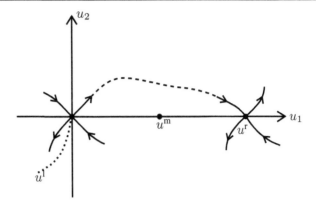

Figure 7.3. *Sketch of the phase portrait of the ODE* (7.20). *The three steady states are indicated as dots (leftmost point label u^l belongs to the origin as indicated by the dotted curve). We look for a heteroclinic orbit starting in $W^u(u^l)$ connecting to (via the dashed trajectory) the stable manifold $W^s(u^r)$.*

The linearization at a steady state u^* is given by

$$w' = \begin{pmatrix} 0 & 1 \\ p - 2(1+p)u_1^* + 3(u_1^*)^2 & -s \end{pmatrix} w =: A(u^*)w. \qquad (7.21)$$

For example, if $u^* = u^l$ then the eigenvalues λ^l of $A(u^l)$ satisfy the equation

$$(\lambda^l)^2 + s\lambda^l - p = 0 \qquad \Rightarrow \qquad \lambda_\pm^l = -\frac{s}{2} \pm \sqrt{\frac{s^2}{4} + p},$$

and it now follows that u^l is a saddle point as $\lambda_-^l < 0 < \lambda_+^l$. Similarly, it can be checked that u^r is a saddle point as well and u^m is completely unstable for $s < 0$ and stable for $s > 0$. It is instructive to sketch the phase portrait as shown in Figure 7.3. Suppose we are interested in left-moving front solutions connecting the states $u = 0$ and $u = 1$. This means looking for a heteroclinic orbit between u^l and u^r, which could potentially exist for some $s < 0$; see Figure 7.3. There are analytical proofs that there exists precisely one s for which there is a trajectory

$$\gamma = \gamma(\xi), \text{ such that } \gamma(-\infty) = (0,0) \text{ and } \gamma(+\infty) = (1,0).$$

In particular, it can be shown that the unstable manifold $W^u(u^l)$ and the stable manifold $W^s(u^r)$ are the same curve in the region $\{u_1 > 0, u_2 > 0\}$; see Figure 7.3. A similar result holds regarding the existence of a unique wave speed $s > 0$ for a right-moving front.

A viewpoint useful for analytical and numerical arguments is the following procedure:

(L1) Pick a codimension one submanifold $\mathcal{M}$ in phase space "between" the two steady states we want to connect; for (7.20) a vertical half-line

$$\mathcal{M} = \{u \in \mathbb{R}^2 : u_1 = \kappa, u_2 > 0\}$$

for some suitable fixed $\kappa \in (0,1)$ works well. Observe that the vector field for (7.20) is tangent to $\mathcal{M}$ if and only if $u_2 = 0$, so if we stay in the positive quadrant this degenerate case does not concern us here; see Figure 7.4.

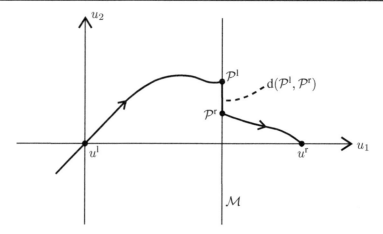

Figure 7.4. *Illustration of the setting of Lin's Method as discussed in (L1)–(L3) in this chapter. In particular, we can use a section $\mathcal{M}$ on which we measure the distance (or Lin gap) between stable and unstable manifolds.*

(L2) Compute the intersection submanifolds of the manifolds $W^u(u^l)$, $W^s(u^r)$ with $\mathcal{M}$:
$$\mathcal{P}^l := \mathcal{M} \cap W^u(u^l), \qquad \mathcal{P}^r := \mathcal{M} \cap W^s(u^r).$$

For (7.20) these are generically just two points. Note that all objects obviously depend upon the choice of the wave speed parameter s, although this is not shown in the notation.

(L3) Define a bifurcation function by

$$\Phi(s) := d(\mathcal{P}^l, \mathcal{P}^r) \tag{7.22}$$

where $d(\cdot, \cdot)$ is a suitable metric measuring the distance on $\mathcal{M}$; for (7.20) we can just take the Euclidean distance between two points; see Figure 7.4.

If we can find a zero $s = s_0$ of the bifurcation function ($\Phi(s_0) = 0$) then we have found a connecting orbit between the two steady states as desired. This idea naturally generalizes into higher dimensions and is also called **Lin's method**, and $\Phi(s)$ is the **Lin gap**. ♦

Another strategy worth being aware of can be useful if there is a conserved quantity available.

Example 7.5. Consider (7.20) in the standing wave case $s = 0$ and pick for simplicity $p = 1/4$. Then we see that the resulting ODEs in the travelling wave frame form a **Hamiltonian system**

$$
\begin{array}{lll}
u_1' = u_2, & & u_1' = -\frac{\partial H}{\partial u_2}(u_1, u_2), \\
u_2' = -u_1\left(1 - u_1\right)\left(u_1 - \frac{1}{4}\right), & \Leftrightarrow & u_2' = \frac{\partial H}{\partial u_1}(u_1, u_2),
\end{array}
\tag{7.23}
$$

where $H(u_1, u_2) := -\frac{1}{2}(u_2)^2 + \frac{1}{8}(u_1)^2 - \frac{5}{12}(u_1)^3 + \frac{1}{4}(u_1)^4$ is the **Hamiltonian function**, or just the **Hamiltonian**. The Hamiltonian is always a **first integral** of the flow since

$$\frac{d}{dt} H(u_1, u_2) = \frac{\partial H}{\partial u_1} \cdot u_1' + \frac{\partial H}{\partial u_2} \cdot u_2' \overset{(7.23)}{=} u_2' \cdot u_1' - u_1' \cdot u_2' = 0,$$

so H is constant along trajectories. Hence, the level curves $\{H(u_1, u_2) = \text{constant}\}$ are trajectories of (7.23). If we want to find a stationary/standing pulse solution of the original PDE (7.1) with the cubic Nagumo nonlinearity, then it suffices to find a homoclinic orbit of the Hamiltonian ODEs (7.23), which is just a bounded level curve of the Hamiltonian function containing the origin. ◆

In summary, we have introduced several important dynamical systems methods to find travelling waves: explicit integration, phase plane analysis, Lin's method, and Hamiltonian structure. However, there are many more, such as the Conley index relating the area to algebraic topology or analytical sub- and super-solution techniques; see references for more details.

Exercise 7.6. (a) Justify the steps to go from (7.10) to the soliton solution (7.12). (b) Derive the existence of the family of travelling waves (7.14) for the sine-Gordon equation. ◊

Exercise 7.7. Consider (7.1) with the cubic nonlinearity $f(u) = u(1 - u)(u - p)$ with $p \in (0, 1)$. Prove that $u \equiv 0, 1$ are locally stable stationary states for this PDE, while $u \equiv p$ is locally unstable for the PDE. Hint: Fourier transform; see also Chapter 8. Remark: The situation is also referred to as **bistability**. Hence, the stability of the travelling wave ODEs does *not* provide information about the actual stability of solutions for the PDE! ◊

Exercise 7.8. (a) Consider Example 7.4 and study, with any direction numerical simulation method of your choice, how the wave speed of a front depends upon the value of $p \in (0, 1)$; see Appendix A for one possible starting point. (b) Can you figure out conceptually how to use numerical continuation as presented in Appendix B to study the wave speed? ◊

Background and Further Reading: The material in this chapter is based upon [186, 248]. Regarding solitons, or solitary waves, we mention just a few basic sources here [3, 4, 143, 164, 428]. The monograph [554] covers many travelling wave problems in the context of parabolic nonlinear PDE; some additional classical papers in the area are [30, 197]. For the Conley index and bistable equations see [412, 507] and for sub- and super-solutions in the bistable case an excellent source is the very general work [105]. The motivation for the Nagumo equation originally arose in neuroscience for modeling propagation of electrical impulses along nerve axons [423]. An excellent description of Lin's method can be found in [345]; for the original source see [380].

Chapter 8

Pushed and Pulled Fronts

Having understood basic existence questions in Chapter 7, we now focus on one important class of nonlinear PDEs, where it is possible to analyze the wave in more detail. We again consider the one-dimensional reaction-diffusion PDE

$$\partial_t u = \partial_x^2 u + f(u), \qquad u = u(x,t) \in \mathbb{R},\ x \in \mathbb{R}, \tag{8.1}$$

and we make the usual ansatz (7.2), i.e., $u(x,t) = u(x - st) = u(\xi)$, to study the travelling wave. In this chapter we will focus on the actual value of the **wave speed** s.

Example 8.1. The main example in this chapter will be the **Fisher-Kolmogorov-Petrovskii-Piskounov (FKPP)** nonlinearity

$$f(u) = u(1 - u). \tag{8.2}$$

This quadratic nonlinearity has fundamentally different properties from the cubic nonlinearity discussed in detail in Chapter 7. For example, we have only two constant steady states $u \equiv 0$ and $u \equiv 1$. The travelling wave frame ODEs are given by

$$\begin{aligned} u_1' &= u_2, \\ u_2' &= -su_2 - u_1(1 - u_1), \end{aligned} \tag{8.3}$$

where $' = \frac{\mathrm{d}}{\mathrm{d}\xi}$ and it is easily checked that $u^{\mathrm{l}} = (0,0)$ and $u^{\mathrm{r}} = (1,0)$ have associated eigenvalues of the linearized system

$$\lambda_\pm^{\mathrm{l}} = \frac{1}{2}\left(\pm\sqrt{s^2 - 4} - s\right), \quad \lambda_\pm^{\mathrm{r}} = \frac{1}{2}\left(\pm\sqrt{s^2 + 4} - s\right). \tag{8.4}$$

Therefore, u^{l} is a **stable node** if $s \geq 2$ and an **unstable source** if $s \leq -2$, while u^{r} is always a hyperbolic saddle; see Figure 8.1. From Figure 8.1, it is apparent that there are infinitely many wave speeds s for which front solutions could exist. ◆

One technique to understand the wave speed better is to try to linearize directly on the level of the PDE (8.1). Suppose u^* is a steady state of (8.1) and also one of the end states of the wave (e.g., $u(\xi) \to u^*$ as $\xi \to +\infty$) we want to analyze. Consider

$$\partial_t w = \partial_x^2 w + \underbrace{(\mathrm{D}_u f)(u^*)}_{=:a^*} w, \qquad w = w(x,t) \in \mathbb{R},\ x \in \mathbb{R}, \tag{8.5}$$

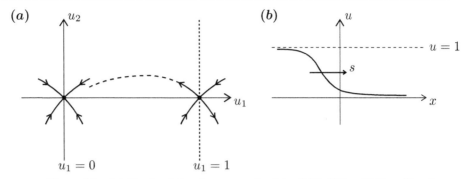

Figure 8.1. *(a) Sketch of the phase portrait of the ODE (8.3). (b) Travelling front solution of the FKPP equation (8.1)–(8.2) showing the pulled front invading the locally unstable steady state $u^* = 0$.*

where linear equations describe the local stability and dynamics near u^*. Note that we currently work on an *unbounded* domain. Hence, it is natural to consider the spatial **Fourier transform**

$$\hat{w}(k,t) := \int_{\mathbb{R}} e^{-ikx} w(x,t) \, dx, \qquad k \in \mathbb{R}, \tag{8.6}$$

as the problem is linear. Applying the Fourier transform to (8.5) yields

$$\partial_t \hat{w} = (ik)^2 \hat{w} + a^* \hat{w} \quad \Rightarrow \quad \hat{w}(t) = e^{((ik)^2 + a^*)t} \hat{w}(0). \tag{8.7}$$

The spatial Fourier modes decay if $\mathrm{Re}[(ik)^2 + a^*] = -k^2 + a^* < 0$ for all $k \in \mathbb{R}$, while a mode $k \in \mathbb{R}$ grows if $-k^2 + a^* > 0$. Hence, we can also check that u^* is stable if all Fourier modes decay to zero.

Example 8.2. Returning to the FKPP equation from Example 8.1, we find

$$\partial_t w = \partial_x^2 w + (D_u f)(u^*) w = \begin{cases} \partial_x^2 w + w & \text{if } u^* = 0, \\ \partial_x^2 w - w & \text{if } u^* = 1, \end{cases} \tag{8.8}$$

so $a^* = \pm 1$ in the notation above and we see that $u^* = 0$ is unstable while $u^* = 1$ is stable. Therefore, we also refer to the FKPP equation as **monostable**, while the Nagumo equation, i.e., (8.1) with $f(u) = u(1-u)(u-a)$ for $a \in (0,1)$, is **bistable** (checking bistability is an easy exercise). For the FKPP equation, the travelling front $u(\xi)$ we want to analyze is a heteroclinic with the following properties (cf. Example 8.1):

$$u(-\infty) = 0, \ u(+\infty) = 1; \text{ if } s < 0, \text{ front moves left},$$
$$u(-\infty) = 1, \ u(+\infty) = 0; \text{ if } s > 0, \text{ front moves right},$$

and $u \geq 0$; see also Figure 8.1. In particular, the front always propagates *into the unstable state*; we also say it *invades* the unstable state. ◆

In the general case, another way to think about the linearized problem is to substitute an ansatz similar to variation-of-constants idea

$$\hat{w}(k,t) = \hat{w}(k,0) e^{-i\omega(k)t},$$

where $\omega = \omega(k)$ will be called the (complex) **frequency** and k the **wave number**, which yields

$$-i\omega = (-k^2 + a^*) \quad \Leftrightarrow \quad \omega = -i(k^2 - a^*). \tag{8.9}$$

Remark: Observe that $\hat{w}(k, 0)$ is just the Fourier transform of the initial condition $w(x, t = 0)$. Furthermore, there is an implicit sign convention as one may equally well use $-\omega$ instead of ω.

Definition 8.3. A relation between wave numbers (or Fourier modes) k and a frequency ω is called a **dispersion relation**.

In general, there is quite some confusion in the literature regarding what one should call *the* dispersion relation. For example, another way to get *a* dispersion relation is to substitute a single Fourier-mode wave ansatz directly into the linearized problem

$$w_k(x, t) = e^{ikx + \sigma t} \quad \Rightarrow \quad \sigma = -k^2 + a^*, \tag{8.10}$$

which is also usually called **dispersion relation**, where $\text{Im}[\omega] = \sigma$. Hence, there is a choice whether we want to look at ω or just its imaginary part.

Remark: The frequency ω of the wave is related to the speed. Indeed, if $\omega(k) = ks(k)$ is real then $e^{i(kx + \omega t)} = e^{ik(x + st)}$ is travelling with **phase velocity** $s = s(k)$. Since different wave numbers may have different speeds, it is possible that an initial wave form **disperses**; this explains the name dispersion relation. Also, note that

$$e^{ik(x + st)} + \text{c.c.} = e^{ik(x + st)} + e^{-ik(x + st)} = 2\cos(k(x - st))$$

is then the real representation of the wave, where c.c. denotes complex conjugate.

The next step is to finally tackle the wave speed problem. Consider the following setup, which is assumed to hold throughout this chapter:

(A1) $u^* = 0$ is an unstable state of (8.1),

(A2) the dispersion relation $\omega(k)$ is a complex-analytic function for $k \in \mathbb{C}$,

(A3a) $w(\cdot, 0) \in C_c^\infty(\mathbb{R}, \mathbb{R})$ (smooth with compact support),

(A3b) $w(x, 0) > 0$ for some $x \in \mathbb{R}$, $w(x, 0) \geq 0$ for all $x \in \mathbb{R}$,

(A4) $x_\kappa(t)$ is a (level) curve in $\mathbb{R} \times [0, \infty)$ such that $w(x_\kappa(t), t) = \kappa$.

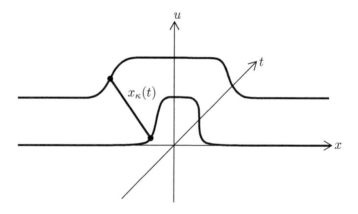

Figure 8.2. *Illustration of wave propagation for a compact or sufficiently localized initial condition in a space-time plot. The level curve $x_\kappa(t)$ can be used to track the speed of the front.*

Observe that the time derivative of the level curve x_κ can be used to track the speed at which waves of an initial compact support spread to the left or right into the unstable state, as shown in Figure 8.2.

Definition 8.4. A **linear spreading speed** $s^* \in (-\infty, +\infty)$ is any value which can be obtained as a well-defined limit of the form

$$s^* = \lim_{t \to +\infty} \frac{\mathrm{d}x_\kappa}{\mathrm{d}t}(t)$$

and can be calculated by just using the linearized equation at the unstable state.

Usually, we expect to have one linear spreading speed to the right and one to the left if a front propagates into an unstable state; see Figure 8.2.

Proposition 8.5. *Assume (A1)–(A4). Then any generic linear spreading speed $s^* \neq 0$ satisfies*

$$s^* = \frac{\mathrm{d}\omega}{\mathrm{d}k}(k^*), \tag{8.11}$$

$$s^* = \frac{\mathrm{Im}[\omega(k^*)]}{\mathrm{Im}[k^*]}, \tag{8.12}$$

where $\omega = \omega(k)$ is the dispersion relation and $k^* \in \mathbb{C}$ is a constant.

It will be made clear in the following proof what the role of $k^* \in \mathbb{C}$ is.

Proof. Using a variation-of-constants ansatz and the inverse Fourier transform, we write the solution of the linearized problem in the original variables as

$$w(x, t) = \frac{1}{2\pi} \int_{\mathbb{R}} \hat{w}(k, 0) e^{\mathrm{i}kx - \mathrm{i}\omega(k)t} \, \mathrm{d}k. \tag{8.13}$$

Consider a travelling wave frame $\xi := x - s^* t$ with some linear spreading speed s^*. Then we obtain from (8.13) that

$$\begin{aligned}
w(\xi, t) &= \frac{1}{2\pi} \int_{\mathbb{R}} \hat{w}(k, 0) e^{\mathrm{i}k(\xi + s^* t) - \mathrm{i}\omega(k)t} \, \mathrm{d}k \\
&= \frac{1}{2\pi} \int_{\mathbb{R}} \hat{w}(k, 0) e^{\mathrm{i}k\xi} e^{(\mathrm{i}ks^* - \mathrm{i}\omega(k))t} \, \mathrm{d}k \\
&= \frac{1}{2\pi} \int_{\mathbb{R}} \underbrace{\hat{w}(k, 0) e^{\mathrm{i}k\xi}}_{=:h(k)} e^{\rho(k)t} \, \mathrm{d}k,
\end{aligned}$$

where $\rho(k) := \mathrm{i}s^* k - \mathrm{i}\omega(k)$. By (A3a), we know that $\hat{w}(k, 0)$ is analytic as a function of $k \in \mathbb{C}$, and since $e^{\mathrm{i}k\xi}$ is analytic it follows that $h(k)$ is analytic. We are interested in a fixed ξ and $t \to +\infty$ limit as we look for linear asymptotic spreading speeds. Then, the main contribution of the integral

$$\frac{1}{2\pi} \int_{\mathbb{R}} h(k) e^{\rho(k)t} \, \mathrm{d}k, \quad \text{as } t \to +\infty, \tag{8.14}$$

can be found by deforming the integration contour into the upper or lower half in $\mathbb{C}$ by **Cauchy's Theorem** [221, p. 110] as $h(k)$ and $\rho(k)$ are analytic. As a matter of fact, we accept that the **method of steepest descents** [50, p. 280–302] shows that the main contribution can be calculated if the integration contour contains a point k^*, where $\rho(k)$ varies least (so-called *saddle points*; here we call k^* the **linear spreading point**), but this just means

$$\frac{\mathrm{d}\rho}{\mathrm{d}k}(k^*) = 0.$$

By **genericity**, we know that the minimum is nondegenerate (any slight perturbation makes it nondegenerate). Furthermore, we see that

$$\frac{d\rho}{dk}(k^*) = -i\frac{d\omega}{dk}(k^*) + is^* = 0, \tag{8.15}$$

from which (8.11) follows. For (8.12), we now evaluate the integral, and the dominant contribution is the integrand involving t evaluated at the saddle point:

$$\frac{1}{2\pi} \int_{\mathbb{R}} h(k)e^{\rho(k)t}\, dk = e^{\rho(k^*)t} + \text{h.o.t.} \quad \text{as } t \to +\infty,$$

where **h.o.t.** denotes **higher-order terms** in the limit. To have a linear spreading speed, a consistency requirement is that the leading-order term neither *grows* nor *decays* as we are in the moving frame of the asymptotic linear spreading speed. This implies

$$\text{Re}[\rho(k^*)] = 0 \quad \Rightarrow \quad \text{Re}[-i\omega(k^*) + is^*k^*] = \text{Im}[\omega(k^*) - s^*k^*] = 0,$$

from which the second part in (8.12) follows easily.   □

Usually, there is more than just one linear spreading speed, e.g., two linear spreading speeds: one for the left-moving front and one for the right-moving front. If we lift the genericity requirement on ρ, i.e., we require less of the structure on the dispersion relation and do not have that the minimum of $\rho(k)$ is nondegenerate, there are a lot more possibilities.

Example 8.6. It is helpful to work out what Proposition 8.5 says about our FKPP example (8.1)–(8.2). The dispersion relation follows from (8.9) with $a^* = 1$, as (8.8) has to be applied for the unstable state $u^* = 0$:

$$\omega(k) = i(-k^2 + 1) \quad \Rightarrow \quad \frac{d\omega}{dk} = -2ik,$$

so $s^* = -2ik^*$ by (8.11). Since s^* has to be real, we conclude that k^* is purely imaginary, say $k^* = i\beta$. Then (8.11) implies

$$2\beta = s^* = \frac{\text{Im}[\omega(k^*)]}{\text{Im}[k^*]} = \frac{\beta^2 + 1}{\beta},$$

so $\beta = \pm 1$ and we have

$$k^* = \pm i, \qquad s^* = -2ik^* = \pm 2.$$

So if we believe that the propagation into the unstable state of the FKPP equation is governed by the behavior of the front near the unstable state, then we would expect that the selected wave speed satisfies $|s| = 2$. ◆

Definition 8.7. A travelling front propagating into an unstable state is called a **pulled front** if the wave speed of the full nonlinear system equals the linear spreading speed near the unstable state. Otherwise, the wave is called a **pushed front**.

Giving a rigorous proof for precise wave speed selection is a nontrivial task and there are many results in the literature. It is a folklore result in applied mathematics that for the FKPP equation there are nonnegative waves for $|s| \geq 2$, and that all *sufficiently rapidly decaying* initial conditions sufficiently close to the shape of the front actually converge to waves with minimal speed $|s| = 2$, i.e., the practically stable nonlinear fronts are pulled fronts for the classical FKPP equation.

Exercise 8.8. Prove the following asymptotics of an integral:

$$\int_0^1 \frac{e^{ikt}}{1+k} dk = -\frac{i}{2t}e^{it} + \frac{i}{t} + \text{h.o.t.} \qquad \text{as } t \to +\infty \qquad (8.16)$$

using integration by parts repeatedly. It is frequently very useful to be able to estimate **Fourier-type integrals** such as (8.16); for more background see [50, 408]. ◊

Exercise 8.9. Consider the FKPP equation with **convection/advection** given by

$$\partial_t u = \partial_x^2 u - \kappa \partial_x u + u(1-u) \qquad (8.17)$$

and calculate the dispersion relation and the linear spreading speed. Fact: the PDE (8.17) does generate pushed or pulled fronts depending on the parameter $\kappa \in \mathbb{R}$. ◊

Exercise 8.10. Consider the **complex Ginzburg-Landau equation (cGL)**

$$\partial_t u = (1 + ic_1)\partial_x^2 u + c_2 u - (1 - ic_3)|u|^2 u, \qquad u = u(x, t) \in \mathbb{C}, \qquad (8.18)$$

where $c_j \in \mathbb{R}$, $j \in \{1, 2, 3\}$. Check for which c_j the steady state $u^* = 0$ is unstable using the dispersion relation. In the unstable case, calculate the linear spreading speed and the spreading point depending upon the c_j's. ◊

Background and Further Reading: The material in this chapter is based upon [543]. Detailed proofs for the full nonlinear case of many equations using sub- and super-solutions can be found in [554]. The FKPP equation [201, 340] is a cornerstone model of mathematical biology [72, 82, 342, 420, 537] but also appears in many other applications [540, 543]. The mathematical literature on monostable equations is vast, and so we can only cite a few interesting sources here [31, 52, 65, 170, 252].

Chapter 9

Sturm-Liouville and Stability of Travelling Waves

Having established some basic techniques and properties regarding existence and speed of travelling waves, it is natural to also look at **stability** of waves.

Example 9.1. For $u = u(x,t)$, $(x,t) \in \mathbb{R} \times [0, +\infty)$, consider the reaction-diffusion PDE

$$\partial_t u = \partial_x^2 u - u + 2u^3. \tag{9.1}$$

Then it can actually be checked that there is a **standing** (speed $s = 0$) **pulse** solution

$$u(x,t) = u(x - st) = u(x - 0 \cdot t) = U(\xi) = \text{sech}(\xi), \tag{9.2}$$

i.e., $U(\xi)$ solves the stationary problem for (9.1) with x replaced by ξ. Following the usual local linearization paradigm, we consider $u(x,t) = U(\xi) + \varepsilon w(\xi, t)$ and obtain

$$\partial_t w = \partial_\xi^2 w + (6U(\xi)^2 - 1)w =: Lw. \tag{9.3}$$

Hence, we might expect that the eigenvalue problem

$$Lw = \lambda w, \qquad \lambda \in \mathbb{C}, \ w \in X, \ w = w(\xi), \tag{9.4}$$

where X is a suitable Banach space, can help us to determine the stability of the pulse. In fact, the example shows that we should study linear operators L, which are defined on an unbounded spatial domain and which depend in general upon the travelling wave coordinate ξ, as we have linearized around the wave profile. ♦

On bounded domains, it was actually sufficient to just look at the eigenvalues. On unbounded domains, we have to consider a more general setup. Let $L : X \to Y$ be a linear operator between two Banach spaces X, Y.

Definition 9.2. If $\mathcal{N}[L - \lambda \, \text{Id}] \neq 0$, then $\lambda \in \mathbb{C}$ is called an **eigenvalue**.

The following definition is only relevant for the infinite-dimensional operator case; see also Figure 9.1.

Definition 9.3. The **essential spectrum** $\sigma_{\text{ess}}(L)$ consists of those $\lambda \in \mathbb{C}$ such that $L - \lambda \, \text{Id}$ is not a Fredholm operator of index zero. The **point spectrum** is the complement, i.e., $\sigma_{\text{pt}}(L) := \sigma(L) - \sigma_{\text{ess}}(L)$, where $\sigma(L)$ denotes the spectrum of L, i.e., $\sigma(L)$ contains those $\lambda \in \mathbb{C}$ such that $L - \lambda \, \text{Id}$ is not invertible.

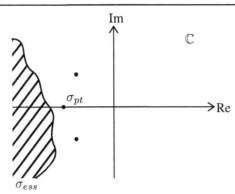

Figure 9.1. *Illustration of a possible spectrum for a linear operator as introduced in Definition 9.3.*

Remark: There are several slightly different definitions of the essential spectrum! For example, some authors drop the index zero condition or use even weaker variants of the Fredholm condition.

As a key classical example consider a **Sturm-Liouville operator**

$$Lw = \partial_\xi^2 w + a_1(\xi)\partial_\xi w + a_0(\xi)w, \qquad w = w(\xi),\ \xi \in \mathbb{R}, \qquad (9.5)$$

where $a_{0,1}(\xi)$ are smooth coefficients decaying exponentially at infinity to asymptotic values

$$\lim_{\xi \to \pm\infty} e^{\nu|\xi|}|a_1(\xi) - a_1^\pm| = 0, \qquad \lim_{\xi \to \pm\infty} e^{\nu|\xi|}|a_0(\xi) - a_0^\pm| = 0 \qquad (9.6)$$

for some $\nu > 0$ and $a_{0,1}^\pm \in \mathbb{R}$. If we would have $\xi \in \Omega$ for some bounded domain/interval Ω, then the results in Chapter 6 imply that L has only point spectrum. However, on an unbounded domain, Sturm-Liouville operators may have essential spectrum. For now, we shall not worry about this problem and look at the point spectrum on unbounded domains; see also Example 10.8. One may also consider the Sturm-Liouville operator (9.5) as a mapping $L : H_\mu^2(\mathbb{R}) \to L_\mu^2(\mathbb{R})$, where the subscript μ indicates that we endow the usual H^2 and L^2 spaces with the **weighted inner product**

$$\langle v, w \rangle_\mu := \int_\mathbb{R} v(x)w(x)\mu(x)\,\mathrm{d}x, \qquad \mu(x) := e^{\int_0^x a_1(y)\,\mathrm{d}y}. \qquad (9.7)$$

Observe that the operator $L : H_\mu^2(\mathbb{R}) \to L_\mu^2(\mathbb{R})$ is self-adjoint; see also Exercise 9.9. Classical **Sturm-Liouville theory** provides the following result.

Theorem 9.4. (*Sturm-Liouville Theorem*; see, e.g., [259, Ch. XI]) *Consider the eigenvalue problem $Lw = \lambda w$ for the Sturm-Liouville operator $L : H^2(\mathbb{R}) \to L^2(\mathbb{R})$ defined in (9.5) with the condition (9.6). The point spectrum $\sigma_{\mathrm{pt}}(L)$ is given by a finite number of eigenvalues such that*

$$\lambda_0 > \lambda_1 > \cdots > \lambda_N, \qquad \lambda_j \in \mathbb{R} \ \ \text{for all } j \in \{0, 1, 2, \ldots, N\}.$$

Furthermore, the eigenfunction e_j for λ_j has j simple zeros, the eigenfunctions are orthonormal in the inner product (9.7), and we have the formula

$$\lambda_0 = \sup_{\|w\|_\mu = 1} \langle Lw, w \rangle_\mu, \qquad (9.8)$$

where the supremum is achieved at $w = e_0$.

Frequently, one also refers to e_0 associated to (9.8) as the **ground state**.

Example 9.5. We continue with Example 9.1. Observe that (9.4) is an eigenvalue problem for a Sturm-Liouville operator (9.5) with

$$a_1(\xi) = 0, \qquad a_0(\xi) = 6(U(\xi)^2 - 1) = 6(\text{sech}^2(\xi) - 1).$$

The condition (9.6) holds with $a_0^\pm = -6$ as $\text{sech}(\xi) = 2/(e^{-\xi} + e^\xi)$ decays exponentially at $\xi = \pm\infty$ and easily with $a_1^\pm = 0$. Next, differentiating the steady state equation

$$0 = \partial_\xi^2 U - U + 2U^3$$

it follows that

$$0 = \partial_\xi^2(\partial_\xi U) - \partial_\xi U + 6U^2 \partial_\xi U = L(\partial_\xi U).$$

So we have $L(\partial_\xi U) = 0 \cdot (\partial_\xi U)$. Since

$$(\partial_\xi U)(\xi) = -\text{sech}(\xi)\tanh(\xi), \qquad \text{recalling that } \tanh(\xi) = \frac{e^\xi - e^{-\xi}}{e^\xi + e^{-\xi}},$$

it is easy to see that $(\partial_\xi U)$ decays exponentially at infinity. We conclude that $\lambda = 0$ is an eigenvalue of L. Furthermore, $(\partial_\xi U)(\xi)$ has precisely one zero at $\xi = 0$ (the maximum of the pulse; see Figure 7.2(b)). So we must have $\lambda_1 = 0$ in the notation of Theorem 9.4. Since Theorem 9.4 also yields the existence of a ground state eigenvalue $\lambda_0 > \lambda_1 = 0$, we see that the point spectrum contains a positive eigenvalue, which implies that the standing pulse solution of (9.1) is unstable. ♦

The goal is to prove a more general theorem about stability for equations of the form

$$\partial_t u = \partial_x^2 u - \mathcal{F}'(u), \qquad u = u(x, t), \ x \in \mathbb{R}, \ \mathcal{F} : \mathbb{R} \to \mathbb{R}, \tag{9.9}$$

and $\mathcal{F}'$ denotes the derivative of the smooth **potential** $\mathcal{F}$. Assume that $\mathcal{F}$ has three critical points at $u = 0, p, 1$ for $p \in (0, \frac{1}{2})$ and $u = 0, 1$ are minima, i.e.,

$$\mathcal{F}''(0) > 0, \qquad \mathcal{F}''(1) > 0$$

just means that $u = 0, 1$ are stable stationary states and we should think of the bistable Nagumo equation with $\mathcal{F}'(u) = -u(1 - u)(u - p)$ as an example. We can apply the phase-plane and Lin-type methods from Chapter 7, or sub- and super-solution arguments, to investigate the existence of travelling waves. We summarize the results here:

Theorem 9.6. *(existence of waves for bistable PDE) Consider the one-dimensional bistable reaction-diffusion equation (9.9) in one space dimension. Then there exist*

(T1) a standing pulse solution $U(\xi) \geq 0$ with $U(\pm\infty) = 0$, and

(T2) a travelling front solution $U(\xi) \geq 0$ with $U(-\infty) = 0$, $U(+\infty) = 1$, $U'(\xi) > 0$ for a unique wave speed $s = s^* > 0$.

Both waves $U(\xi)$ and their derivatives $(\partial_\xi U)(\xi)$ decay exponentially to their asymptotic limits as $\xi \to \pm\infty$.

Proof. (Sketch; see also [30, 105]) The idea is to use the geometric insights from Chapter 7. Indeed, we can use Lin's method to set up a root-finding problem to determine the unique heteroclinic orbit that yields the positive monotone travelling front. We can adapt the Hamiltonian argument from Example 7.5 to get the homoclinic orbit representing the standing pulse. The exponential decay at the end states just follows from the local linearization at the hyperbolic saddle points of the travelling wave ODEs in coordinates $(u_1, u_2) = (u, \partial_\xi u) \in \mathbb{R}^2$ associated to (9.9); see Figure 7.3.   □

Theorem 9.7. *(stability of waves for bistable PDE) Consider the same setup as in Theorem 9.6. Denote the point spectra for the standing pulse and the travelling front by $\sigma_{\mathrm{pt}}^{\mathrm{pulse}}$ and $\sigma_{\mathrm{pt}}^{\mathrm{front}}$, respectively, and let $z = a + \mathrm{i}b \in \mathbb{C}$; then*

$$\sigma_{\mathrm{pt}}^{\mathrm{pulse}} \cap \{a > 0, b = 0\} \neq \{\}, \tag{9.10}$$

$$\sigma_{\mathrm{pt}}^{\mathrm{front}} \subset \{a^* < a < 0, b = 0\} \cup \{0\} \tag{9.11}$$

for some constant $a^* < 0$.

Proof. We start with the standing pulse denoted by $U(\xi)$, which satisfies

$$\partial_\xi^2 U = \mathcal{F}'(U). \tag{9.12}$$

The relevant eigenvalue problem is

$$Lw = \lambda w, \quad L = \partial_\xi^2 - \mathcal{F}''(U),$$

so L is a Sturm-Liouville operator with $a_0(\xi) = -\mathcal{F}''(U(\xi))$ and $a_1(\xi) = 0$. As in Example 9.5, we differentiate (9.12) and obtain

$$0 = \partial_\xi^2(\partial_\xi U) - \mathcal{F}''(U)(\partial_\xi U) = L(\partial_\xi U) \tag{9.13}$$

so $(\partial_\xi U)(\xi)$ is an eigenfunction with eigenvalue $\lambda = 0$. Since $(\partial_\xi U)(\xi)$ vanishes precisely once for $\xi \in \mathbb{R}$ by its phase-plane construction, we conclude from Theorem 9.4 that $\lambda_1 = 0$ and the ground state has a positive eigenvalue $\lambda_0 > 0$, which proves (9.10). To show (9.11), we use again the notation $U(\xi)$ for the travelling front, which solves

$$0 = \partial_\xi^2 U + s^* \partial_\xi U - \mathcal{F}'(U). \tag{9.14}$$

The relevant eigenvalue problem is

$$Lw = \lambda w, \quad L = \partial_\xi^2 + s^* \partial_\xi - \mathcal{F}''(U),$$

so L is a Sturm-Liouville operator with $a_0(\xi) = -\mathcal{F}''(U(\xi))$ and $a_1(\xi) = s^*$. As before, direct differentiation of (9.14) shows that $\partial_\xi U$ is an eigenfunction with eigenvalue $\lambda = 0$. However, $U(\xi)$ is monotone by Theorem 9.6(T2), so it follows that $\partial_\xi U$ is the ground state. Therefore, $\lambda_0 = 0$ and Theorem 9.4 implies that all other eigenvalues must be real and negative. The lower bound in (9.11) follows also from Theorem 9.4, as the number of eigenvalues is finite.   □

Remark: The eigenvalue $\lambda = 0$ always occurs due to **translation invariance**, i.e., if $U(\xi)$ is a travelling wave so is $U(\xi + \xi_0)$ for any $\xi_0 \in \mathbb{R}$; see also Chapter 10. More generally, zero eigenvalues may arise due to other symmetries of the problem as well; cf. Chapter 32.

Basically, Theorem 9.7 is a first step towards the folklore statement that, in bistable one-component reaction diffusion equations in one space dimension, the family of monotone fronts is locally stable while the family of pulses is unstable. In particular, we have just shown one part of **spectral stability**, i.e., we analyzed the point spectrum; for methods to treat the essential spectrum see Chapter 10. However, a generalization of the stability result for fronts and pulses fails miserably in multicomponent systems, as in the next example.

Example 9.8. Consider a version of the FitzHugh-Nagumo equation

$$\begin{aligned}
\partial_t u &= \partial_x^2 u + u(1 - u)(u - 0.1) - v, \\
\partial_t v &= \varepsilon(u - v)
\end{aligned} \tag{9.15}$$

for $x \in \mathbb{R}$. It is a known that for $0 < \varepsilon \ll 1$ sufficiently small there exists a travelling pulse solution for (9.15), which is locally asymptotically stable. However, the proof of this result is difficult; see references as well as Chapter 36. ◆

The last example shows that we cannot, despite the elegance, rely on studying just one-dimensional cases based upon Sturm-Liouville theory.

Exercise 9.9. Prove that the Sturm-Liouville operator L with smooth coefficient functions (9.5) is **self-adjoint** in H_μ^2, i.e., $\langle Lu, v \rangle_\mu = \langle u, L^*v \rangle_\mu$ with **adjoint operator** $L^* = L$. ◊

Exercise 9.10. Consider (9.9) and define the **energy** as $E(u) := \frac{1}{2}(\partial_x u)^2 + \mathcal{F}(u)$. What is the relation between the energy and the Hamiltonian function of Example 7.5? ◊

Exercise 9.11. Consider the eigenvalue problem associated to **Hill's equation**

$$Lw := \partial_\xi^2 w + a_0(\xi)w = \lambda w \qquad \text{on } H_{\text{per}}^2([0, \pi]),$$

where $a_0 : \mathbb{R} \to \mathbb{R}$ is π-periodic. (a) Consider the **Bloch-wave decomposition** via a change of variables $w = e^{i\mu\xi}u$ to prove that the spectrum of L is real. (b) What can you say about the location of the spectrum? Hint: This is more difficult and requires some classical ODE Floquet theory (see, e.g., background references below). ◊

Background and Further Reading: This chapter is mainly based upon the monograph [318]. Example 9.1 can be found in [317]. An important survey of travelling wave stability results is [486] and we also refer to [489] for an early work on stability of waves. It should also be noted that under certain additional conditions one can conclude from spectral stability also nonlinear stability [263, 486, 318]. For example, the travelling front in Theorem 9.7 is not only spectrally stable but nonlinearly stable; i.e., small perturbations of the front asymptotically converge to a translate of the front. For classical Sturm-Liouville theory a good source is [559]. For the existence proof of the FitzHugh-Nagumo [202, 423] pulse consider [84, 185, 305, 346] and for stability see [301, 566]. The FitzHugh-Nagumo model has spawned quite a number of similar PDEs [45, 196]. For more on Hill's equation see [390], and we refer to [106, 529] for introductions to Floquet theory.

Chapter 10

Exponential Dichotomies and the Evans Function

For the results in Chapter 9 we relied crucially on classical Sturm-Liouville theory to understand the spectrum. In this chapter, we develop the basics of a more general theory for equations of the form

$$\partial_t u = \mathcal{A}(\partial_x)u + f(u), \qquad u = u(x,t),\ x \in \mathbb{R}, \tag{10.1}$$

where $\mathcal{A}(\cdot)$ is a polynomial of its argument, $\mathcal{A}(\partial_x) : X \to X$ is a linear operator on a suitable Banach space X, u can be vector-valued, and f is a mapping representing the nonlinear terms.

Example 10.1. A key example to keep in mind for (10.1) is reaction-diffusion systems

$$\partial_t u = D\Delta u + f(u), \qquad u = u(x,t) \in \mathbb{R}^d, \tag{10.2}$$

where D is a diagonal matrix with positive entries. ♦

Let $U(\xi) = u(x - st)$ be a travelling wave solution for (10.1). Then the first step to investigate (linear) stability of the wave is to consider the operator

$$L := \mathcal{A}(\partial_\xi) + s\partial_\xi + [(\mathrm{D}_u f)(U(\xi))] \tag{10.3}$$

and study the eigenvalue problem $Lw = \lambda w$.

Example 10.2. For (10.2) we find that

$$Lw = D\partial_\xi^2 w + s\partial_\xi w + (\mathrm{D}_u f)(U(\xi))w.$$

The eigenvalue problem $Lw = \lambda w$ can be written as a first-order system using $\frac{\mathrm{d}w}{\mathrm{d}\xi} = \tilde{w}$:

$$\begin{pmatrix} w' \\ \tilde{w}' \end{pmatrix} = \begin{pmatrix} 0 & \mathrm{Id} \\ D^{-1}[\lambda - (\mathrm{D}_u f)(U(\xi))] & -sD^{-1} \end{pmatrix} \begin{pmatrix} w \\ \tilde{w} \end{pmatrix}, \tag{10.4}$$

where $' = \frac{\mathrm{d}}{\mathrm{d}\xi}$. Note that the eigenvalue problem for $v := (w, \tilde{w}) \in \mathbb{R}^{2d}$ has the structure

$$v' = A(\xi; \lambda)v = (\tilde{A}(\xi) + B(\lambda))v \tag{10.5}$$

for matrix-valued functions $\tilde{A}$ and B, which follow directly from (10.4). ♦

Since we work in one spatial dimension, we can always reduce to an operator L which is a first-order differential operator acting on a space of more function components, as illustrated in Example 10.2. We still use L to denote this operator; i.e., the rewriting of (10.3) to a first-order ODE system is understood. This spectrum of L can be studied by considering

$$Lw = \lambda w \qquad \Leftrightarrow \qquad (L - \lambda \, \mathrm{Id})w = 0,$$

which is equivalent to a first-order ODE system of the form

$$\frac{\mathrm{d}v}{\mathrm{d}\xi} = A(\xi; \lambda)v \tag{10.6}$$

for some matrix-valued function $A(\xi; \lambda)$ and for some $v \in \mathbb{C}^n$; note that it is useful to directly consider (10.6) over $\mathbb{C}^n$, as $A(\xi; \lambda)$ can be a complex matrix. Hence, we have to study the family of linear operators

$$\mathcal{L}(\lambda)v := \frac{\mathrm{d}v}{\mathrm{d}\xi} - A(\cdot; \lambda)v.$$

We are going to work in spaces with additional uniform continuity requirements and let

$$X = C^1_{\mathrm{unif}}(\mathbb{R}, \mathbb{C}^n), \quad Y = C^0_{\mathrm{unif}}(\mathbb{R}, \mathbb{C}^n), \quad \mathcal{L} = \mathcal{L}(\lambda) : X \to Y$$

so that $\mathcal{L}$ is a closed and densely defined operator. X, Y are Banach spaces under their usual supremum norms. We assume that the matrix-valued function $A \in \mathbb{C}^{n \times n}$ decomposes as

$$A(\xi; \lambda) = \tilde{A}(\xi) + B(\lambda), \tag{10.7}$$

where $\tilde{A}, B$ are smooth functions of their arguments; frequently one just has $B(\lambda) = \lambda B$ for a constant matrix B. The spectral properties of $\mathcal{L}$ can obviously be studied by looking at the ODEs

$$v' = A(\xi; \lambda)v, \tag{10.8}$$

which is a **nonautonomous** linear system, as A depends upon the spatial-dynamics "time" variable ξ. In particular, we have to look for those λ where $\mathcal{L}$ is not invertible.

Definition 10.3. Denote by $\phi = \phi(\xi, \zeta)$ the **fundamental solution** (or **propagator**) for the system (10.8), i.e.,

$$v(\xi) = \phi(\xi, 0)v(0)$$

solves (10.8), $\phi(\xi, \xi) = \mathrm{Id}$ for all $\xi \in \mathbb{R}^n$, and $\phi(\xi, \chi)\phi(\chi, \zeta) = \phi(\xi, \zeta)$ holds for all $\xi, \chi, \zeta \in \mathbb{R}^n$.

Example 10.4. If the linear system is autonomous,

$$v' = A(\lambda)v, \qquad v \in \mathbb{C}^n, \tag{10.9}$$

then we could easily solve (10.9) using the matrix exponential, and solutions can be classified depending on their behavior near the steady-state $v \equiv 0$. For example, if the steady state is hyperbolic for some λ, then there is a splitting

$$\mathbb{C}^n = E^{\mathrm{s}}(0; \lambda) \oplus E^{\mathrm{u}}(0; \lambda) = \mathcal{R}[P_0^s(\lambda)] \oplus \mathcal{N}[P_0^s(\lambda)]$$

where P_0^s is the **spectral projection** onto the stable eigenspace $E^{\mathrm{s}}(0; \lambda)$ for $v \equiv 0$. Note also that the stable and unstable subspaces are invariant under the propagator $\phi(\xi, \zeta) = \mathrm{e}^{(\xi-\zeta)A(\lambda)}$ and solutions decay in forward time in the stable space and in backward time in the unstable space. ♦

The right concept to generalize the splitting from the last example is given in the next definition.

Definition 10.5. Let $I = \mathbb{R}_0^+ = [0, +\infty)$, $\mathbb{R}_0^- = (-\infty, 0]$ or $\mathbb{R}$ and fix $\lambda \in \mathbb{C}$. The ODE (10.8) has an **exponential dichotomy** on I if there exist constants $K > 0$, $\kappa_s < 0 < \kappa_u$ and a continuous family of projectors $P(\xi)$ for $\xi \in I$ such that for $\xi, \zeta \in I$ the following hold:

(D1) $\|\phi(\xi, \zeta)P(\zeta)\| \le K\, e^{\kappa_s(\xi - \zeta)}$ for $\xi \ge \zeta$,

(D2) $\|\phi(\xi, \zeta)[\mathrm{Id} - P(\zeta)]\| \le K\, e^{\kappa_u(\xi - \zeta)}$ for $\xi \le \zeta$,

(D3) projections commute with evolution, i.e., $\phi(\xi, \zeta)P(\zeta) = P(\xi)\phi(\xi, \zeta)$.

The ξ-independent dimension of the nullspace $\mathcal{N}[P(\xi)]$ is also called the **Morse index**. If the exponential dichotomies hold on $\mathbb{R}_0^+$ and $\mathbb{R}_0^-$, the associated Morse indices are denoted by $m_+(\lambda)$ and $m_-(\lambda)$.

Remark: It can be shown that $\mathcal{L}$ is a Fredholm operator if and only if it has exponential dichotomies on $\mathbb{R}_0^+$ and $\mathbb{R}_0^-$. Furthermore, then $m_-(\lambda) - m_+(\lambda)$ is equal to the Fredholm index; see Chapter 4.

Theorem 10.6. *(Palmer's Theorem [444, 445]) The following characterization of spectrum and resolvent set of L holds:*

(T1) $\lambda \notin \sigma(L)$ *if and only if (10.6) has an exponential dichotomy on $\mathbb{R}$;*

(T2) $\lambda \in \sigma_{\mathrm{pt}}(L)$ *if and only if (10.6) has an exponential dichotomy on $\mathbb{R}_0^+$ and $\mathbb{R}_0^-$ with the same Morse index and $\dim(\mathcal{N}[\mathcal{L}(\lambda)]) > 0$;*

(T3) *if (T2) holds, then $\mathcal{N}[P_-(0; \lambda)] \cap \mathcal{R}[P_+(0; \lambda)] \cong \mathcal{N}[\mathcal{L}(\lambda)]$, where $P_\pm(\xi; \lambda)$ are the projections for the exponential dichotomies on $\mathbb{R}^\pm$;*

(T4) $\lambda \in \sigma_{\mathrm{ess}}(L)$ *if we are not in the situation (T1) or (T2).*

Suppose we are interested in applying Palmer's Theorem 10.6 to the case of a travelling front $U(\xi)$ with

$$\lim_{\xi \to \pm\infty} U(\xi) = U_\pm^*,$$

with homogeneous endstates $U_\pm^* \in \mathbb{R}^d$. Consider (10.6) with $A(\xi; \lambda)$ and define

$$A_\pm(\lambda) := \lim_{\xi \to \pm\infty} A(\xi; \lambda). \tag{10.10}$$

It turns out that the asymptotic matrices $A_\pm(\lambda)$ can be used to characterize the exponential dichotomies and hence, by Palmer's Theorem 10.6, also the spectrum.

Theorem 10.7. *(end states and wave stability [128, 486]) Fix $\lambda \in \mathbb{C}$ and consider a travelling front. Then the following results hold:*

(T1) *The ODE (10.6) has an exponential dichotomy on $\mathbb{R}_0^+$ if and only if the matrix A_+ is hyperbolic. In this case $m_+(\lambda) = \dim E_+^u(\lambda)$, where $E_+^u(\lambda)$ is the unstable eigenspace of $A_+(\lambda)$.*

(T2) *The statement (T1) holds with $\mathbb{R}_0^+$ replaced by $\mathbb{R}_0^-$ and $A_+(\lambda)$ replaced by $A_-(\lambda)$.*

(T3) *The ODE (10.6) has an exponential dichotomy on $\mathbb{R}$ if and only if it has exponential dichotomies on $\mathbb{R}_0^+$ and $\mathbb{R}_0^-$ with projections $P_\pm(\xi; \lambda)$ such that $\mathcal{N}[P_-(0; \lambda)] \oplus \mathcal{R}[P_+(0; \lambda)] = \mathbb{C}^n$.*

Example 10.8. Here we use the previous results to gain more insight on stability for reaction-diffusion systems. Consider the bistable Nagumo equation

$$\partial_t u = \partial_x^2 u + u(1-u)(u-p), \qquad p \in (0, 1/2), \; x \in \mathbb{R}, \; u = u(x,t), \qquad (10.11)$$

with a right-moving travelling front solution $U(\xi)$, $\xi = x - st$ with $s > 0$, as discussed in Chapter 7 with end states $U_-^* \equiv 0$ and $U_+^* \equiv 1$. A direct calculation using the formulas (10.4)–(10.5) yields

$$A(\xi; \lambda) = \begin{pmatrix} 0 & 1 \\ \lambda - f'(U(\xi)) & -s \end{pmatrix} = \begin{pmatrix} 0 & 1 \\ \lambda + 3U(\xi)^2 - U(\xi)(2 + 2p) + p & -s \end{pmatrix}$$

so that the asymptotic matrices are

$$A_-(\lambda) = \begin{pmatrix} 0 & 1 \\ \lambda + p & -s \end{pmatrix}, \quad A_+(\lambda) = \begin{pmatrix} 0 & 1 \\ \lambda + 1 - p & -s \end{pmatrix}.$$

Suppose we are interested in determining the essential spectrum; i.e., we look at point spectrum later. By Palmer's Theorem 10.6 and Theorem 10.7, λ is in the essential spectrum if either (I) at least one of the matrices $A_\pm(\lambda)$ is not hyperbolic, or (II) both are hyperbolic but there are exponential dichotomies on $\mathbb{R}_0^+$ and $\mathbb{R}_0^-$ with differing Morse indices $m_+(\lambda) \neq m_-(\lambda)$. To check these conditions, it is helpful to compute trace and determinant:

$$\det(A_\pm(\lambda)) = -\lambda \pm p - U_\pm^*, \qquad \operatorname{tr}(A_\pm(\lambda)) = -s < 0.$$

The trace is real and negative. Since the trace is the sum of the eigenvalues $\mu_1^\pm, \mu_2^\pm$ of $A_\pm(\lambda)$ it follows that if complex conjugate eigenvalues $\mu_1 = \overline{\mu_2}$ occurred then those have negative real part, so those eigenvalues must be contained in the left-half plane. Hence, we focus on real eigenvalues $\mu_1^\pm, \mu_2^\pm \in \mathbb{R}$, which implies $\lambda \in \mathbb{R}$ as $\det(A_\pm(\lambda)) = \mu_1^\pm \mu_2^\pm$. By using the trace-determinant analysis from Exercise 2.13 it follows that hyperbolicity with real eigenvalues fails only when the determinant vanishes, which happens when

$$-\lambda \pm p - U_\pm^* = 0 \quad \Leftrightarrow \quad -\lambda - p = 0 \text{ or } -\lambda + p - 1 = 0,$$

and since $p \in (0, 1/2)$ by assumption, this can only happen when $\lambda < 0$. Therefore, we have checked that (I) can occur only for $\lambda < 0$. Next, pick $\lambda \geq 0$ and consider the case when $A_\pm(\lambda)$ are hyperbolic, but then one easily computes that the Morse indices coincide checking (II). This implies that for fixed p we have

$$\sigma_{\mathrm{ess}}(L) \subset \{\lambda \in \mathbb{C} : \operatorname{Re}(\lambda) \leq \lambda_b < 0\}$$

for some fixed negative λ_b. Therefore, linear instability can only arise via the point spectrum. However, the point spectrum has already been analyzed in Theorem 9.7, which implies for the Nagumo equation that the front is spectrally stable and the pulse is unstable; in fact, similar arguments hold for many classes of one-component PDEs with bistable nonlinearity. ♦

The last example shows that it is often quite easy to show that the essential spectrum is contained in the left-half plane. However, we still lack a general tool for the point spectrum. Palmer's Theorem 10.6(T3) implies that one possibility is to look at the intersection

$$\mathcal{N}[P_-(0; \lambda)] \cap \mathcal{R}[P_+(0; \lambda)] \qquad (10.12)$$

and check when it is nonempty to characterize the point spectrum. This means looking at bounded solutions as those lying in the intersection (10.12). Indeed, $\mathcal{R}[P_+(0; \lambda)]$ consists of all initial conditions $v(0)$ with solutions $v(\xi)$ for (10.8), which decay exponentially as $\xi \to +\infty$, while $\mathcal{N}[P_-(0; \lambda)]$ consists of all initial conditions $v(0)$ with solutions $v(\xi)$, which decay exponentially as $\xi \to -\infty$; see Figure 10.1 and also compare this to the weighted space in the Sturm-Liouville problems in Chapter 9.

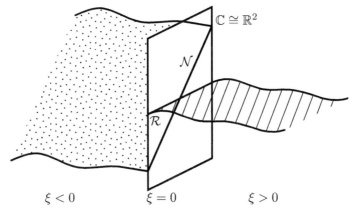

$\xi < 0 \qquad\qquad \xi = 0 \qquad\qquad \xi > 0$

Figure 10.1. *Sketch of one possible intersection scenario for* (10.12). *This also illustrates the geometric intuition behind Palmer's Theorem 10.6(T3).*

Suppose the essential spectrum is in the left-half plane and let Ω denote the connected component of $\mathbb{C} \setminus \sigma_{\mathrm{ess}}$ intersecting the right-half plane. One very nice device to study the intersection (10.12) for $\lambda \in \Omega$ is to consider bases

$$
\begin{aligned}
v_1(\lambda), \dots, v_k(\lambda) \qquad & \text{of } \mathcal{N}[P_-(0; \lambda)], \\
v_{k+1}(\lambda), \dots, v_n(\lambda) \qquad & \text{of } \mathcal{R}[P_+(0; \lambda)],
\end{aligned}
$$

where $k = \dim(\mathcal{N}[P_-(0; \lambda)])$ is the Morse index; it can be shown that the Morse index does not change in Ω and the bases can be chosen to depend analytically [319] upon λ.

Definition 10.9. The **Evans function** is defined as

$$
E(\lambda) := \det[v_1(\lambda), \dots, v_n(\lambda)]. \tag{10.13}
$$

Theorem 10.10. *(Evans function properties) The Evans function $E(\lambda)$ is analytic for $\lambda \in \Omega$ and $E(\lambda) = 0$ if and only if λ is an eigenvalue of L.*

Proof. If λ is an eigenvalue, then (10.12) is nonempty. Therefore, the bases are linearly dependent and the determinant (10.13) vanishes. The converse is equally simple. Analyticity of the Evans function follows from the analyticity of the bases. $\qquad\square$

In general, it is difficult to compute the Evans function explicitly, except for certain special cases and integrable/conservative systems. However, it provides a useful abstract theoretical as well as practical numerical tool.

Example 10.11. To see the difficulty in the computation of the Evans function for reaction-diffusion systems consider the case from Example 9.1 given by

$$
\partial_t u = \partial_x^2 u - u + 2u^3 \tag{10.14}
$$

with standing pulse solution $U(\xi) = \text{sech}(\xi)$. Then we must eventually understand the exponential dichotomy properties of the linear system

$$v' = A(\xi; \lambda)v = \begin{pmatrix} 0 & 1 \\ 1 + \lambda - 6U(\xi)^2 & 0 \end{pmatrix},$$

which can be turned into an autonomous system via setting $\frac{d\xi}{d\tau} = 1$ using the new (dummy) time variable; note $\xi(\tau) = \tau$ if we assume $\xi(0) = 0$. However, in the autonomous case we still have the nonlinear term $U(\xi) = \text{sech}(\xi)$ in the problem. Hence, we are still left with understanding a nonlinear ODE, which is a highly nontrivial task. ◆

Example 10.12. The previous example can be generalized. For a d-component reaction-diffusion system (10.2) with a nonlinearity $f : \mathbb{R}^d \to \mathbb{R}^d$, we are facing the problem

$$\begin{pmatrix} \dot{u} \\ \dot{\tilde{u}} \end{pmatrix} = \begin{pmatrix} \tilde{u} \\ D^{-1}(-s\tilde{u} - f(u)) \end{pmatrix}, \tag{10.15}$$

$$\begin{pmatrix} \dot{w} \\ \dot{\tilde{w}} \end{pmatrix} = \begin{pmatrix} 0 & \text{Id} \\ D^{-1}[\lambda - (\mathbf{D}_u f)(u)] & -sD^{-1} \end{pmatrix} \begin{pmatrix} w \\ \tilde{w} \end{pmatrix}, \tag{10.16}$$

$$\dot{\xi} = 1, \tag{10.17}$$

which is an autonomous, fully coupled, $(4d + 1)$-dimensional, nonlinear system of ODEs. The $2d$ equations (10.15) come from the existence problem in the travelling wave frame, the $2d$ equations (10.16) come from the eigenvalue problem to determine stability, and $\dot{\xi} = \frac{d\xi}{d\tau}$ makes the system autonomous but represents another difficulty. Indeed, the ξ equation shows directly that there are no steady states, so we are always dealing, in some sense, with trying to understand transients. ◆

However, for some systems with a special structure, one may actually find the Evans function explicitly. Here we just quote a result to illustrate this fact.

Example 10.13. Consider the following coupled hierarchy of **nonlinear Schrödinger equations (NLS)** given by

$$\begin{aligned} \partial_t u &= \mathrm{i} \left(\partial_x^2 u - 2u - 2u^2 v \right), \\ \partial_t v &= \mathrm{i} \left(\partial_x^2 v - 2v - 2v^2 u \right), \end{aligned} \tag{10.18}$$

where $u = u(x, t), v = v(x, t) \in \mathbb{C}$. Then it turns out that a highly nontrivial explicit calculation yields for a so-called stationary soliton solution the Evans function

$$E(\lambda) = 8a(\lambda)^2 b(\lambda)^2 \sqrt{\lambda - \mathrm{i}} \sqrt{\lambda + \mathrm{i}},$$

where there are explicit formulas for a, b given by

$$a(\lambda) = \frac{\mathrm{e}^{\mathrm{i}\pi/4}\sqrt{\lambda - \mathrm{i}} - 1}{\mathrm{e}^{\mathrm{i}\pi/4}\sqrt{\lambda - \mathrm{i}} + 1}, \quad b(\lambda) = \frac{\mathrm{e}^{-\mathrm{i}\pi/4}\sqrt{\lambda + \mathrm{i}} - 1}{\mathrm{e}^{-\mathrm{i}\pi/4}\sqrt{\lambda + \mathrm{i}} + 1}.$$

Now one has an explicit formula, but then a new problem arises as $E(\pm\mathrm{i}) = 0$, so one has to deal with spectrum on the imaginary axis, which, a priori, indicates neither stability nor instability; frequently symmetries play a key role in understanding this spectrum, as discussed in Chapter 32. ◆

A fundamental step has been discussed here, as we have converted a full PDE problem for waves in one-dimensional spatial domains into systems of ODEs, which elegantly links dynamics of ODEs to existence and stability of travelling wave patterns of spatially extended systems posed on the spatial domain $\mathbb{R}$.

Exercise 10.14. Consider the reaction-diffusion system (10.2) and suppose it has a travelling pulse or a travelling front solution $U(\xi)$. Show that $\lambda = 0$ is always in the spectrum with associated eigenfunction $U'(\xi)$. $\Diamond$

Exercise 10.15. Consider the PDE (9.1) from Example 9.1. Calculate the Evans function for the standing pulse solution explicitly. Hint: Use hypergeometric series. $\Diamond$

Exercise 10.16. Prove that for a homogeneous rest state $U(\xi) \equiv U^* \in \mathbb{R}^d$ of (10.1) the point spectrum is empty. This illustrates a fundamental difference between spectra for problems on bounded domains and those for problems on unbounded domains. $\Diamond$

Background and Further Reading: The exposition here summarizes several key aspects from the review [486] with the last example from [317]. The idea of looking at exponential dichotomies [443, 444, 445] has found many applications in differential equations and dynamical systems; here we can just mention a few [114, 299, 337, 483, 410]. A key paper developing, and naming, the Evans function is [7]. The original work of Evans [184] was actually focused on trying to understand the stability of waves in FitzHugh-Nagumo type reaction-diffusion systems and associated nerve impulse equations. We refer to [487] for the relation between spectra on bounded and unbounded domains for waves. Furthermore, the Evans function approach has been used successfully in the context of solitary waves [451, 452] as well as shock waves [224, 578]. The setup for shock waves in hyperbolic conservation laws is the topic of Chapter 11.

Chapter 11

Characteristics and Shocks

Another key class of PDEs with a direct relation to dynamical systems are **first-order PDEs**, e.g.,

$$F(\nabla u, u, y) = 0, \qquad u = u(y) \in \mathbb{R}, \ y \in \Omega \subset \mathbb{R}^N, \tag{11.1}$$

where F is assumed to be smooth and $u = g : \partial\Omega \to \mathbb{R}$ on $\partial\Omega$.

Example 11.1. A typical example is **Burgers' equation**

$$u_t + u\partial_x u = 0 \qquad \text{or} \qquad u_t + \frac{1}{2}\partial_x(u^2) = 0 \tag{11.2}$$

where $\Omega = \mathbb{R} \times [0, +\infty)$, $y = (x, t)$, $\partial\Omega = \mathbb{R} \times \{0\}$. ♦

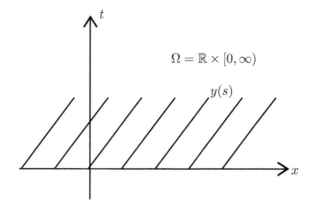

Figure 11.1. *Basic sketch of the idea of the method of characteristics. The space-time domain is dissected by curves $y = y(s)$, which propagate the solution from $t = 0$ into the domain via the characteristic ODEs as stated in Theorem 11.2 for a very general setting.*

It is possible to study (11.1) via the **method of characteristics**. Let $s \in \mathbb{R}$, and consider dissecting Ω via curves

$$y = y(s), \quad z(s) := u(y(s)), \qquad p(s) = \nabla u(y(s)),$$

and denote $\mathrm{d}/\mathrm{d}s = {}'$; see Figure 11.1. One also refers to $y(s)$ as a **characteristic curve**. The following result is still remarkable despite being very classical.

Theorem 11.2. *(method of characteristics) Let $u \in C^2(\Omega)$ solve (11.1). Assume y solves the ODE*

$$y' = \nabla_p F(p, z, y); \tag{11.3}$$

then, for each s such that $y(s) \in \Omega$, we have that p, z solve

$$p' = -\nabla_y F(p, z, y) - \partial_z F(p, z, y)p, \tag{11.4}$$

$$z' = \nabla_p F(p, z, y) \cdot p. \tag{11.5}$$

Definition 11.3. The ODEs (11.3)–(11.5) are called **characteristic ODEs**. A curve $y(s) \in \Omega$ is called the **projected characteristic**. Obviously these ODEs provide a direct link between dynamical systems and first-order PDEs.

Proof. (of Theorem 11.2) Differentiating p with respect to s yields

$$p_i' = \sum_{j=1}^{N} y' \partial_{y_i y_j} u(y) \qquad \text{for } i \in \{1, 2, \dots, N\}. \tag{11.6}$$

Differentiating the first-order PDE (11.1) with respect to y_i gives

$$\sum_{j=1}^{N} \partial_{p_j} F(\nabla u, u, y) \, \partial_{y_i y_j} u + \partial_z F(\nabla u, u, y) \, \partial_{y_i} u + \partial_{y_i} F(\nabla u, u, y) = 0. \tag{11.7}$$

Since (11.3) holds by assumption, we may evaluate (11.7) at $y = y(s)$ and insert the resulting expression and (11.3) into (11.6) to obtain

$$p_i' = -\partial_{y_i} F(p, z, y) - \partial_z F(p, z, y)p_i,$$

which eliminates the second-order derivatives in (11.6) and already yields (11.4). To obtain (11.5), we simply differentiate the definition of z. $\qquad\square$

Example 11.4. (Example 11.1 continued) Consider (11.2) with given initial condition

$$u_0(x) = \begin{cases} 1 & \text{if } x \leq 0, \\ 1 - x & \text{if } x \in (0, 1), \\ 0 & \text{if } x \geq 1. \end{cases} \tag{11.8}$$

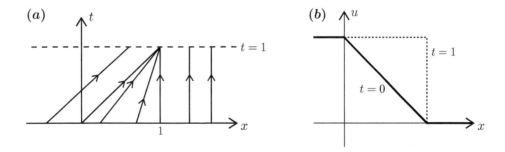

Figure 11.2. *(a) Method of characteristics applied to Burgers' equation (11.2) with initial condition (11.8). Observe that the characteristics cross at time $t = 1$ leading to shock formation. (b) Illustration of the difference of the solution at $t = 0$ and $t = 1$ showing a sharpening of the profile to a vertical front at the shock time.*

The method of characteristics yields $z' = 0$ and $y' = (1, z)^\top$. Therefore, the solution is constant along the projected characteristic

$$y(s) = (u_0(x_0)s + x_0, s) \qquad \text{for each } x_0 \in \mathbb{R},$$

as shown in Figure 11.2. Observe that the solution

$$u(x, t) = \begin{cases} 1 & \text{if } x \le t, t \in [0, 1], \\ 1 - x & \text{if } t \le x \le 1, t \in [0, 1], \\ 0 & \text{if } x \ge 1, t \in [0, 1] \end{cases}$$

is only easy to define for $t \le 1$. We shall develop a way to deal with the situation also for $t > 1$ in what follows. ◆

The last example shows a general problem for scalar **conservation laws**

$$\partial_t u + \partial_x f(u) = 0, \qquad x \in \mathbb{R}, \ t \in (0, \infty) \tag{11.9}$$

with certain initial data $u(x, 0) = u_0(x)$, i.e., characteristics may cross, or characteristics may not entirely fill the domain as discussed in Exercise 11.11. The natural way out seems to be to consider weak solutions $u \in L^\infty(\mathbb{R} \times (0, \infty))$, which are required to satisfy

$$\int_0^\infty \int_{\mathbb{R}} u \partial_t \phi + f(u) \partial_x \phi \ dx \ dt + \int_{\mathbb{R}} u_0(x) \phi(x, 0) \ dx = 0$$

for all test functions $\phi \in C_c^\infty$. However, one may check that there are infinitely many weak solutions in certain cases, and one has to impose additional conditions to single out a solution; see also Figure 11.2.

Definition 11.5. Fix an open region $\mathcal{V}$ in $\mathbb{R} \times (0, \infty)$, and consider a smooth curve $\Gamma = \{x = s(t)\} \subset \mathcal{V}$ dissecting $\mathcal{V}$ into two domains $\mathcal{V}_l$ and $\mathcal{V}_r$. Consider a solution u to (11.9) such that u_l and u_r are the two limits in two domains approaching a point $x \in \Gamma$. Then u satisfies the **Rankine-Hugoniot condition** if

$$(u_l - u_r)\frac{ds}{dt} = f(u_l) - f(u_r) \tag{11.10}$$

for each point $x \in \Gamma$.

Example 11.6. (Example (11.4) continued) Let us set $S := \{s(t) = \frac{1}{2}(1 + t)\}$ for $t \ge 1$, and

$$u(x, t) = \begin{cases} 0 & \text{if } s(t) < x, \\ 1 & \text{if } s(t) > x \end{cases} \tag{11.11}$$

is easily checked to define a weak solution. Furthermore, this weak solution satisfies the Rankine-Hugoniot condition since $u_l = 1$, $u_r = 0$, $f(u_l) = \frac{1}{2}$, $f(u_r) = 0$ so that we have

$$(u_l - u_r)\frac{ds}{dt} = (1 - 0)\frac{1}{2} = \frac{1}{2} - 0 = f(u_l) - f(u_r).$$

The solution $u(x, t)$ given by (11.11) is an example of a **shock** with a curve S of discontinuities. ◆

Unfortunately, even the Rankine-Hugoniot condition is sometimes not enough to discriminate between weak solutions, as shown in Exercise 11.11.

Definition 11.7. A solution satisfies the **Lax entropy condition** along a curve of discontinuities $S = \{x = s(t)\}$ if

$$f'(u_l) > \frac{ds}{dt} > f'(u_r).$$

If in addition the Rankine-Hugoniot condition holds, then the curve S is called a **shock (curve)**. Furthermore, we say that u satisfies the **Oleinik entropy condition** if

$$\frac{f(w) - f(u_r)}{w - u_r} \geq \frac{f(u_l) - f(u_r)}{u_l - u_r}$$

for all w between u_l and u_r.

The Rankine-Hugoniot and entropy conditions are the classical tools to limit the solution space. One may prove [507, 367] that Oleinik entropy, Rankine-Hugoniot, and *convexity of f* imply the existence of a unique weak solution, which can be identified as the limit of the **viscous regularization**

$$\partial_t u + \partial_x f(u) = \varepsilon \partial_x^2 u \tag{11.12}$$

as $\varepsilon \to 0$. This naturally links shocks to travelling wave problems as discussed in Chapters 7–10. However, convexity of f is a very strong assumption and it is not immediately evident why one should select the regularization (11.12). Here we shall briefly consider the **diffusive-dispersive regularization**

$$\partial_t u + \partial_x f(u) = \varepsilon_1 \partial_x^2 u + \varepsilon_2 \partial_x^3 u, \tag{11.13}$$

which for the choice $f(u) = u^3$ is also known as the **modified Korteweg-deVries-Burgers' (mKdVB) equation**. The third-order derivative represents the dispersive part, and it is balanced with the diffusive part if we select

$$\varepsilon_1 = \varepsilon, \qquad \varepsilon_2 = p^{-1}\varepsilon^2,$$

where $p > 0$ is a parameter with $p = \mathcal{O}(1)$ as $\varepsilon \to 0$.

Definition 11.8. The conservation law (11.9) together with the initial condition

$$u_0(x) = \begin{cases} u_l & \text{if } x < 0, \\ u_r & \text{if } x \geq 0 \end{cases}$$

is referred to as a **Riemann problem**.

Theorem 11.9. *(mKdVB front existence) There exists a travelling front solution to* (11.13) *which satisfies the Rankine-Hugoniot condition but not the Lax (and hence also not the Oleinik) entropy criterion as $\varepsilon \to 0$ for a Riemann problem.*

Proof. (Sketch [262]) We consider a scaled travelling wave frame ansatz

$$u(x,t) = u(\xi), \qquad \xi := \frac{x - st}{\varepsilon},$$

where s is already fixed since we want to satisfy the Rankine-Hugoniot condition. As boundary data, suppose we have indeed a jump discontinuity already with u_l and u_r as discussed above. Therefore, the travelling front should satisfy

$$\lim_{\xi \to -\infty} u(\xi) = u_l, \qquad \lim_{\xi \to +\infty} u(\xi) = u_r, \qquad \lim_{\xi \to \pm\infty} u'(\xi) = 0 = \lim_{\xi \to \pm\infty} u''(\xi),$$

where $' = d/d\xi$. The mKdVB equation in the travelling wave frame is

$$-su' + (u^3)' = pu'' + u'''.$$

We may integrate once and obtain

$$-su + (u^3) + C = pu' + u'' \tag{11.14}$$

for some constant $C \in \mathbb{R}$. Rewriting (11.14) as a first-order system and analyzing via phase-plane methods, it is very similar to the classical Nagumo case studied in Example 7.4. One finds three equilibria and under certain conditions two of them are saddles connected by a heteroclinic orbit. In this situation one finds, after quite a few calculations, the shock/wave speed as

$$s = u_l^2 - \frac{\sqrt{2}}{3} p u_l + \frac{2}{9} p^2.$$

In addition, one needs the conditions

$$u_l > \frac{2p\sqrt{2}}{3}, \qquad u_r = -u_l + \frac{\sqrt{2}}{3} p \tag{11.15}$$

to have a heteroclinic saddle-to-saddle connection. In particular, once p is fixed, then u_l determines u_r. We want to show that this wave/shock is *nonclassical*. First, note that the limit characteristic speeds (for $\varepsilon = 0$) on each side of the wave can be found using the method of characteristics since $y' = (1, 3z^2)^\top$, so that the speeds are $3u_l^2$ and $3u_r^2$. Next, we pick some $u_l > 0$; then we also have

$$s = u_l^2 - \frac{\sqrt{2}}{3} p u_l + \frac{2}{9} p^2 \overset{(11.15)}{\leq} -\frac{2}{9} p^2 + u_l^2 \leq u_l^2 < 3u_l^2.$$

Similarly we can find $s < 3u_r^2$ using the second relation in (11.15). This implies that characteristics outside the shock in the limit $\varepsilon \to 0$ are faster than the separating curve (also referred to as an **undercompressive shock**). In this case, it is easy to see that the Lax and Oleinik entropy conditions cannot hold. □

We have now finished introducing some basics of waves in various contexts, the latest being shocks in hyperbolic conservation laws and their regularizations. Of course, one may wonder which other patterns are possible. Chapters 12–13 provide one key tool to analyze pattern formation.

Exercise 11.10. Consider the **Hamilton-Jacobi equations**

$$\partial_t u + H(\nabla u, x) = 0, \qquad H : \mathbb{R}^{2N} \to \mathbb{R}. \tag{11.16}$$

Use Theorem 11.2 to derive the characteristic ODEs for (11.16). Can you conclude that the function H is a **Hamiltonian** (cf. Chapter 27 and/or Example 7.5)? ◇

Exercise 11.11. Consider Burgers' equation (11.2) with initial data

$$u_0(x) = \begin{cases} 0 & \text{if } x \leq 0, \\ 1 & \text{if } x > 0. \end{cases}$$

(a) Use the method of characteristics and check that it fails to define the solution in the cone $\{(x, t) \in \mathbb{R} \times (0, \infty) : 0 < x < t\}$. Show that

$$u(x, t) = \begin{cases} 1 & \text{if } x > t, \\ \frac{x}{t} & \text{if } 0 < x < t, \\ 0 & \text{if } x < 0 \end{cases} \tag{11.17}$$

is a weak solution which satisfies the Oleinik entropy condition; it is called a **rarefaction (wave)**. (b) Find a weak solution different from (11.17) on an unbounded open set that does not satisfy the Oleinik entropy condition. ◊

Exercise 11.12. Consider the following system of conservation laws:

$$\partial_t u + \partial_x(v^{-2}) = 0,$$
$$\partial_t v - \partial_x u = 0, \tag{11.18}$$

which is a simplified model for isotropic gas dynamics. (a) Show that if $v > 0$, then (11.18) is a **hyperbolic system of conservation laws**, i.e., it is of the form

$$\partial_t U + A(U)\partial_x U = 0$$

for a matrix $A = A(U)$ having no eigenvalues with zero real part. (b) Diagonalize $A(U)$ by a coordinate change and sketch the two fields of characteristic curves for (11.18). ◊

Background and Further Reading: The first part of this chapter mainly follows [186]. The diffusive-dispersive regularization approach is taken from [262]. For a broad general background for conservation laws we refer to [138, 361, 367, 507], for more on multidimensional/system cases consider [51, 68, 238, 271, 391], and for some classic works see [130, 274, 358, 359]. Of course, one may also consider stability questions for shocks; see, e.g., [210, 382, 383, 577] and cf. Chapters 9–10. Nonclassical shocks can be obtained actually in quite a wide variety of systems [366]. Recall that the KdV equation already appeared in Chapter 7 and the mKdVB equation is just one modification in a long list of variants that have been considered; here we can just mention a few more classical KdV references [362, 413, 414, 569], but there are many other works for all KdV variants, mostly focusing on finding classes of special solutions.

Chapter 12

Onset of Patterns and Multiple Scales

In this chapter, and in Chapter 13, we are going to discuss how to approximate many time-dependent PDEs near an instability using just a few universal PDEs. These simple PDEs are called **amplitude equations** or **modulation equations**. They can be viewed from a dynamical systems standpoint also as normal forms; see the remarks following Example 2.5. We study a concrete model problem to motivate the development of amplitude equations on a formal level. Consider the **Swift-Hohenberg equation**

$$\partial_t u = [p - (\Delta + 1)^2]u - u^3, \quad u = u(x, y, t) \in \mathbb{R}, \quad (x, y) \in \mathbb{R}^2, \qquad (12.1)$$

where $p \in \mathbb{R}$ is a parameter and Δ is the Laplacian; note that we deviate from our convention to use x as the spatial variable, as using (x, y) as coordinates is going to simplify the notation later on. The Swift-Hohenberg equation is an idealized model of convective instabilities in fluid dynamics. Note carefully that the domain is $\Omega = \mathbb{R}^2$. Therefore, we cannot expect to capture the dynamical effects for bifurcations of steady states using the point spectrum as we did in Chapters 5–6, and the essential spectrum plays a crucial role; cf. Exercise 10.16. Suppose we are interested in the stability of $u \equiv 0$ and linearize (12.1) around $u \equiv 0$; then we obtain

$$\partial_t w = pw - (\Delta + 1)^2 w, \qquad w = w(x, y, t). \qquad (12.2)$$

Substituting the dynamics of an individual Fourier mode,

$$w_k(x, y, t) = e^{\sigma t + ik \cdot (x,y)^\top} + \text{c.c.}, \qquad k = (k_x, k_y)^\top \in \mathbb{R}^2,$$

where c.c. denotes complex conjugate and k is a **wave vector**, into (12.2) yields the following dispersion relation (check it!):

$$\sigma = p - (k^2 - 1)^2, \qquad k^2 := \|k\|^2. \qquad (12.3)$$

If $p < 0$, then $\sigma < 0$, so all modes decay. When $p = 0$, then the critical modes, which are no longer linearly stable, occur for wave numbers with $\|k\| = 1$. When $p > 0$, a whole band of wave numbers is linearly unstable and could occur in a **pattern** bifurcating from the homogeneous solution; see Figure 12.1.

Let us assume we only look for a simple "stripe" pattern at the bifurcation $p = 0$ of the form

$$u(x, y) = A e^{ix} + \overline{A} e^{-ix} = 2 \operatorname{Re}(A) \cos(x) - 2 \operatorname{Im}(A) \sin(x) \qquad (12.4)$$

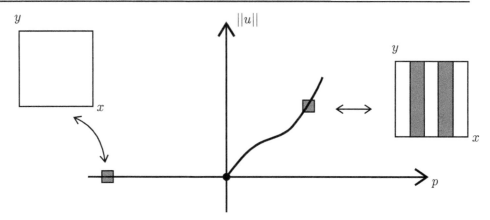

Figure 12.1. *Sketch of one possible basic bifurcation structure we are interested in for the Swift-Hohenberg equation (12.1). For $p < 0$ we have a locally stable homogeneous branch (the zero solution on this branch is marked by a small grey square; the zoom shows the level set of u), while for $p > 0$ one option would be to consider the bifurcation to stripes (the nonzero solution on the new branch is marked by small grey square; the zoom shows the level set of u).*

with **amplitude** A; see Figure 12.2(d). A formal perturbation ansatz based at the critical wave number for the pattern (12.4) is given by

$$k = (1 + \hat{k}_x) \begin{pmatrix} 1 \\ 0 \end{pmatrix} + \hat{k}_y \begin{pmatrix} 0 \\ 1 \end{pmatrix}.$$

If we also let $p = \varepsilon^2 \hat{p}$, and substitute the wave vector perturbation into the dispersion relation, we obtain

$$\sigma = \varepsilon^2 \hat{p} - (2\hat{k}_x + (\hat{k}_x)^2 + (\hat{k}_y)^2)^2. \tag{12.5}$$

Observe carefully that both directions x and y contribute to the same order in (12.5) if we set $\hat{k}_x \sim \varepsilon$, $\hat{k}_y \sim \sqrt{\varepsilon}$, and $\sigma \sim \varepsilon^2$, where $h_1(\varepsilon) \sim h_2(\varepsilon)$ means that $\lim_{\varepsilon \to 0} h_1(\varepsilon)/h_2(\varepsilon) = C \neq 0$ for a constant C and functions h_1, h_2. Our calculation provides a formal way to look at the size of the unstable band of wavenumbers; see also Figure 12.2(b)–(c).

The main idea is that very close to the bifurcation point, near the onset of the pattern, the amplitude A in (12.4) is not stationary but will be slowly modulated. The assumption of slow modulation of the amplitude also leads to the names **amplitude equation** and **modulation equation**, which are the PDEs we are going to derive below. In this chapter, the derivation will be formal, while a rigorous justification will be provided in Chapter 13.

We have several scales involved in the problem: the quickly varying **carrier wave** e^{ix} and the slowly varying amplitude (or **envelope**) A; see Figure 12.2(d). Motivated by the dispersion relation, we consider the slow/small variables

$$T := \varepsilon^2 t, \quad X := \varepsilon x, \quad Y := \sqrt{\varepsilon} y$$

and the fast/large variables $\tilde{x} := x, \tilde{t} := t$ for the ansatz

$$u(x, y, t) = u(\tilde{x}, \tilde{y}, \tilde{t}; X, Y, T) = A(X, Y, T)e^{i\tilde{x}} + \text{c.c.}, \tag{12.6}$$

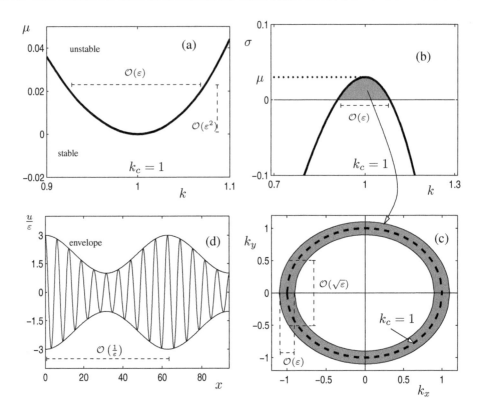

Figure 12.2. *Illustration for the Swift-Hohenberg equation* (12.1). *(a) Curve of neutral stability* $\mu = (k^2 - 1)^2$ *delimiting the regions of stable and unstable modes. (b) Spectrum* σ *for* $\mu = 0.03$ *fixed. The shaded areas are unstable modes that are presented in (c) in the space of wave vectors to illustrate the relevant scalings. (d) Sketch of a potential pattern* (12.4). *Figure adapted from [346] and reprinted with permission from Springer Nature.*

which is also called a **multiple scales** or **two-scale** ansatz. Furthermore, we consider a regime near onset with $p = \varepsilon^2 \hat{p}$ and note that the chain rule formally prescribes the relations

$$\partial_x = \partial_{\tilde{x}} + \varepsilon \partial_X, \quad \partial_y = \partial_{\tilde{y}} + \sqrt{\varepsilon}\partial_Y, \quad \partial_t = \partial_{\tilde{t}} + \varepsilon^2 \partial_T. \tag{12.7}$$

Substituting everything into the Swift-Hohenberg equation (12.1) and collecting terms leads to

$$\varepsilon^2 \partial_T u = \varepsilon^2 \hat{p} u - \big[(1 + \partial_{\tilde{x}}^2)^2 + 2\varepsilon(\partial_{\tilde{x}}^2 + 1)(2\partial_{\tilde{x}X} + \partial_Y^2) \tag{12.8}$$
$$+ 2\varepsilon^2 \partial_X^2 (\partial_{\tilde{x}}^2 + 1) + \varepsilon^2 (2\partial_{\tilde{x}X} + \partial_Y^2)^2 + \cdots\big] u - u^3,$$

where u does not depend upon $\tilde{y}$ and $\tilde{t}$. It is now convenient to drop the tildes and revert to the notation x for $\tilde{x}$. A classical ansatz to study equations such as (12.8) is to use an **asymptotic expansion**

$$u = u_0 + \varepsilon u_1 + \varepsilon^2 u_2 + \varepsilon^3 u_3 + \cdots, \tag{12.9}$$

where each $u_j = u_j(x, X, Y, T)$ does not depend upon ε and $u_0 \equiv 0$ as we **perturb** near the trivial solution.

Remark: One refers to $\{\varepsilon^j\}_{j=0}^\infty$ as an asymptotic sequence, i.e., $\lim_{\varepsilon \to 0} \varepsilon^{j+1}/\varepsilon^j = 0$ for all $j \in \mathbb{N}_0$. In principle, other asymptotic sequences depending upon ε could work as well. Usually this involves some trial and error for each problem at hand; for a lot more on multiscale and perturbation problems in PDE dynamics see Chapters 33–36.

Inserting (12.9) into (12.8), we can collect terms at different orders of ε and obtain for the first two orders

$$\text{at } \mathcal{O}(\varepsilon): \qquad 0 = (\partial_x^2 + 1)^2 u_1 =: L(u_1),$$
$$\text{at } \mathcal{O}(\varepsilon^2): \quad L(u_2) = -2(\partial_x^2 + 1)(2\partial_{xX} + \partial_Y^2)u_1,$$

where standard **(Landau) order notation** is used, i.e., $h_1(\varepsilon) = \mathcal{O}(h_2(\varepsilon))$ indicates $\limsup_{\varepsilon \to 0} |h_1(\varepsilon)/h_2(\varepsilon)| < \infty$. We now apply a key procedure from asymptotics and perturbation methods. To find $u_1 = u_1(x, X, Y, T)$ we solve the equations order-by-order so we start with

$$0 = (\partial_x^2 + 1)^2 u_1. \tag{12.10}$$

Using the Fourier transform in the x component with Fourier variable $\xi \in \mathbb{R}$, we find

$$0 = [(\mathrm{i}\xi)^4 + 2(\mathrm{i}\xi)^2 + 1]\hat{u}_1(\xi, X, Y, T) = (\xi^2 - 1)^2 \hat{u}_1(\xi, X, Y, T). \tag{12.11}$$

One possibility is that $\xi = \pm 1$ so in the x component u_1 is just a linear combination of the Fourier modes $\mathrm{e}^{\pm \mathrm{i}x}$, and so we let

$$u_1(x, X, Y, T) = \tilde{A}(X, Y, T)\mathrm{e}^{\mathrm{i}x} + \text{c.c.}, \tag{12.12}$$

where the amplitude $\tilde{A}(X, Y, T)$ naturally incorporates the trivial solution $u_1 \equiv 0$ if $\tilde{A} \equiv 0$. Note that $\tilde{A} = A/\varepsilon$ in comparison to the amplitude A in the ansatz (12.4); again, we shall drop the tilde from now on, so if we derive an equation for A, it is understood that variations of A on the scale $\mathcal{O}(1)$ become small modulations on the original scale.

For the order $\mathcal{O}(\varepsilon^2)$, we observe that it does not provide additional information on u_1 as we find the equation $L(u_2) = 0$; we may absorb u_2 into u_1 via $\tilde{u}_1 = u_1 + \varepsilon u_2$ and drop the tilde. We now assume that the amplitude from solving $L(u_2) = 0$ is zero, i.e., $u_2 \equiv 0$. The interesting part occurs at order $\mathcal{O}(\varepsilon^3)$, which is given by

$$L(u_3) = -\partial_T u_1 + [\hat{p} - 2\partial_X^2(\partial_x^2 + 1) - (2\partial_{xX} + \partial_Y^2)^2]u_1 - u_1^3. \tag{12.13}$$

If we assume that u_3 is bounded as $x \to \pm\infty$ and the expansion is uniformly valid in space-time, then the coefficients in front of $\mathrm{e}^{\pm \mathrm{i}x}$ of the right-hand side of (12.13) must vanish. Instead of directly substituting (12.12) into (12.13) a few preliminary calculations are helpful:

$$\partial_T u_1 = (\partial_T A)\mathrm{e}^{\mathrm{i}x} + (\partial_T \overline{A})\mathrm{e}^{-\mathrm{i}x},$$
$$(\partial_x^2 + 1)u_1 = (\mathrm{i}^2 + 1)A\mathrm{e}^{\mathrm{i}x} + ((-\mathrm{i})^2 + 1)\overline{A}\mathrm{e}^{-\mathrm{i}x} = 0,$$
$$(2\partial_{xX} + \partial_Y^2)u_1 = (2\mathrm{i}\partial_X A + \partial_Y^2 A)\mathrm{e}^{\mathrm{i}x} + (-2\mathrm{i}\partial_X \overline{A} + \partial_Y^2 \overline{A})\mathrm{e}^{-\mathrm{i}x}.$$

So we already see that one term is going to vanish. Furthermore, we have to differentiate the last expression once more, and also calculate the cubic term, which yields

$$(2\partial_{xX} + \partial_Y^2)^2 u_1 = (4\mathrm{i}^2 \partial_X^2 A + 4\mathrm{i}\partial_{YYX}A + \partial_Y^4 A)\mathrm{e}^{\mathrm{i}x}$$
$$+ (4\mathrm{i}^2 \partial_X^2 \overline{A} - 4\mathrm{i}\partial_{YYX}\overline{A} + \partial_Y^4 \overline{A})\mathrm{e}^{-\mathrm{i}x}$$
$$= -4\mathrm{e}^{\mathrm{i}x}\left(\partial_X - \frac{\mathrm{i}}{2}\partial_Y^2\right)^2 A - 4\mathrm{e}^{-\mathrm{i}x}\left(\partial_X + \frac{\mathrm{i}}{2}\partial_Y^2\right)^2 \overline{A},$$
$$u_1^3 = (A\mathrm{e}^{\mathrm{i}x} + \overline{A}\mathrm{e}^{-\mathrm{i}x})^3 = 3|A|^2 A\mathrm{e}^{\mathrm{i}x} + 3|A|^2 \overline{A}\mathrm{e}^{-\mathrm{i}x} + \cdots,$$

which shows that the coefficient of e^{ix} is just the complex conjugate of the coefficient for e^{-ix} so there is only one amplitude equation to satisfy. Indeed, with the preparations inserting (12.12) into (12.13) easily yields

$$\partial_T A = \hat{p} A - 3A|A|^2 + 4\left(\partial_X - \frac{i}{2}\partial_Y^2\right)^2 A. \tag{12.14}$$

There is a slight simplification of the previous equation if we rescale:

$$\hat{X} := X/2, \qquad \hat{Y} := Y/\sqrt{2}, \qquad \hat{A} := \sqrt{3}A.$$

Using this scaling, and dropping all the hats from the variables in equation (12.14), finally results in the classical version of the **Newell-Whitehead-Segel equation**

$$\partial_T A = \hat{p} A - A|A|^2 + \left(\partial_X - \frac{i}{2}\partial_Y^2\right)^2 A. \tag{12.15}$$

Using this equation, we can now study parameter variations of $\hat{p}$; note that $\varepsilon^2 \hat{p} = p$ so $p = \mathcal{O}(\varepsilon^2)$ corresponds to $\hat{p} = \mathcal{O}(1)$. Another amplitude equation is obtained if we neglect the Y dependence of the amplitude in (12.14) so that

$$\partial_T A = \hat{p} A - 3A|A|^2 + 4\partial_X^2 A, \tag{12.16}$$

which in this context is called the **(real) Ginzburg-Landau equation**; see also Chapter 13. Note carefully that (12.15) and (12.16) again have the cubic nonlinearity, which we have become accustomed to when dealing with problems involving bifurcations and travelling waves.

It can be shown that the particular form of the Swift-Hohenberg equation we used as a starting point does not matter as much as one would think: other pattern-forming problems lead to similar (classes of) amplitude equations. Hence, an amplitude equation can also be viewed as a normal form or universality class in the **weakly nonlinear** regime near instability; see also Exercise 12.2. So far our calculations have been formal. We still need to address the question, in what sense do the Newell-Whitehead-Segel and/or the Ginzburg-Landau equations really approximate true solutions of the original system we started with?

Exercise 12.1. Consider the Swift-Hohenberg equation (12.1) and replace $\Delta + 1$ by $\Delta + k_c$ for some $k_c > 0$. What changes in the derivation of the amplitude equation? $\Diamond$

Exercise 12.2. Consider an abstract dispersion relation of the form $\sigma = p - (k^2 - 1)^2 + \mathcal{O}([k^2 - 1]^3)$ as $\|k\| \to 1$. Use the formal methods in this chapter to show that the linear part of the amplitude equation (12.15) is already prescribed by this dispersion relation. $\Diamond$

Exercise 12.3. Substitute $A = R_0 e^{iqX}$ into (12.15) to determine a relationship between p, q and R_0. Explain why this is related to the occurrence of solutions of the form $u = R_0 e^{i(1+\varepsilon q)x}$; this is called a **roll solution**. $\Diamond$

Background and Further Reading: This chapter mostly follows [283], where a lot more details on pattern formation and amplitude equations can be found. The classical physics-oriented survey of the area is [135] and other sources with similar flavor are [134, 462]. The Swift-Hohenberg equation [519] is a standard model problem [119] in the area of amplitude equations and has also very interesting dynamics away from the weakly nonlinear regime [76, 384]. For detailed surveys of the Ginzburg-Landau equation (GLE) see [19, 370, 405]. An English translation of the derivation by Ginzburg and Landau is also available [233]. Of course, one still has to study the GLE and understand its dynamics, e.g., waves [544], which we have already studied in quite some detail in Chapters 7–10 for the bistable case, which also occurs for GLE. Another interesting direction is studying the GLE attractor [162, 406]; see also Chapters 17–18.

Chapter 13

Validity of Amplitude Equations

In Chapter 12 we saw how to derive amplitude equations from the Swift-Hohenberg equation. Here we provide a proof of our formal approximations. As before, we consider as a starting point the **Swift-Hohenberg equation**

$$\partial_t u = \mathcal{L}_p u - u^3, \qquad \mathcal{L}_p := p \operatorname{Id} - (\partial_x^2 + 1)^2, \tag{13.1}$$

where $p \in \mathbb{R}$ is a parameter and $u = u(x,t)$ for $x \in \mathbb{R}$; observe that we restrict our attention here to the one-dimensional case to simplify the exposition. For the scalings

$$p = \varepsilon^2, \qquad T = \varepsilon^2 t, \qquad X = \varepsilon x$$

and the formal ansatz

$$u_A(x,t) := \varepsilon(A(X,T)e^{ix} + \text{c.c.}) \tag{13.2}$$

we already know from Chapter 12 that the amplitude $A = A(X,T)$ is expected to satisfy the **(real) Ginzburg-Landau equation**

$$\partial_T A = A - 3A|A|^2 + 4\partial_X^2 A. \tag{13.3}$$

We always assume that initial conditions for (13.1) and (13.3) are in C_b^4 so that the solutions are also in this space [120]. The natural question is, in what sense does u_A approximate u?

Theorem 13.1. *(amplitude equation validity) Let u denote the solution of (13.1). For each $T_0 > 0$ and $\kappa > 0$ there exist $\varepsilon_0, K > 0$ such that for all $\varepsilon \in (0, \varepsilon_0)$ the following holds: If*

$$|u(x,0) - u_A(x,0)| \le \kappa \varepsilon^2 \quad \forall x \in \mathbb{R},$$

then we have the estimate

$$|u(x,t) - u_A(x,t)| < K\varepsilon^2 \qquad \forall (x,t) \in \mathbb{R} \times [0, T_0/\varepsilon^2], \tag{13.4}$$

where u_A is given by (13.2) and A solves (13.3).

Remark: Note that u_A and u are $\mathcal{O}(\varepsilon)$ near criticality so the error is one order higher in Theorem 13.1. Furthermore, it is natural that we only get a finite-time estimate, as the Ginzburg-Landau equation only approximates the Swift-Hohenberg equation. However, the closer we get to criticality as $\varepsilon \to 0$, the better the approximation becomes.

Before proving the result, we state a helpful lemma and motivate the strategy.

Lemma 13.2. *For the linearized operator $\mathcal{L}_0$ at criticality we have*

$$-\mathcal{L}_0(B(\varepsilon x)\mathrm{e}^{kix}) = [(1-k^2)^2 B + \varepsilon 4ik(1-k^2)B'$$
$$+\varepsilon^2(2-6k^2)B'' + \varepsilon^3 4ikB''' + \varepsilon^4 B'''']\mathrm{e}^{ikx}.$$

The proof is left as an exercise in calculating derivatives. Using Lemma 13.2 it is easy to see that substituting u_A into (13.1) yields

$$\varepsilon^3 \mathrm{e}^{ix}\partial_T A = [\varepsilon^3 \mathrm{e}^{ix}A + 4\varepsilon^3 \mathrm{e}^{ix}\partial_X^2 A - 3\varepsilon^3 \mathrm{e}^{ix}A|A|^2] + \varepsilon^3 \mathrm{e}^{3ix}A^3 + \mathcal{O}(\varepsilon^4) \quad (13.5)$$

with the complex conjugate terms understood on both sides. Therefore, the residual error is of order $\mathcal{O}(\varepsilon^3)$ and given by the term $\varepsilon^3 \mathrm{e}^{3ix}A^3$. Upon integrating the equation up to time T_0/ε^2, a total error of $\mathcal{O}(\varepsilon)$ seems to remain and this is not good enough. The idea of the following proof is to use a modified approximation.

Proof. (of Theorem 13.1) The improved approximation is defined as

$$v_A(x,t) = \varepsilon A(X,T)\mathrm{e}^{ix} - \frac{\varepsilon^3}{64}A(X,T)^3 \mathrm{e}^{3ix} + \text{c.c.}$$

The idea is to study the error by looking at

$$R(x,t) := \frac{u(x,t) - v_A(x,t)}{\varepsilon^2}.$$

If we can show that

$$\|R(x,t)\|_\infty = \sup_{x\in\mathbb{R}} |R(x,t)|$$

is bounded for $t \in [0, T_0/\varepsilon^2]$ by a constant independent of ε, then we have

$$|u(x,t) - u_A(x,t)| = |\varepsilon^2 R(x,t) - \varepsilon^3(A^3 \mathrm{e}^{3ix} + \text{c.c.})/64| = \mathcal{O}(\varepsilon^2)$$

for $(x,t) \in \mathbb{R} \times [0, T_0/\varepsilon^2]$ and the result follows. Hence, it remains to analyze $R(x,t)$. We calculate

$$\varepsilon^2 \partial_t R = \partial_t u - \partial_t v_A$$
$$= \underbrace{\partial_t u - \mathcal{L}_0 u - \varepsilon^2 u + u^3}_{=0} + \mathcal{L}_0 u + \varepsilon^2 u - u^3 - \partial_t v_A$$
$$= \mathcal{L}_0(\varepsilon^2 R + v_A) + \varepsilon^2(\varepsilon^2 R + v_A) - (\varepsilon^2 R + v_A)^3 - \partial_t v_A$$
$$= \varepsilon^2 \mathcal{L}_0 R + \varepsilon^2 R(\varepsilon^2 - 3v_A^2) + \varepsilon^3(-\varepsilon^3 R^3 - 3\varepsilon R^2 v_A)$$
$$\quad - \partial_t v_A - v_A^3 + \varepsilon^2 v_A + \mathcal{L}_0 v_A.$$

Upon grouping terms and dividing through by ε^2 we find

$$\partial_t R = \mathcal{L}_0 R + \varepsilon^2 a(x,t;\varepsilon)R + \varepsilon^3 b(x,t,R;\varepsilon) + \varepsilon^2 r(x,t;\varepsilon), \quad (13.6)$$

where $a(x,t;\varepsilon) = 1 - 3(v_A/\varepsilon)^2$, $b(x,t,R;\varepsilon) = -\varepsilon R^3 - 3(v_A/\varepsilon)R^2$, and $r(x,t;\varepsilon) = [-\partial_t v_A - v_A^3 + \varepsilon^2 v_A + \mathcal{L}_0 v_A]/\varepsilon^4$. The last term is the interesting one since

$$-\varepsilon^4 r(x,t;\varepsilon) = \partial_t v_A + v_A^3 - \varepsilon^2 v_A - \mathcal{L}_0 v_A$$
$$= \varepsilon^3 \left[(\partial_T A - 4\partial_X^2 A - A)\mathrm{e}^{ix} - (1-3^2)^2 \frac{1}{64}A^3 \mathrm{e}^{3ix} + \text{c.c.}\right]$$
$$+ \varepsilon^3(A\mathrm{e}^{ix} + \text{c.c.})^3 + \mathcal{O}(\varepsilon^4)$$

by applying Lemma 13.2. Indeed, the $\mathcal{O}(\varepsilon^3)$ term in the last expression completely vanishes (which explains the choice of prefactor $1/64$), and this shows that $r(x, t; \varepsilon)$ is bounded over $\mathbb{R} \times [0, \infty) \times (0, \varepsilon_0)$. Furthermore, we have for our given $\kappa > 0$ that

$$|R(x, 0)| \leq 2\kappa$$

if ε_0 is chosen sufficiently small. To study (13.6) we are going to use some basic semigroup theory; see Chapter 14. In particular, we drop the x variable in the notation and write (13.6) as

$$\partial_t R = \mathcal{L}_0 R + h(t), \qquad R(0) = R_0. \tag{13.7}$$

The linear part $\partial_t R = \mathcal{L}_0 R$ is solved by a uniformly bounded strongly continuous semigroup $e^{t\mathcal{L}_0}$ when the equation is considered on $C_b(\mathbb{R})$ (bounded continuous functions) with the usual supremum norm $\| \cdot \|_\infty$; see Lemma 13.3. Hence we may write the solution of (13.7), respectively (13.6), via the **variation-of-constants formula** or **Duhamel's formula**,

$$R(t) = e^{t\mathcal{L}_0} R(0) + \int_0^t e^{(t-s)\mathcal{L}_0} h(s) \, ds \tag{13.8}$$

$$= e^{t\mathcal{L}_0} R(0) + \varepsilon^2 \int_0^t e^{(t-s)\mathcal{L}_0} [a(s; \varepsilon) R(s) + \varepsilon b(s, R(s); \varepsilon) + r(s; \varepsilon)] \, ds.$$

For a given $\delta > 0$ we have $\|b(s, R(s); \varepsilon)\|_\infty \leq K_b$ for all R with $\|R(s)\|_\infty \leq \delta$ and $\varepsilon \in (0, \varepsilon_0)$. Since the semigroup is uniformly bounded we have $\|e^{s\mathcal{L}_0}\|_\infty \leq K$ for some $K > 0$. We can also ensure that $\|r(s; \varepsilon)\|_\infty \leq K$ by the argument above about the boundedness of r and clearly $|a(s; \varepsilon)| \leq K$. Then one estimates (13.8) and obtains

$$\|R(t)\|_\infty \leq 2K\kappa + \int_0^t \varepsilon^2 K^2 \|R(s)\|_\infty \, ds + \varepsilon^2 t K (\varepsilon K_b + K) \tag{13.9}$$

as long as $\|R(s)\|_\infty \leq \delta$ holds. **Gronwall's inequality** (see Exercise 13.6) applied to (13.9) yields

$$\|R(t)\|_\infty \leq [2K\kappa + T_0 K(\varepsilon K_b + K)] e^{\varepsilon^2 K^2 t}, \tag{13.10}$$

and for $t \leq T_0/\varepsilon^2$ we can easily see that, upon making ε small so that $\varepsilon K_b \leq K$, we find that $\|R(t)\|_\infty$ is bounded independent of ε. $\qquad \square$

Lemma 13.3. *([335, Lem. 2.3]) The semigroup $e^{t\mathcal{L}_0} : C_b(\mathbb{R}) \to C_b(\mathbb{R})$ is strongly continuous and uniformly bounded.*

We shall not prove this result here but refer to the discussion in Example 14.12 below. The main idea of the proof of Theorem 13.1 was to insert a clever approximation to remove higher-order correction terms. This strategy is not limited to the Ginzburg-Landau equation. Consider the **sine-Gordon equation**

$$\partial_t^2 u = \partial_x^2 u - \sin u, \qquad u = u(x, t), \ x \in \mathbb{R}. \tag{13.11}$$

Then it is shown in Exercise 13.7 that the dispersion relation for the linearized sine-Gordon equation is given by $\omega^2 = k^2 + 1$.

Definition 13.4. For a given dispersion relation $\omega(k)$, the **group velocity** ν is defined by $\nu := \frac{d\omega}{dk}$.

The solutions we are interested in for the sine-Gordon equation are of the form

$$u_A(x,t) = \varepsilon A(\varepsilon t, \varepsilon(x - \nu t))e^{i(kx-\omega t)} + \text{c.c.} \tag{13.12}$$

It can be shown using a formal calculation via the method of multiple scales as in Chapter 12 that the amplitude A satisfies the **nonlinear Schrödinger equation (NLS)**

$$2i\omega\partial_T A = (\nu^2 - 1)\partial_X^2 A + \frac{1}{2}A|A|^2, \qquad A = A(X,T). \tag{13.13}$$

Since we are dealing with hyperbolic problems, it is natural to consider $(u, \partial_t u) \in H^1(\mathbb{R}) \times L^2(\mathbb{R})$ and measure the approximation in this space.

Theorem 13.5. *(amplitude equation validity [335]) Let u denote the solution of (13.11) and let u_A be given by (13.12), where A is a solution of (13.13) with $\partial_T^{n_T}\partial_X^{n_X} A \in C([0,T], L^2(\mathbb{R}))$ for all $n_X + n_T \leq 2$. Then, for each $T_0 > 0$, $T_0 \leq T$, and $\kappa > 0$ there exist $\varepsilon_0, K > 0$ such that for all $\varepsilon \in (0, \varepsilon_0)$ the following holds: If*

$$\|u(\cdot,0) - u_A(\cdot,0)\|_{H^1(\mathbb{R})\times L^2(\mathbb{R})} \leq \kappa\varepsilon^{3/2},$$

then we have the estimate

$$\|u(\cdot,t) - u_A(\cdot,t)\|_{H^1(\mathbb{R})\times L^2(\mathbb{R})} < K\varepsilon^{3/2} \qquad \forall t \in [0, T_0/\varepsilon^2]. \tag{13.14}$$

Theorem 13.5 can be proved in a similar way as Theorem 13.1. One has to consider the linearized semigroup on a different Banach space. Furthermore, a new improved approximation has to be used as discussed in Exercise 13.8. In summary, we now know that, over a long time scale, amplitude equations can be regarded as normal forms. Furthermore, we found that it should be helpful to study semigroup methods in more detail, so we introduce this material in the next chapter.

Exercise 13.6. Prove **Gronwall's inequality**, i.e., show that for a given time interval $(a, b) \subset \mathbb{R}$ and continuous functions α, β, g with $\beta \geq 0$ the inequality

$$g(t) \leq \alpha(t) + \int_a^t \beta(s)g(s)\,\mathrm{d}s, \qquad \forall t \in (a, b)$$

implies that

$$g(t) \leq \alpha(t) + \int_a^t \alpha(s)\beta(s)\exp\left(\int_s^t \beta(r)\,\mathrm{d}r\right)\mathrm{d}s$$

for $t \in (a, b)$. Show that if in addition $\alpha(t)$ nondecreasing, then

$$g(t) \leq \alpha(t)\exp\left(\int_a^t \beta(r)\,\mathrm{d}r\right),$$

which is the version we used above. $\Diamond$

Exercise 13.7. Show that the dispersion relation for the linearized sine-Gordon equation is given by $\omega^2 = k^2 + 1$ using the ansatz $e^{i(kx-\omega t)}$. $\Diamond$

Exercise 13.8. Show that the improved approximation for the sine-Gordon equation leading to the NLS equation is given by

$$v_A(x,t) = \varepsilon A(X,T)e^{i(kx-\omega t)} - \frac{1}{54k^2 - 54\omega^2 + 6}\varepsilon^3 A(X,T)e^{3i(kx-\omega t)} + \text{c.c.},$$

where $T = \varepsilon^2 t$ and $X = \varepsilon(x - \nu t)$. $\Diamond$

Background and Further Reading: This chapter is mainly based upon [335], which developed the idea of using improved approximations to get rid of certain error terms; see also [407, 494]. Other approaches to proving validity of amplitude equations can be found in [120, 542]. For more on the sine-Gordon equation see [46, 136]. The GLE and NLS are amplitude equations for many different problems, e.g., the NLS is an amplitude equation for certain water wave problems [354]. The NLS is also one of the paradigmatic problems in nonlinear PDE dynamics, and we can only mention just a few references here [95, 147, 169, 203, 235, 326, 330, 424, 438, 518]; see also Chapter 27. For a lot more on amplitude equations we refer to the monograph [495]. There is also a lot of amplitude equation theory for stochastic PDEs available [61].

Chapter 14

Semigroups and Sectorial Operators

We have already seen in Chapter 13 that it will be convenient to have the language and some basic results from semigroup theory available. Here we provide a broader introduction to the topic, also with a view towards more general classes of PDEs.

Definition 14.1. Let X be a Banach space. A **(strongly continuous) semigroup** on X is a family of continuous linear operators $\{S(t)\}_{t \geq 0}$ on X such that

(D1) $S(0) = \mathrm{Id}$,

(D2) $S(t)S(s) = S(t+s)$ for $t, s \geq 0$,

(D3) $\|S(t)u - u\|_X \to 0$ as $t \searrow 0$ for every $u \in X$.

All semigroups we consider will be strongly continuous so we omit this prefix. However, Definition 14.1 can be modified by requiring additional regularity; e.g., $S(t)$ is an **analytic semigroup** if $t \mapsto S(t)u$ is real analytic for $t \in (0, +\infty)$ and every $u \in X$.

Definition 14.2. The **infinitesimal generator** A of a semigroup $S(t) : X \to X$ is defined by

$$Au := \lim_{t \searrow 0} \frac{1}{t} \left(S(t)u - u \right)$$

with domain $\mathcal{D}(A)$ consisting of all $u \in X$, where the limit exists. If the generator can be identified, then we write $S(t) = e^{tA}$.

An important example is the linear ODE $\frac{du}{dt} = Au$, where the matrix $A \in \mathbb{R}^{d \times d}$ is easily seen to be the infinitesimal generator of the flow $\phi(u_0, t) = u(t)$, which also defines a semigroup $S(t)$.

Example 14.3. The natural example from the context of PDEs is the heat equation

$$\partial_t u = \Delta u, \qquad (x, t) \in \Omega \times [0, +\infty), \ u = u(x, t) \in \mathbb{R}, \tag{14.1}$$

say for $\Omega = [0, 1]$ and with Dirichlet boundary conditions for simplicity. We shall see below that $A = \Delta$ is the infinitesimal generator for an analytic semigroup $e^{t\Delta} : L^2(\Omega) \to L^2(\Omega)$ with $\mathcal{D}(\Delta) = H_0^1(\Omega) \cap H^2(\Omega)$. In fact, Δ is also a closed and densely defined operator on $L^2(\Omega)$. ◆

The next definition is useful to set up an abstract framework to check which operators generate (analytic) semigroups.

Definition 14.4. A linear operator $A : X \to X$ is called **sectorial** if it is closed, densely defined, and for some $\theta \in (\frac{\pi}{2}, \pi)$, $M \geq 1$, $a \in \mathbb{R}$, the sector

$$S_{a,\theta} := \{\lambda \in \mathbb{C} : |\arg(\lambda - a)| \leq \theta, \lambda \neq a\}$$

is contained in the **resolvent set** $\rho(A) := \mathbb{C} - \sigma(A)$, where $\sigma(A)$ is the spectrum. Furthermore, we have the bound

$$\|(\lambda \, \text{Id} - A)^{-1}\|_{\mathcal{L}(X,X)} \leq \frac{M}{|\lambda - a|} \tag{14.2}$$

for all $\lambda \in S_{a,\theta}$, where $\|L\|_{\mathcal{L}(X,X)} := \sup\{\|Lv\|_X : v \in X, \|v\|_X = 1\}$ is the usual **operator norm** on the space $\mathcal{L}(X, X)$ of linear operators.

We refer to Figure 14.1 for an illustration of the sector $S_{a,\theta}$. From now on, we shall just abbreviate $\|\cdot\|_X = \|\cdot\|$ and $\|\cdot\|_{\mathcal{L}(X,X)} = \|\cdot\|$ if it is clear from the arguments which norm is considered.

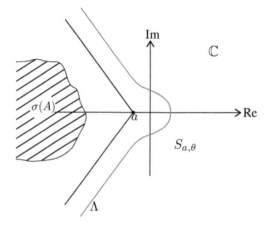

Figure 14.1. *Sketch to visualize Definition 14.4 for a sectorial operator A. The spectrum $\sigma(A)$ has to be contained in a certain wedge/cone in $\mathbb{C}$.*

Remark: Slightly different versions of the definition of sectorial operators exist, particularly regarding sign conventions.

Example 14.5. Continuing Example 14.3 with $\Delta = \partial_x^2$ on $\Omega = [0, 1]$, we know from Chapter 6 that the eigenvalues of ∂_x^2 are $\lambda_n = -(\pi n)^2$ with $n \in \{1, 2, 3, \ldots\}$. So ∂_x^2 is a sectorial operator on $L^2(\Omega)$, with

$$\mathcal{D}(\partial_x^2) = \{u \in L^2(\Omega) : \partial_x^2 u \in L^2(\Omega)\} = H_0^1(\Omega) \cap H^2(\Omega), \tag{14.3}$$

where we could, e.g., take $a = 0$ and any $\theta \in (\frac{\pi}{2}, \pi)$ in Definition 14.4. In fact, the Laplacian considered with suitable boundary conditions on a sufficiently regular domain $\Omega \subset \mathbb{R}^N$ and many other elliptic operators turn out to be sectorial; cf. Chapter 3. ◆

Theorem 14.6. *(sectorial generator $\Rightarrow$ analytic semigroup [263, Sec. 1.3]) If A is sectorial, then A is the infinitesimal generator of an analytic semigroup e^{tA} with the representation formula*

$$e^{tA} = \frac{1}{2\pi i} \int_\Lambda (\lambda \, \text{Id} - A)^{-1} e^{\lambda t} \, d\lambda, \tag{14.4}$$

where Λ is a contour in $\rho(A)$ with $\arg(\lambda) \to \pm\theta^*$ as $|\lambda| \to +\infty$ for $\theta^* \in (\frac{\pi}{2}, \pi)$. Furthermore, if $\mathrm{Re}(\lambda) < a$ for $\lambda \in \sigma(A)$, then

$$\|\mathrm{e}^{tA}\| \leq K\mathrm{e}^{at} \quad \text{for } t > 0 \tag{14.5}$$

and some $K > 0$.

See also Figure 14.1 for an illustration. The **Dunford integral** (14.4) involving the **resolvent** $(\lambda\,\mathrm{Id} - A)^{-1}$ of A is frequently key to working with semigroups, as it is often more convenient to work with resolvents.

Remark: Theorem 14.6 starts from a linear operator A and yields a semigroup. The **Hille-Yosida Theorem** provides an if-and-only-if characterization, when a given closed linear operator A generates a semigroup. Basically, it states that the requirements are that the domain $\mathcal{D}(A)$ is dense in X and the resolvent $(\lambda\,\mathrm{Id} - A)^{-1}$ has to satisfy the bound

$$\|(\lambda\,\mathrm{Id} - A)^{-n}\| \leq \frac{\text{const.}}{(\lambda - \omega)^n} \qquad \forall n \in \mathbb{N}, \lambda > \omega, \lambda \in \rho(A).$$

Another important result in a similar spirit is the **Lumer-Phillips Theorem**, which requires some decay/dissipation of A and surjectivity of the resolvent; we refer to the references for more details.

Example 14.7. We continue with Example 14.5. Let $u_0 = u(x, 0)$ be the initial value. Since the operator $\Delta = \partial_x^2$ is sectorial, Theorem 14.6 implies that it generates the analytic semigroup $\mathrm{e}^{t\Delta}$. Therefore, the heat equation (14.1) is solved by

$$u(t) = \mathrm{e}^{t\Delta}u_0. \tag{14.6}$$

However, by separation of variables we can solve the heat equation by

$$u(x, t) = \sum_{n=1}^{\infty} \mathrm{e}^{\lambda_n t} \langle e_n, u_0 \rangle_{L^2(\Omega)} e_n(x),$$

where $e_n(x) = \sqrt{2}\sin(n\pi x)$ is the eigenfunction for λ_n. This gives concrete formulas to work with A and e^{tA} by their action on basis functions. For example, we have

$$(\lambda\,\mathrm{Id} - \Delta)^{-1}v = \sum_{n=1}^{\infty} (\lambda - \lambda_n)^{-1} \langle e_n, v \rangle_{L^2(\Omega)} e_n(x)$$

for $v \in L^2(\Omega)$. Another example defining $(-\Delta)^\alpha$ and computing $\mathcal{D}((-\Delta)^\alpha)$ for $\alpha > 0$ is considered in Exercise 14.13. ♦

One may generalize the previous observations about $(-\Delta)^\alpha$.

Definition 14.8. Suppose A is a sectorial operator and $\mathrm{Re}(\sigma(A)) < 0$. Then for any $\alpha > 0$ define

$$(-A)^{-\alpha} := \frac{1}{\Gamma(\alpha)} \int_0^\infty t^{\alpha-1}\mathrm{e}^{tA}\,\mathrm{d}t,$$

where $\Gamma(\alpha)$ denotes the **gamma function** (recall $\Gamma(n) = (n-1)!$ for $n \in \mathbb{N}$ and $\Gamma(\alpha) = \int_0^\infty x^{\alpha-1}\mathrm{e}^{-x}\,\mathrm{d}x$ for $\alpha \in \mathbb{C} \setminus \{0, -1, -2, \ldots\}$).

Theorem 14.9. *(fractional powers of operators [263, Thm. 1.4.2]) If A is sectorial with $\mathrm{Re}(\sigma(A)) < 0$, then for any $\alpha > 0$ the operator $(-A)^{-\alpha}$ is bounded and injective.*

The last theorem shows why it is sometimes nicer to analyze A via its negative powers. Furthermore, one can also use a similar idea with positive exponents to define spaces, which turns out to be very suitable for many dynamical problems.

Definition 14.10. Let A be sectorial on X and fix $a \in \mathbb{R}$ such that $A_1 := A + a$ Id satisfies $\text{Re}(\sigma(A_1)) < 0$. For $\alpha \geq 0$, define the **fractional power space** $X^\alpha := \mathcal{D}(A_1^\alpha)$ with the norm $\|x\|_\alpha := \|(-A_1)^\alpha x\|_X$ for $x \in X^\alpha$.

It is easy to see that X^α are Banach spaces. The main reason to use the spaces X^α is that they provide well-defined domains for operators. Furthermore, there are **embedding theorems** for $X = L^p(\Omega)$ which show, under suitable conditions, that X^α is a subset of smooth functions $C^k(\Omega)$ or inside some Sobolev space $W^{k,q}(\Omega)$ for some values k, q. As before, we shall not worry too much about regularity assumptions and just select the domain, the boundary and initial data, and those equations which are sufficiently smooth, potentially giving up sharp/optimal regularity results in the process.

The next step is to move from the linear problem $\frac{du}{dt} = Au$ to the inhomogeneous problem

$$\frac{du}{dt} = Au + f(t), \qquad u \in X,\ f : (0, T) \to X,\ u(0) = u_0. \tag{14.7}$$

If A generates a strongly continuous semigroup e^{tA}, then we may always consider a **mild solution** to (14.7) given by

$$u(t) = e^{tA} u_0 + \int_0^t e^{(t-s)A} f(s)\, ds. \tag{14.8}$$

Theorem 14.11. *(existence of classical solutions [263, Thm. 3.2.2]) Consider* (14.7) *and assume that A is a sectorial operator. Furthermore, suppose $f : (0, T) \to X$ is continuous and $\int_0^\rho \|f(t)\|\, dt < +\infty$ for some $\rho > 0$. Then there is a unique strong (classical) solution of* (14.7) *which coincides with the mild solution* (14.8).

The Laplacian and other classical elliptic operators on L^2 are key examples in semigroup theory for PDEs. However, the framework is obviously not limited to this setup as the next example illustrates.

Example 14.12. Consider the linear evolution equation

$$\partial_t u = \mathcal{L}_0 u := -(1 + \partial_x^2)^2 u, \qquad u = u(x,t),\ x \in \mathbb{R}, \tag{14.9}$$

arising in the context of the **Swift–Hohenberg equation**; see also Chapter 13. In fact, it is relatively easy to check that

$$e^{t\mathcal{L}_0} : C_b(\mathbb{R}) \to C_b(\mathbb{R})$$

is a strongly continuous semigroup. Lemma 13.3 states that the semigroup is uniformly bounded. Although we do not give a full proof, let us motivate this fact formally. The dispersion relation (12.3) yields that at criticality the spectrum touches the imaginary axis precisely at zero and the rest of the (essential) spectrum is contained in the left-half complex plane. Hence, it is natural to consider Definition 14.4 with a sector $\mathcal{S}_{0,\theta}$. It turns out that $\theta = 2\pi/3$ is a good choice for the angle; see Figure 14.1. Motivated by (14.2), we aim to estimate the resolvent. Consider $\omega := (-\lambda)^{1/2}$, $\text{Im}(\omega) > 0$, and the operator splitting

$$\mathcal{L}_0 v - \lambda v = -(1 + \partial_x^2 + \omega)(1 + \partial_x^2 - \omega)v = h$$

for some h. This yields the solution

$$v = (\mathcal{L}_0 - \lambda)^{-1}h = -[G_{\beta+} \circ G_{\beta-}]h, \quad \beta_{\pm} = (-1 \pm \omega)^{1/2}, \; \mathrm{Re}(\beta_{\pm}) > 0,$$

where the operator

$$[G_\beta h](x) = -\int_{\mathbb{R}} \frac{1}{2\beta} e^{-\beta|x-\xi|} h(\xi) \, d\xi \tag{14.10}$$

can be derived using Fourier transforms; see Exercise 14.15. Therefore, we obtain

$$\|(\mathcal{L}_0 - \lambda \, \mathrm{Id})^{-1}\| \leq \frac{1}{|\beta_+| \mathrm{Re}(\beta_+)} \frac{1}{|\beta_-| \mathrm{Re}(\beta_-)}$$

$$\leq \frac{1}{\mathrm{Re}(\beta_+)\mathrm{Re}(\beta_-)} \frac{1}{\sqrt{|1+\lambda|}}. \tag{14.11}$$

From this inequality it is not too difficult to obtain that

$$\|(\mathcal{L}_0 - \lambda \, \mathrm{Id})^{-1}\| \leq \frac{M}{|\lambda|} \tag{14.12}$$

for all $\lambda \in \mathcal{S}_{0,\theta}$. Then we may conclude sectoriality and employ the Dunford integral (14.4) to see that the semigroup is uniformly bounded. ♦

Having introduced some basic results in semigroup theory, it is probably most important to always remember that there are key relations between the semigroup, the generator, and the associated resolvent. These ideas are going to appear in many PDE dynamics problems.

Exercise 14.13. Consider the one-dimensional Laplacian $\Delta = \partial_x^2$ with Dirichlet boundary conditions on $\Omega = [0,1]$ from Example 14.7 and define

$$(-\Delta)^\alpha v = \sum_{n=1}^{\infty} (-\lambda_n)^\alpha \, \langle e_n, v \rangle_{L^2(\Omega)} \, e_n(x)$$

for $v \in L^2(\Omega)$, where λ_n denotes the eigenvalue of Δ with eigenfunction e_n. Prove that

$$\mathcal{D}((-\Delta)^\alpha) = \left\{ v \in L^2(\Omega) : \sum_{n=1}^{\infty} \lambda_n^{2\alpha} \langle e_n, v \rangle_{L^2(\Omega)}^2 < +\infty \right\}. \tag{14.13}$$

Furthermore, prove that $\mathcal{D}((-\Delta)^{1/2}) = H_0^1(\Omega)$. Note that this exercise generalizes to a domain $\Omega \subset \mathbb{R}^N$, $N \geq 1$. ◊

Exercise 14.14. Show that if $-A$ is positive definite and self-adjoint then so is $(-A)^\alpha$ for all $\alpha > 0$. ◊

Exercise 14.15. Consider Example 14.12 and formally solve the equation

$$(1 + \omega + \partial_x^2)v = h, \quad v = v(x), \; x \in \mathbb{R},$$

using the Fourier transform; this calculation essentially yields (14.10). Furthermore, prove that (14.11) implies (14.12) for $\lambda \in \mathcal{S}_{0,\frac{2}{3}\pi}$. ◊

Background and Further Reading: This chapter is mainly based upon the introduction to analytic semigroups in [263]. Other excellent books focusing more on semigroup theory are [179, 180, 448]; classical sources for semigroups, functional analysis, the Hille-Yosida Theorem, and related topics are [131, 266, 474, 567]. For more relations to spectral theory and nonlinear problems we also refer to [42, 96, 107]. The uniform boundedness of the linearized Swift-Hohenberg equation at criticality is taken from [335]. Sharp regularity estimates for many parabolic-type semigroup problems can be found in [387]. For further variations on the semigroup theory and related problems, one may consider [190, 251, 538]. There is also an important role of semigroups in the theory of stochastic PDEs [464, 465] and there are deep links to fundamental topics in probability theory [148, 193].

Chapter 15

Dissipation and Absorbing Sets

In Chapter 14, we saw that semigroups arising from generators with spectra in the left half of the complex plane are globally contracting; see equation (14.5) in Theorem 14.6. More generally, dissipation refers to the frequent situation in PDEs that the dynamics contracts or reduces to a subset of phase space.

Example 15.1. Prototypical examples for dissipative systems are reaction-diffusion equations such as

$$\partial_t u = \Delta u + f(u), \quad (x, t) \in \Omega \times [0, +\infty), \ u = u(x, t) \quad (15.1)$$

for a suitable sufficiently smooth nonlinearity $f : \mathbb{R} \to \mathbb{R}$ on a compact domain $\Omega \subset \mathbb{R}^N$ with smooth boundary and Dirichlet boundary conditions $u(x) = 0$ for $x \in \partial\Omega$. To obtain dissipativity, one has to require that the nonlinearity f satisfies certain additional conditions. ◆

Let X be a Banach space and suppose we may view the solution $u(x, t)$ of the PDE with initial condition $u_0 \in X$ as a **semiflow** (or **nonlinear semigroup**)

$$S(t)u_0 = u(x, t), \qquad S(t) : X \to X.$$

In particular, $S(t)$ satisfies $S(0) = \text{Id}$, S is continuous in t and in the argument u_0, and $S(t + s) = S(t)S(s)$ for $t, s \geq 0$; see also Definition 14.1. There are several different notions of dissipativity. For us, the following will suffice.

Definition 15.2. $S(t)$ is called **(bounded) dissipative** if there exists a bounded set $\mathcal{B} \subset X$, such that for each bounded set $\mathcal{Y}$ there exists a time $t_{\mathcal{Y}}$ such that $S(t)\mathcal{Y} \subset \mathcal{B}$ for all $t \geq t_{\mathcal{Y}}$.

The previous statement can be rephrased by saying that there exists a bounded **absorbing set** for the dynamics.

Remark: Frequently one also finds explicit requirements on the dynamics such as a **dissipative estimate**

$$\|S(t)u_0\|_X \leq Q(\|u_0\|_X)\mathrm{e}^{-\alpha t} + K$$

for some constants $K, \alpha > 0$ and a monotone increasing function Q.

The natural question to ask is, which PDEs are actually dissipative?

Example 15.3. (Example 15.1 continued) Consider the reaction-diffusion equation (15.1) and make the additional assumptions

$$-K - \alpha_1 |v|^p \leq f(v)v \leq K - \alpha_2 |v|^p, \qquad f'(v) \leq K_d, \quad (15.2)$$

for all $v \in \mathbb{R}$ and some constants $K, K_d, \alpha_{1,2} > 0$. It also helps to simplify the calculations to require $f(0) = 0$. A typical example is the case

$$f(v) = v - v^3.$$

Indeed, in this case we have

$$-|v|^4 \leq f(v)v = v^2 - v^4 \leq K - \frac{1}{2}|v|^4$$

and the upper bound on the derivative is trivial to show. Under the assumptions (15.2) it is not difficult to see that $S(t)u_0 = u(x,t)$ yields a semiflow with

$$u \in C(0, T; L^2(\Omega)) \cap L^2(0, T; \mathcal{D}((-\Delta)^{1/2})), \tag{15.3}$$

and we know from Exercise 14.13 that $\mathcal{D}((-\Delta)^{1/2}) = H_0^1(\Omega)$. Furthermore, we expect from Theorem 14.11 that solutions should be classical ones for many cases of the nonlinearity f. ◆

A main strategy to prove the existence of a bounded absorbing set is to consider a PDE on a suitable Banach space X on which the differential operators appearing in the equation are at least densely defined and which is "tractable" analytically. For the reaction-diffusion equation (15.1) one excellent guess is to take $X = L^2(\Omega)$.

Theorem 15.4. *(absorbing set in L^2) Suppose* (15.1) *satisfies* (15.2). *Then there exist constants $K_1, K_2 > 0$ and a time $t_{u_0} > 0$ such that*

$$\|u(t)\|_{L^2(\Omega)} \leq K_1 \tag{15.4}$$

for all $t \geq t_{u_0}$. Furthermore, we have the bound

$$\int_t^{t+1} \|u(s)\|_{H_0^1(\Omega)}^2 \, \mathrm{d}s \leq K_2 \tag{15.5}$$

for all $t \geq t_{u_0}$.

Proof. The domain Ω will be dropped as a subscript in the spatial norms on L^2 and H_0^1 throughout the proof. We write (15.1) in the form

$$\frac{\mathrm{d}u}{\mathrm{d}t} - \Delta u = f(u), \qquad u = u(t) = u(x,t), \tag{15.6}$$

and take the L^2-inner product with u, which yields

$$\frac{1}{2}\frac{\mathrm{d}}{\mathrm{d}t}\|u\|_{L^2}^2 + \|u\|_{H_0^1}^2 = \int_\Omega f(u)u \, \mathrm{d}x \leq \int_\Omega K - \alpha_2|u|^p \, \mathrm{d}x, \tag{15.7}$$

where we used (15.2) in the last inequality. In the next step, we use a version of the **Poincaré inequality** given by

$$\|u\|_{L^2} \leq \lambda^{-1/2}\|\nabla u\|_{L^2} = \lambda^{-1/2}\|u\|_{H_0^1},$$

where $\lambda > 0$ is the smallest eigenvalue of $-\Delta$ on Ω; see also Theorem 3.9. Using this inequality and dropping the negative term on the right-hand side in (15.7) leads to

$$\frac{\mathrm{d}}{\mathrm{d}t}\|u\|_{L^2}^2 + 2\lambda\|u\|_{L^2}^2 \leq K \tag{15.8}$$

for some constant $K > 0$. By Gronwall's inequality we may conclude that

$$\|u\|_{L^2}^2 \le e^{-2\lambda t} \|u_0\|_{L^2}^2 + \int_0^t e^{-2\lambda t + 2\lambda s} K \, ds$$

$$= e^{-2\lambda t} \|u_0\|_{L^2}^2 + K e^{-2\lambda t} \int_0^t e^{2\lambda s} \, ds \le e^{-2\lambda t} \|u_0\|_{L^2}^2 + \frac{K}{2\lambda}.$$

The first result (15.4) now follows. The second result (15.5) is left as Exercise 15.9. $\square$

Theorem 15.4 is a typical first step to check, whether there is a possibility that the long-term dynamics of a PDE is finite-dimensional. The second estimate (15.5) helps to construct a better absorbing set in H_0^1. The idea is to multiply (15.6) by Δu and integrate, which yields

$$\frac{1}{2} \frac{d}{dt} \|u\|_{H_0^1}^2 + \|\Delta u\|_{L^2}^2 \le K \|u\|_{H_0^1}^2 \tag{15.9}$$

for some constant $K > 0$, by using the upper bound of the derivative f' from (15.2). Integrating the last inequality between s and t, we get

$$\|u(t)\|_{H_0^1}^2 \le 2K \int_s^t \|u(r)\|_{H_0^1}^2 \, dr + \|u(s)\|_{H_0^1}^2.$$

Integrating again, now with respect to s and between $t - 1$ and t (with $t \ge 1$), yields

$$\|u(t)\|_{H_0^1}^2 \le 2K \int_{t-1}^t \int_s^t \|u(r)\|_{H_0^1}^2 \, dr \, ds + \int_{t-1}^t \|u(s)\|_{H_0^1}^2 \, ds$$

$$\le (2K + 1) \int_{t-1}^t \|u(s)\|_{H_0^1}^2 \, ds \le K_2(1 + 2K),$$

where the second inequality follows from Fubini's Theorem and the last inequality follows from (15.5). Note that the last calculation was formal since we required regularity of $u(t)$ that we may not have, but this can be made rigorous either by proving directly that solutions are regular enough or by using a standard Galerkin approximation for parabolic PDEs [186, Sec. 7.1], which we do not detail here. With the H_0^1 estimate we get the following.

Theorem 15.5. *(absorbing set in H_0^1) Suppose* (15.1) *satisfies* (15.2) *and $u(x, t)$ is sufficiently smooth. Then there exists an absorbing set in H_0^1.*

Corollary 15.6. *Suppose* (15.1) *satisfies* (15.2) *and $u(x, t)$ is sufficiently smooth. Then there exists a compact absorbing set in L^2.*

Proof. Using Theorem 15.5 and assuming a **compact embedding** of H^1 into L^2, the result would follow. Hence, it remains to justify the embedding. The standard **Sobolev Embedding Theorem** [6, 70, 186, 308, 348] (or **Rellich-Kondrachov Theorem**) states that for a bounded domain $\Omega \subset \mathbb{R}^N$ with C^1-boundary, $k > l$, and $k - N/p > l - N/q$, we have

$$W^{k,p}(\Omega) \subset\subset W^{l,q}(\Omega),$$

i.e., $W^{k,p}(\Omega) =: W^{k,p}$ is **compactly embedded** in $W^{l,q}(\Omega) =: W^{l,q}$, which just means that $\| \cdot \|_{W^{l,q}} \le C \| \cdot \|_{W^{k,p}}$ uniformly for some constant $C > 0$, and each bounded sequence in $W^{k,p}$ is precompact in $W^{l,q}$. So if we select $p = 2$, $k = 1$, $l = 0$, $q = 2$, this gives $W^{1,2}(\Omega) = H^1(\Omega) \subset\subset L^q(\Omega)$. $\square$

Reaction-diffusion equations are a good benchmark scenario for other classes of dissipative PDEs.

Example 15.7. Another classical, albeit technically more involved, example is the **(incompressible) Navier-Stokes equation**

$$\partial_t u - \nu \Delta u + (u \cdot \nabla)u + \nabla p = f(x,t), \qquad \nabla \cdot u = 0, \qquad (15.10)$$

where $u = u(x,t)$ represents the velocity of a fluid in a domain $\Omega \subset \mathbb{R}^N$ ($N = 2, 3$), $\nu > 0$ controls the strength of the viscosity, f is a forcing term, $\nabla \cdot u = 0$ is the incompressibility condition, and $p = p(x,t)$ is the pressure. One may actually use incompressibility to eliminate the pressure term and, with quite a few calculations, end up with

$$\frac{du}{dt} + \nu A u + B(u,u) = f, \qquad (15.11)$$

where A is a linear operator and B represents the (quadratic!) nonlinearity of the Navier-Stokes equations. A simple domain to consider is $\Omega = [0,L] \times [0,L] \subset \mathbb{R}^2$ with further simplifications

$$\int_\Omega u_0(x)\,\mathrm{d}x = 0, \qquad \int_\Omega f(x,t)\,\mathrm{d}t = 0$$

for the initial condition and the forcing term. For this case, the existence, uniqueness, and regularity theory for the Navier-Stokes equations is complicated but still possible to carry out via classical techniques. Then one may carry out similar arguments to show that (15.11) in the two-dimensional periodic domain Ω generates a semiflow on

$$L^2_{\mathrm{NS}} := \left\{ u \in L^2(\Omega) : \int_\Omega u(x)\,\mathrm{d}x = 0, \nabla \cdot u = 0, u \text{ periodic} \right\}$$

as well as on the suitable Sobolev space analogue

$$H^1_{\mathrm{NS}} := \left\{ u \in H^1(\Omega) : \int_\Omega u(x)\,\mathrm{d}x = 0, \nabla \cdot u = 0, u \text{ periodic} \right\}.$$

In this setup, similar arguments, as demonstrated above for the reaction-diffusion case, can be used to establish the existence of absorbing sets in L^2_{NS} as well as in H^1_{NS}. ◆

Although it is nice to have a suitable compact absorbing set, we still have to analyze the dynamics inside this set, which will be the topic of Chapters 16–19. First, we aim towards identifying geometric concepts to define a lower-dimensional effective phase space, and then we are going to provide upper bounds on the dimensions of the attractors of PDEs.

Exercise 15.8. A semiflow $S(t) : X \to X$ is called **point dissipative** if there exists a bounded set $\mathcal{B}$ such that for every initial condition u_0 there exists a time t_{u_0} such that $S(t)u_0 \subset \mathcal{B}$ for all $t \geq t_{u_0}$. Prove that if $X = \mathbb{R}^d$, then point dissipativity implies bounded dissipativity. Hint: Use the Heine-Borel Theorem. ◊

Exercise 15.9. Prove the inequality (15.5) from Theorem 15.4. ◊

Exercise 15.10. Prove the inequality (15.9) under the assumptions that u is sufficiently smooth and f' is bounded above. ◊

Background and Further Reading: The presentation here is mainly based upon [477]. Other excellent sources are [35, 254, 411, 528]. For reaction-diffusion equations on unbounded domains see, e.g., [173, 556], for nonautonomous PDEs consider [94], and for general views on reaction-diffusion equations see [200, 245, 507]. The Navier-Stokes equations are a vast topic, essentially forming a research area by itself; here we can just refer to a few interesting sources in this area to get started [109, 124, 161, 192, 204, 215, 220, 508, 527]. One should also mention that there has been a long struggle with the global regularity for the Navier-Stokes equations in $\mathbb{R}^3$ [77, 191, 338].

Chapter 16

Nonlinear Saddles and Invariant Manifolds

Having obtained absorbing sets for certain classes of PDEs in Chapter 15, the next step will be to understand the geometry of phase space inside an absorbing set. In particular, we would like to extend the geometric viewpoint of ODEs from Chapter 2 to the PDE setting. In this chapter, we study the class of PDEs

$$\partial_t u + Au = f(u), \qquad u_0 \in H, \tag{16.1}$$

where $-A : H \to H$ is a linear, negative, self-adjoint, and sectorial operator on a Hilbert space H and f is a sufficiently smooth globally Lipschitz continuous nonlinearity. From Chapter 14 it follows that there exists a well-defined solution

$$u \in C^0(0, T; H) \cap L^2(0, T; \mathcal{D}(A^{1/2})), \quad u(t) = S(t)u_0, \tag{16.2}$$

where $S(t) : H \to H$ is a semiflow. Furthermore, we require that f is globally bounded in H. In Chapter 15 we showed that certain classes of PDEs have global bounded absorbing sets, and if this fact has been shown, then we can just cut off the nonlinearity outside some large ball. Therefore, the requirement on f is less restrictive than one might think at first.

The final goal of our analysis here and in Chapter 17 is to establish that certain PDEs of the form (16.1) essentially reduce to finite-dimensional ODE problems.

Definition 16.1. Consider a semiflow $S(t) : H \to H$ and define an **inertial manifold** $\mathcal{M}$ as a finite-dimensional, exponentially attracting, and invariant (sufficiently) smooth manifold for $S(t)$.

Note that **invariance** means here that $S(t)\mathcal{M} \subset \mathcal{M}$ for $t \geq 0$, i.e., formally one should say **positively invariant**; see Figure 16.1.

Instead of tackling the inertial manifold problem directly, we are going to focus on local stable and unstable manifolds for a saddle point of

$$\partial_t u + Bu = F(t, u) \tag{16.3}$$

for a nonlinearity F and a self-adjoint sectorial operator $B : H \to H$. One approach is to study the related inhomogeneous linear problem

$$\partial_t u + Bu = h(t) \tag{16.4}$$

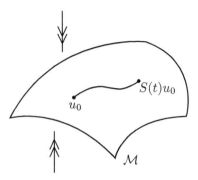

Figure 16.1. *Sketch of a positively invariant manifold $\mathcal{M}$ with a trajectory $u(t) = S(t)u_0$ starting in $\mathcal{M}$ and also remaining in $\mathcal{M}$. We illustrated the situation in which $\mathcal{M}$ is attracting in the normal directions, and positive invariance means that $S(t)u_0 = u(t)$ is not going to leave $\mathcal{M}$ for $t \geq 0$.*

for a given smooth forcing $h(t)$. Suppose H is split into an orthogonal sum

$$H = H_+ \oplus H_- \tag{16.5}$$

and denote the associated projections by $P_+ : H \to H_+$ and $P_- : H \to H_-$. Furthermore, assume $B = \mathrm{diag}(B_+, B_-)$ is split accordingly, i.e., for $\theta > 0$ we have

$$\begin{aligned}
\langle B_+ u, u \rangle_H &\leq -\theta \|u\|_H^2 \quad &\text{for } u \in H_+ \cap \mathcal{D}(B), \\
\langle B_- u, u \rangle_H &\geq \theta \|u\|_H^2 \quad &\text{for } u \in H_- \cap \mathcal{D}(B).
\end{aligned} \tag{16.6}$$

Note that (16.4) is equivalent to

$$\begin{aligned}
\partial_t u_+ + B_+ u_+ &= h_+(t), \\
\partial_t u_- + B_- u_- &= h_-(t),
\end{aligned} \tag{16.7}$$

where $u_\pm = P_\pm u$ and $h_\pm = P_\pm h$; recall that we have encountered a similar projection splitting already in Chapter 4. For $h \equiv 0$, it follows that (16.7) is solved by

$$u_+(t) = \mathrm{e}^{-tB_+} u_+(0), \qquad u_-(t) = \mathrm{e}^{-tB_-} u_-(0).$$

Due to (16.6) we may then conclude that

$$\begin{aligned}
\|\mathrm{e}^{-tB_+}\|_{\mathcal{L}(H,H)} &\leq \mathrm{e}^{\theta t} \quad &\text{for } t \leq 0, \\
\|\mathrm{e}^{-tB_-}\|_{\mathcal{L}(H,H)} &\leq \mathrm{e}^{-\theta t} \quad &\text{for } t \geq 0,
\end{aligned} \tag{16.8}$$

which just means that $u \equiv 0$ is a saddle point for the linear system with unstable eigenspace H_+ and stable eigenspace H_-; see also Chapter 1 for the finite-dimensional saddle case and Chapter 10 for exponential dichotomies of ODEs. It is instructive to see that we may bound $u_\pm$ in the case of nonzero forcing $h \neq 0$. Consider u_+ and observe that Theorem 14.11 implies

$$u_+(t) = \mathrm{e}^{(t_0 - t)B_+} u_+(t_0) + \int_{t_0}^{t} \mathrm{e}^{(s-t)B_+} h_+(s) \, \mathrm{d}s.$$

However, we know that e^{-tB_+} contracts on H_+ as $t \to -\infty$ so, if we let $t_0 \to \infty$, we have

$$u_+(t) = -\int_{t}^{\infty} \mathrm{e}^{(s-t)B_+} h_+(s) \, \mathrm{d}s. \tag{16.9}$$

Suppose that $h \in C_b(\mathbb{R}, H)$ so h is bounded and continuous with values in H. Then it is easy to see that

$$\|u_+\|_{C_b(\mathbb{R},H)} \leq \frac{K}{\theta}\|h_+\|_{C_b(\mathbb{R},H)}$$

for some constant K and for $\theta > 0$ as introduced above. A similar estimate also holds for u_-. Therefore, it follows that

$$\|u\|_{C_b(\mathbb{R},H)} \leq \frac{K}{\theta}\|h\|_{C_b(\mathbb{R},H)}$$

for some generic constant $K > 0$ independent of θ. It turns out that working in the different norm

$$\|u\|^2_{L^2(\mathbb{R},H)} := \int_{\mathbb{R}} \|u(t)\|^2_H \, dt,$$

we can determine a sharp value of K.

Lemma 16.2. *([576, Lem. 2.2]) Consider the inhomogeneous problem* (16.4); *suppose* $h \in L^2(\mathbb{R}, H)$ *and* $h \neq 0$. *Then*

$$\|u\|_{L^2(\mathbb{R},H)} \leq \frac{1}{\theta}\|h\|_{L^2(\mathbb{R},H)},$$

where $\theta > 0$ *is the constant from* (16.6). *In particular, the solution operator* $R : L^2(\mathbb{R}, H) \to L^2(\mathbb{R}, H)$ *of* (16.4) *has norm bounded by* $1/\theta$.

We return to the full problem (16.3) and assume that F is globally Lipschitz,

$$\|F(t,u) - F(t,v)\|_H \leq \kappa\|u - v\|_H, \tag{16.10}$$

uniformly for $t \in \mathbb{R}$. Furthermore, we assume that the nonlinear problem has a hyperbolic saddle point at $u \equiv 0$, i.e., the linear operator B still satisfies (16.8) and

$$F(t,0) \equiv 0.$$

The next step is to prove an analogue of Theorem 2.11 showing persistence of the linear spaces $H_\pm$ as local stable and unstable manifolds. However, the construction turns out to be more technical than for the finite-dimensional case.

Definition 16.3. Fix $\tau \in \mathbb{R}$. The **unstable set** $\mathcal{M}_+(\tau) \subset H$ consists of all $u_\tau \in H$ such that there exists a backward trajectory $u(t)$ with $t \leq \tau$ with

$$u(\tau) = u_\tau, \qquad \|u\|_{L^2((-\infty,\tau],H)} < \infty.$$

Similarly, the **stable set** $\mathcal{M}_-(\tau) \subset H$ consists of all $u_\tau \in H$ such that there exists a forward trajectory $u(t)$ with $t \geq \tau$ with

$$u(\tau) = u_\tau, \qquad \|u\|_{L^2([\tau,+\infty),H)} < \infty.$$

Essentially, the sets $\mathcal{M}_+(\tau)$ and $\mathcal{M}_-(\tau)$ turn out to be the unstable and stable manifolds of the saddle point at $u \equiv 0$. The time τ is somewhat arbitrary, and we shall mostly work with $\tau = 0$ to simplify the notation. However, we keep in mind that the following arguments work in more generality, in fact, even uniformly in τ. The key point is that the sets $\mathcal{M}_\pm = \mathcal{M}_\pm(0)$ turn out to be (Lipschitz) manifolds.

Theorem 16.4. *(PDE stable/unstable manifolds) Suppose the **spectral gap condition***

$$\theta > \kappa \qquad (16.11)$$

holds where θ controls the linear contraction/expansion rates in (16.8) and κ is the Lipschitz constant of the nonlinearity F in (16.10). Then $\mathcal{M}_{\pm}$ are Lipschitz manifolds, i.e.,

$$\mathcal{M}_{\pm} = \{u_{\pm} + M_{\pm}(u_{\pm}),\ u_{\pm} \in H_{\pm}\},$$

where the maps $M_{\pm} : H_{\pm} \to H_{\mp}$ satisfy

$$\|M_{\pm}(v_1) - M_{\pm}(v_2)\|_{H_{\mp}} \leq K\|v_1 - v_2\|_{H_{\pm}}$$

for some constant $K > 0$.

Proof. (Sketch) We only consider $\mathcal{M}_{+}$ since the same arguments can be adapted to $\mathcal{M}_{-}$. Suppose we could verify that for every $u_0 \in H_{+}$ there exists a unique solution $u \in L^2((-\infty, 0], H)$ to

$$\partial_t u + Bu = F(t, u), \quad (P_{+}u)(0) = u_0, \qquad (16.12)$$

and that u depends in a Lipschitz continuous way on u_0. In this scenario, one may simply define

$$M_{+}(u_0) := (P_{-}u)(0). \qquad (16.13)$$

This mapping simply sends the element u_0 to the suitable stable set via solving (16.12); see Figure 16.2.

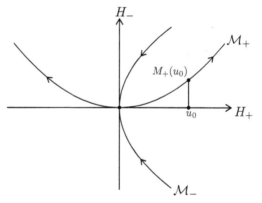

Figure 16.2. *Sketch of the construction of the mapping M_{+} given in (16.13) via solving the auxiliary problem (16.12). Compare this to the situation of the classical finite-dimensional Stable-Unstable Manifold Theorem 2.11.*

Hence it remains to solve (16.12). A natural idea is to measure the deviation from the linear problem and introduce for $t \leq 0$

$$w(t) := u(t) - v(t), \qquad v(t) := e^{-tB_{+}}u_0.$$

In particular, w then solves

$$\partial_t w + Bw = F(t, w + e^{-tB_{+}}u_0), \quad (P_{+}w)(0) = 0. \qquad (16.14)$$

One may extend (16.14) to an equivalent equation for $t \in \mathbb{R}$ by defining

$$\tilde{F}(t, u_0, w) := \begin{cases} F(t, w + v(t)) & \text{for } t < 0, \\ 0 & \text{for } t \geq 0, \end{cases}$$

and consider the problem to be

$$\partial_t w + Bw = \tilde{F}(t, u_0, w), \qquad w \in L^2(\mathbb{R}, H); \qquad (16.15)$$

see also Exercise 16.5. To solve (16.15) one considers the equivalent fixed-point problem

$$w = R \circ \tilde{F}(\cdot, u_0, w), \qquad (16.16)$$

where R is the solution operator defined in Lemma 16.2. It is natural to try to solve (16.16) using the Banach fixed-point theorem on the Banach space $L^2(\mathbb{R}, H)$. A key step is to derive the estimate

$$\|\tilde{F}(t, u_{0,1}, w_1) - \tilde{F}(t, u_{0,2}, w_2)\|_{L^2(\mathbb{R}, H)} \leq \kappa \|w_1 + v_1 - w_2 - v_2\|_{L^2((-\infty, 0], H)}$$
$$\leq \kappa \left(\|w_1 - w_2\|_{L^2(\mathbb{R}, H)} \qquad (16.17) \right.$$
$$\left. + \theta^{-1} \|u_{0,1} - u_{0,2}\|_H \right),$$

where Lipschitz continuity of F was used in the first inequality and the second inequality follows by using the fact

$$\|v\|_{L^2((-\infty, 0], H)} \leq \frac{1}{\theta} \|u_0\|_H,$$

which follows from the definition of v and expansion properties of B_+ considered in (16.8). Since $F(t, 0) \equiv 0$, one observes by taking $w_2 = 0 = u_{0,2}$ in (16.17) that $\tilde{F}(\cdot, u_0, w) \in L^2(\mathbb{R}, H)$ if $u_0 \in H_+$ and $w \in L^2(\mathbb{R}, H)$. Therefore, (16.16) is a well-defined mapping on $L^2(\mathbb{R}, H)$. In addition, Lemma 16.2 yields that the operator norm of R is bounded by $1/\theta$, which implies in combination with (16.17) that the mapping (16.16) is a contraction if $\kappa/\theta < 1$. Since this is precisely the spectral gap condition (16.11), the existence of M_+ follows. The Lipschitz continuity is discussed in Exercise 16.6. $\qquad \square$

Lastly, we note that a little bit of extra work shows that solutions in M_+ indeed decay in backward time:

$$\|u(t)\|_H \leq K e^{\delta t} \|u_0\|_H, \qquad t \leq 0,$$

for $|\delta| < \theta - \kappa$, $\delta > 0$. Hence, we could also formally write $\mathcal{M}_+ = W^u(0)$ as the unstable manifold, and similarly for the stable manifold $\mathcal{M}_- = W^s(0)$. In the vast majority of geometric dynamics approaches, a stable/unstable manifold result for a hyperbolic saddle point is the fundamental basis. Then, this result is used for a more detailed analysis of phase space.

Exercise 16.5. Prove that solving (16.14) is equivalent to (16.15). $\Diamond$

Exercise 16.6. Show that the mapping M_+ is Lipschitz. $\Diamond$

Exercise 16.7. Construct a nontrivial (and nonlinear!) example for (16.1), which satisfies all the assumptions used in this chapter. $\Diamond$

Background and Further Reading: The main line of argument follows the notes in [576]. The Stable-Unstable Manifold Theorem for hyperbolic equilibria of ODEs is discussed in many sources; see, e.g., [71, 247, 320, 325, 529]. The theory can be extended very substantially to cover entire manifolds with a hyperbolic splitting between the tangent and normal dynamics, i.e., requiring **normal hyperbolicity** [194, 269, 346, 534, 562]. The two main proof techniques for stable/unstable and normally hyperbolic manifolds are the functional iteration (or Lyapunov-Perron [388, 456]) method and the graph transform. The case of center manifolds includes nonhyperbolic situations and is already very subtle for ODEs [85, 503]. There are many important resources developing invariant manifold theory for PDEs; see, for example, the material and references in [35, 47, 48, 545].

Chapter 17

Spectral Gap and Inertial Manifolds

We continue the topic from Chapter 16, studying PDEs

$$\partial_t u + Au = f(u), \qquad u_0 \in H, \tag{17.1}$$

where $-A : H \to H$ is a linear, negative, self-adjoint, and sectorial operator on a Hilbert space H and f is a sufficiently smooth globally Lipschitz nonlinearity. Recall also that (17.1) generates a semiflow $S(t) : H \to H$ as defined in (16.2) and that we required that f is globally bounded in H. In this chapter, we want to establish the existence of an **inertial manifold** $\mathcal{M}$ for (17.1) to demonstrate that the dynamics is low-dimensional; see Definition 16.1.

The **Hilbert-Schmidt Theorem** [472, p. 268] states that the compact, self-adjoint, bounded operator A has a complete orthonormal system in H,

$$Ae_j = \lambda_j e_j, \qquad 0 < \lambda_1 \le \lambda_2 \le \cdots,$$

with eigenfunctions e_j and eigenvalues λ_j. Therefore, we have

$$v = \sum_{j=1}^{\infty} v_j e_j, \qquad v_j := \langle v, e_j \rangle_H$$

for every $v \in H$. A natural idea to construct a low-dimensional inertial manifold is simply to project functions onto the first few (Fourier) modes

$$P_J v := \sum_{j=1}^{J} v_j e_j.$$

Also setting $Q_J := \mathrm{Id} - P_J$ leads us naturally to define the linear spaces

$$H_+ := P_J H, \qquad H_- := Q_J H, \qquad H = H_- \oplus H_+.$$

As discussed in a similar setting in Chapter 16, it is helpful to present the PDE (17.1) as a system

$$\partial_t u_+ + Au_+ = f_+(u_+ + u_-), \tag{17.2}$$
$$\partial_t u_- + Au_- = f_-(u_+ + u_-), \tag{17.3}$$

where $f_+ := P_J F$, $f_- := Q_J F$, $u_+ := P_J u$, and $u_- := Q_J u$. By construction we have

$$\begin{aligned} \langle Av, v \rangle_H &\leq \lambda_J \|v\|_H^2 && \text{for } v \in H_+ \cap \mathcal{D}(A), \\ \langle Av, v \rangle_H &\geq \lambda_{J+1} \|v\|_H^2 && \text{for } v \in H_- \cap \mathcal{D}(A). \end{aligned} \tag{17.4}$$

The crucial idea is to construct a mapping $\Phi : H_+ \to H_-$ such that the inertial manifold is given by

$$\mathcal{M} := \{u_+ + \Phi(u_+),\ u_+ \in H_+\}, \qquad u_- = \Phi(u_+),$$

so $\mathcal{M}$ is parametrized by a finite-dimensional set of variables and the dynamics on $\mathcal{M}$ reduces to the ODEs

$$\partial_t u_+ + A u_+ = f_+(u_+ + \Phi(u_+)), \tag{17.5}$$

which is also called the **inertial form**.

Example 17.1. We may view the situation in analogy to multiple time-scale systems. A **fast-slow system** is given by

$$\begin{aligned} \varepsilon \frac{dx}{d\tau} &= f(x, y, \varepsilon), \\ \frac{dy}{d\tau} &= g(x, y, \varepsilon), \end{aligned} \tag{17.6}$$

where $\tau \in \mathbb{R}$ is the time variable, $(x, y) \in \mathbb{R}^{d_x + d_y} =: \mathbb{R}^d$, the maps f, g are sufficiently smooth, and $\varepsilon > 0$ is assumed to be small. The variables $x \in \mathbb{R}^{d_x}$ are **fast** while the variables $y \in \mathbb{R}^{d_y}$ are **slow**. The set

$$\mathcal{C}_0 := \{(x, y) \in \mathbb{R}^{d_x + d_y} : f(x, y, 0) = 0\} \tag{17.7}$$

is called the **critical manifold**. $\mathcal{C}_0$ is called **normally hyperbolic** if the matrix

$$\mathrm{D}_x f(x, y, 0)|_{(x,y) \in \mathcal{C}_0} \in \mathbb{R}^{d_x \times d_x}$$

has no eigenvalues with zero real parts. In the normally hyperbolic case **Fenichel's Theorem** (see background references below) states that for sufficiently small ε there exists a perturbed invariant manifold $\mathcal{C}_\varepsilon$, called a **slow manifold**. The manifold $\mathcal{C}_\varepsilon$, as well as the dynamics on $\mathcal{C}_\varepsilon$, converges to $\mathcal{C}_0$ as $\varepsilon \to 0$; see also Figure 17.1.

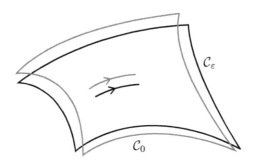

Figure 17.1. *Critical manifold $\mathcal{C}_0$ defined in (17.7) on which the slow dynamics (17.8) (single grey arrow) takes place. The full system dynamics for (17.6) can then be approximated by the slow manifold $\mathcal{C}_\varepsilon$ (black) if we have normal hyperbolicity.*

In particular, if $\mathcal{C}_\varepsilon = \{x = h_\varepsilon(y)\}$ for some map $h_\varepsilon : \mathbb{R}^{d_x} \to \mathbb{R}^{d_y}$, then the effective slow dynamics on $\mathcal{C}_\varepsilon$ is just given by

$$\frac{dy}{d\tau} = g(h_\varepsilon(y), y). \tag{17.8}$$

Note the similarity of (17.8) to the PDE case discussed above leading to equation (17.5). The map Φ above is the direct analogue of the parametrization h_ε here. However, for our PDE setting we also required exponential attraction for the inertial manifold $\mathcal{M}$, which would correspond to requiring attraction of $\mathcal{C}_\varepsilon$ for $0 \leq \varepsilon \ll 1$ with respect to the fast directions; see Chapter 36 for more on fast-slow systems. $\blacklozenge$

The last example shows that it is helpful to think of $\mathcal{M}$ as an attracting slow manifold where the fast variables u_- decay very quickly and the effective long-term dynamics is given by the slow variables u_+.

Theorem 17.2. *(existence of inertial manifolds) Consider* (17.1) *under the assumptions stated above and suppose furthermore that for some J there exists a spectral gap*

$$\lambda_{J+1} - \lambda_J > 2\kappa, \tag{17.9}$$

where $\kappa > 0$ is the Lipschitz constant of f. Then there exists a J-dimensional inertial manifold $\mathcal{M}$ defined via $\Phi : H_+ \to H_-$. In particular, for each u_0 there exists $v_0 \in \mathcal{M}$ such that

$$\|S(t)u_0 - S(t)v_0\|_H \leq Ke^{-\lambda_J t}\|u_0 - v_0\|_H$$

for some constant $K > 0$.

Proof. (Sketch [576]) The idea is to use the previous work from Chapter 16 on invariant manifolds for nonlinear saddles and to obtain $\mathcal{M}$ as a certain unstable manifold. We define

$$B := A - \frac{\lambda_J + \lambda_{J+1}}{2}\mathrm{Id}$$

and observe that

$$\begin{aligned}\langle B_+ u, u\rangle_H &\leq -\theta\|u\|_H^2 &\quad \text{for } u \in H_+ \cap \mathcal{D}(A), \\ \langle B_- u, u\rangle_H &\geq \theta\|u\|_H^2 &\quad \text{for } u \in H_- \cap \mathcal{D}(A)\end{aligned} \tag{17.10}$$

hold due to (17.4) with $\theta = \frac{\lambda_{J+1} - \lambda_J}{2}$. For simplicity we assume $f(0) \equiv 0$. Define

$$\tilde{u}(t) := e^{\alpha t}u(t), \qquad \alpha := \frac{\lambda_J + \lambda_{J+1}}{2}.$$

This transforms (17.1) to

$$\partial_t \tilde{u} + B\tilde{u} = F(t, \tilde{u}), \qquad F(t, \tilde{u}) := e^{\alpha t}f(e^{-\alpha t}\tilde{u}). \tag{17.11}$$

It can be checked that F is also Lipschitz continuous with the same Lipschitz constant $\kappa > 0$. We want to apply Theorem 16.4, which holds for equations of the form (17.11). The assumptions about A and the definition of B easily lead to the correct contraction and expansion rates (16.6) for B. The spectral gap condition in Theorem 16.4,

$$\theta > \kappa,$$

is immediately guaranteed by (17.9). Furthermore, $F(t, 0) \equiv 0$ since $f(0) \equiv 0$. Hence, the existence of an unstable manifold $\mathcal{M}_+$ at $\tilde{u} = 0$ for (17.11) follows. One may then check that $\mathcal{M}_+$ is indeed invariant under the semiflow $S(t)$ of the original problem (17.1). It requires quite a bit of extra work to then also show that trajectories must track it exponentially in forward time; the idea is to use that $e^{-\alpha t}\tilde{u}(t) = u(t)$ by construction so $\mathcal{M}_+ = \mathcal{M}$ is indeed the inertial manifold we wanted to construct. $\square$

Remark: Theorem 17.2 can be generalized in various directions. For example, the smoothness of $\mathcal{M}$ can be improved in many cases. There are also other common alternative proof techniques, e.g., a method based upon **invariant cones** and the **squeezing property**. See background references below for more details.

Example 17.3. A classical example for the existence of inertial manifolds is dissipative reaction-diffusion equations, such as

$$\partial_t u - \partial_x^2 u = f(u), \quad (x,t) \in \Omega \times [0,+\infty), \; u = u(x,t), \tag{17.12}$$

for a suitable sufficiently smooth nonlinearity $f : \mathbb{R} \to \mathbb{R}$ on a bounded interval $\Omega \subset \mathbb{R}^1$ with Dirichlet boundary conditions $u(x) = 0$ for $x \in \partial\Omega$. We have seen in Chapter 15 that under certain assumptions on the nonlinearity an absorbing set exists. In such a case, we may cut off the nonlinearity and assume that f is indeed globally Lipschitz. Since $A = -\partial_x^2$, we know from Example 6.3 that the eigenvalues satisfy

$$\lambda_J = K J^2 \qquad \text{as } J \to +\infty \tag{17.13}$$

for some constant $K > 0$ depending on the length of the interval; in fact, (17.13) is a version of **Weyl's law**

$$\lambda_J \sim K J^2 \qquad \text{as } J \to +\infty,$$

which holds for many other situations involving the Laplacian. We find from (17.13) that

$$\lambda_{J+1} - \lambda_J = K(J+1)^2 - K J^2 = K(2J+1) > 2\kappa$$

for some sufficiently large J. Therefore, the spectral gap condition (17.9) holds and Theorem 17.2 implies the existence of an inertial manifold. ◆

It should be noted that, although it is theoretically very important to know that a certain PDE is effectively finite-dimensional, it is not always immediately useful in practical applications. For example, the dimension of the inertial form ODE system (17.5) could be extremely large or the mapping Φ is difficult to compute; see also Chapters 18–19.

Example 17.4. Consider the **Kuramoto-Sivashinsky equation**

$$\partial_t u + \partial_x^4 u + 2p\partial_x^2 u + \partial_x(u^2) = 0 \tag{17.14}$$

with a parameter $p \in \mathbb{R}$ posed on an interval $\Omega = [0,\pi]$, $(x,t) \in \Omega \times [0,\infty)$, $u = u(x,t)$ and boundary conditions

$$u(x) = 0, \quad (\partial_x^2 u)(x) = 0 \quad \text{for } x \in \partial\Omega = \{0,\pi\}.$$

One may show that (17.14) is well posed and dissipative on $H := L^2(\Omega)$ with an absorbing ball in a suitable space. However, Theorem 17.2 is not directly applicable as the nonlinearity $f(u) = -\partial_x(u^2)$ is not a map from H to H. ◆

The last example shows the need to slightly generalize Theorem 17.2. Consider (17.1) on the Hilbert space H and define

$$H^s := \mathcal{D}(A^s), \; s \in \mathbb{R}, \qquad \|v\|_{H^s}^2 = \sum_{j=1}^{\infty} \lambda_j^{2s} \langle v, e_j \rangle_H^2.$$

Note that for $s > 0$ the definition of H^s is just Definition 14.10 of fractional operator norms, while for $s < 0$ one just takes H^s as the completion of H with respect to the norm $\| \cdot \|_{H^s}$; furthermore $H^0 = H$.

Remark: One frequently finds the definition $H^s := \mathcal{D}(A^{s/2})$ with $\|v\|_{H^s}^2 = \sum_{j=1}^{\infty} \lambda_j^s \langle v, e_j \rangle_H^2$ in the literature, so one just has to keep track of the factor of 2 to match the different definitions.

Suppose now f is a Lipschitz map from H^{α_1} to H^{α_2},

$$\|f(u) - f(v)\|_{H^{\alpha_1}} \le \kappa \|u - v\|_{H^{\alpha_2}} \qquad (17.15)$$

for $\alpha_1 < \alpha_2$ and $u, v \in H^{\alpha_2}$.

Theorem 17.5. *(existence of inertial manifolds) Consider* (17.1) *with the assumptions stated above and suppose f satisfies* (17.15) *for $\alpha_2 = 0$ and some $\alpha_1 \in (-2, 0]$. Furthermore, suppose the spectral gap condition*

$$\frac{\lambda_{J+1} - \lambda_J}{\lambda_{J+1}^{-\alpha_1/2} + \lambda_J^{-\alpha_1/2}} > \kappa \qquad (17.16)$$

holds; then there exists a Lipschitz inertial manifold.

The exercises serve as a guide for how to apply the last result to the Kuramoto-Sivashinsky equation to prove the existence of an inertial manifold.

Exercise 17.6. Consider Example 17.4 and show that $A := \partial_x^4 + 2p\partial_x^2 + p^2 + 1$ is positive and self-adjoint on $\mathcal{D}(A)$. Rewrite (17.14) in the form (17.1) using A, i.e., compute f. ◊

Exercise 17.7. Continue with Exercise 17.6 and check that A has eigenvalues $\lambda_J = (J^2 - p)^2 + 1$ and use this result to verify that the spectral gap condition (17.16) must hold for sufficiently large J. ◊

Exercise 17.8. Continue with Exercises 17.6-17.7 and prove that f is Lipschitz with $\alpha_1 = -1/2$ and $\alpha_2 = 0$ once the nonlinearity f has been cut off properly. ◊

Background and Further Reading: In this chapter, we mainly followed the notes [576]. Detailed important sources on inertial manifold theory are [125, 477, 528]. For more on variations of the concept and original papers, we refer to a few further interesting sources [115, 126, 205, 206, 393, 397, 535]. An overview of finite-dimensional multiple time scale dynamics can be found in [346]; see also [195, 302] and Chapters 33–36. The Kuramoto-Sivashinsky equation is another benchmark model in PDE dynamics [285, 329, 403]; inertial/invariant manifolds for this PDE are one well-studied theme [20, 300, 434].

Chapter 18

Attractors and the Variational Equation

Consider the dissipative semiflow $S(t) : X \to X$ on a Banach space X for your favorite PDE; see Chapter 14 for examples and definitions. In Chapter 15, we discussed absorbing sets as global invariant sets containing the long-time dynamics and extended this idea in Chapters 16–17 to inertial manifolds. Here we also consider PDEs with an absorbing set $\mathcal{B}$ but instead focus on the notion of an **attractor**.

Definition 18.1. An invariant set $\mathcal{A} \subset X$ is called an **attractor** if there exists an open neighborhood $\mathcal{N}$ such that for every $u_0 \in \mathcal{N}$ we have

$$\inf_{\tilde{u} \in \mathcal{A}} \|S(t)u_0 - \tilde{u}\|_X \to 0 \qquad \text{as } t \to \infty.$$

The largest open set $\mathcal{N}$ satisfying this property is called the **basin of attraction**. If we have

$$\mathcal{A} = \bigcap_{t>0} S(t)\overline{\mathcal{B}},$$

then $\mathcal{A}$ is called a **maximal attractor**.

Remark: Even for relatively simple cases, there are other options to define an attractor—not only the very classical Definition 18.1. For example, in Figure 18.1(a) a maximal attractor is shown for a planar ODE. However, one never really sees large parts of the maximal attractor in practice. Hence, one could also use a measure-theoretic viewpoint. Suppose the absorbing set $\mathcal{B}$ can be endowed with a finite measure. Then we could define a statistical attractor, also called a **Milnor attractor**, as the smallest closed set containing ω-limit sets of orbits of almost all points in $\mathcal{B}$; see also Figure 18.1(b). Here we shall continue with Definition 18.1, but it is important to remember that this definition may incorporate larger sets than we might like for certain practical applications.

Our key example in this chapter will be the (**damped**) **sine-Gordon equation**

$$\partial_t^2 u + \partial_t u = \Delta u - \sin(u), \qquad u = u(x,t), \tag{18.1}$$

with $(x,t) \in \Omega \times [0, \infty)$ for a square domain $\Omega = [0, \pi]^2 \subset \mathbb{R}^2$, with Dirichlet boundary conditions on $\partial\Omega$, and initial conditions

$$u_0(x) = u(x,0), \qquad u_1(x) = \partial_t u(x,0).$$

One may then prove that the sine-Gordon equation generates a semiflow [528, p. 189–190]

$$S(t) : H_0^1(\Omega) \times L^2(\Omega) \to H_0^1(\Omega) \times L^2(\Omega), \qquad S(t)\begin{pmatrix} u_0 \\ u_1 \end{pmatrix} = \begin{pmatrix} u(t) \\ \frac{du}{dt}(t) \end{pmatrix}.$$

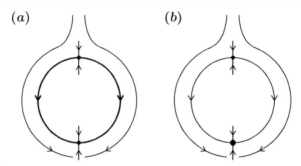

Figure 18.1. *Difference between two attractor definitions sketched for a simple ODE phase portrait. (a) Definition via maximal invariant sets, so that we get an entire circle as the attractor, including two steady states and the two heteroclinic orbits between them. (b) In the Milnor/statistical attractor sense, we only obtain the locally stable steady state at the bottom marked by a bigger dot as the attractor.*

Furthermore, an absorbing set can be constructed similarly to the techniques in Chapter 15, and together with further technical steps, one may then establish the existence of an attractor.

Theorem 18.2. *(damped sine-Gordon attractor [528, p. 201]) The (damped) sine-Gordon equation* (18.1) *has a compact, connected, maximal attractor $\mathcal{A}$ in $X = H_0^1(\Omega) \times L^2(\Omega)$.*

As nice as Theorem 18.2 is, it is not very quantitative. To make the estimate at least a little bit more quantitative, i.e., to get a grip on the "dimension" of $\mathcal{A}$, we first rewrite the semiflow $S(t)$ slightly as $S_\varepsilon(t)$ by setting

$$v = \left(u, \frac{\mathrm{d}u}{\mathrm{d}t} + \varepsilon u \right)^\top \tag{18.2}$$

for $\varepsilon \in (0, \varepsilon_0]$. Actually it turns out that one needs

$$\varepsilon_0 = \frac{1}{4} \tag{18.3}$$

for our purposes here. Using (18.2) we obtain the equation

$$\frac{\mathrm{d}v}{\mathrm{d}t} = A_\varepsilon v - f(v) =: F(v), \tag{18.4}$$

where the simple rewriting and shifting just implies that

$$A_\varepsilon = \begin{pmatrix} -\varepsilon \mathrm{Id} & \mathrm{Id} \\ \Delta + \varepsilon(1-\varepsilon)\mathrm{Id} & (\varepsilon - 1)\mathrm{Id} \end{pmatrix}, \qquad f(v) = \begin{pmatrix} 0 \\ \sin(u) \end{pmatrix}.$$

It is relatively easy to see that $S_\varepsilon(t)$ is still a well-defined semiflow on X. So if $v_\mathcal{A}(t)$ is a solution of (18.4) lying in the attractor $\mathcal{A}$, then it is quite natural to consider the **variational equation**

$$\frac{\mathrm{d}V}{\mathrm{d}t} = \mathrm{D}_v F(v_\mathcal{A}(t))V, \qquad V(0) =: \eta \in X. \tag{18.5}$$

If we can control this nonautonomous system along the attractor *uniformly*, then there is certainly hope of getting an estimate for a suitable notion of "dimension"; see Figure 18.2.

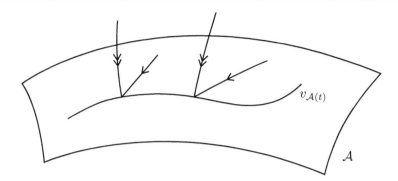

Figure 18.2. *Illustration of the variational equation* (18.5) *for an attractor* $\mathcal{A}$. *We linearize along a trajectory* $v_{\mathcal{A}(t)}$. *Then there are generically several directions of different strengths pushing towards the attractor.*

To capture these effects, we are going to use the fact that $X = H_0^1(\Omega) \times L^2(\Omega)$ is actually a Hilbert space with the inner product formed by the sum of the inner products in $H_0^1(\Omega)$ and $L^2(\Omega)$ so that it is possible to define an **exterior product** $\wedge$ and an associated norm/volume $\|\cdot\|_{\wedge^m X}$; see Exercise 18.5. Let $\eta_1, \ldots, \eta_m$ be initial conditions for (18.5) with associated solutions $V_1, \ldots, V_m$. The important formula we just state here is

$$\|V_1(t) \wedge \cdots \wedge V_m(t)\|_{\wedge^m X} = \|\eta_1(t) \wedge \cdots \wedge \eta_m(t)\|_{\wedge^m X} \tag{18.6}$$
$$\cdot \exp\left[\int_0^t \operatorname{Tr} D_v F(S_\varepsilon(s)v_0) \circ Q_m(s)\mathrm{d}s\right],$$

where $Q_m(t)$ is the orthogonal projection in X onto the space spanned by $V_1(t), \ldots, V_m(t)$. Observe that (18.6) is just a variation-of-constants type formula for the linearized growth or decay of an initial volume along the attractor.

Example 18.3. Just consider the (almost) trivial ODE case

$$\frac{\mathrm{d}V}{\mathrm{d}t} = \begin{pmatrix} -1 & 0 \\ 0 & -2 \end{pmatrix} V, \qquad V(0) = \eta \in \mathbb{R}^2 \tag{18.7}$$

with global attractor $V = 0$. The linearized problem along $V = 0$ obviously just coincides with (18.7). Let us pick two initial conditions $\eta_1 = (1, 0)^\top$ and $\eta_2 = (0, 1)^\top$; then we have two solutions

$$V_1(t) = (\mathrm{e}^{-t}, 0)^\top \qquad \text{and} \qquad V_2(t) = (0, \mathrm{e}^{-2t})^\top.$$

Therefore, we can evaluate the left-hand side in (18.6) explicitly as $V_1 = \mathrm{e}^{-t}(1, 0)^\top$ and $V_2 = \mathrm{e}^{-2t}(0, 1)^\top$; the wedge product definition from standard linear algebra says

$$\|V_1(t) \wedge V_2(t)\|_{\wedge^2 \mathbb{R}^2}^2 = \det((\langle V_i(t), V_j(t)\rangle)_{1 \leq i,j \leq 2} = \mathrm{e}^{-t}\mathrm{e}^{-2t},$$

which yields the time evolution of the area of the rectangle generated by the two basis vectors under the flow of (18.7). ◆

Theorem 18.4. *(variational equation estimate)* *For the (damped) sine-Gordon equation* (18.1)*, the variational equation along the attractor admits the growth bound*

$$\operatorname{Tr} D_v F(v_{\mathcal{A}}(t)) \circ Q_m(t) \leq -\frac{m}{4}\varepsilon + \frac{1}{\varepsilon}\sum_{j=1}^m \lambda_j^{-1},$$

where λ_j *are the eigenvalues of the negative Laplacian on* Ω.

Proof. (Sketch [528]) Let $\Phi_j = (\xi_j, \zeta_j)$ for $j = 1, \ldots, m$ denote an orthonormal basis of $Q_m X$; obviously all objects are time dependent here but we shall later on provide uniform estimates over the attractor, and so we just drop the time dependence t for now from the notation. We have

$$\operatorname{Tr} D_v F(v_\mathcal{A}) \circ Q_m = \sum_{j=1}^{\infty} \langle D_v F(v_\mathcal{A}) \circ Q_m \Phi_j, \Phi_j \rangle_X$$

$$= \sum_{j=1}^{m} \langle D_v F(v_\mathcal{A}) \Phi_j, \Phi_j \rangle_X. \tag{18.8}$$

Each individual inner product can be estimated. Indeed, we have

$$\langle D_v F(v_\mathcal{A}) \Phi_j, \Phi_j \rangle_X = \langle A_\varepsilon \Phi_j, \Phi_j \rangle_X - \langle \cos(v_\mathcal{A}) \xi_j, \zeta_j \rangle_{L^2}, \tag{18.9}$$

where the term $\cos(v_\mathcal{A})$ just arises from the differentiation of the nonlinearity $\sin(u)$ in (18.1). For the first term in (18.9) ones finds

$$\langle A_\varepsilon \Phi_j, \Phi_j \rangle_X = -\varepsilon \|\xi_j\|_{H_0^1}^2 + (\varepsilon - 1)\|\zeta_j\|_{L^2}^2 + \varepsilon(1 - \varepsilon)\langle \xi_j, \zeta_j \rangle_{L^2}$$

$$\leq -\frac{\varepsilon}{2} \left(\|\xi_j\|_{H_0^1}^2 + \|\zeta_j\|_{L^2}^2 \right), \tag{18.10}$$

where the last inequality is Exercise 18.6(a). The second term is relatively easy since cos is uniformly bounded so that

$$-\langle \cos(v_\mathcal{A}) \xi_j, \zeta_j \rangle_{L^2} \leq |\cos(v_\mathcal{A})| \|\xi_j\|_{L^2} \|\zeta_j\|_{L^2} \leq \|\xi_j\|_{L^2} \|\zeta_j\|_{L^2}.$$

Therefore, using the previous estimates and **Young's inequality** in the variant with $\varepsilon > 0$ (i.e., $ab \leq a^2/\varepsilon + \varepsilon b^2/4$), we have

$$\langle D_v F(v_\mathcal{A}) \Phi_j, \Phi_j \rangle_X \leq -\frac{\varepsilon}{2} \left(\|\xi_j\|_{H_0^1}^2 + \|\zeta_j\|_{L^2}^2 \right) + \|\xi_j\|_{L^2} \|\zeta_j\|_{L^2}$$

$$\leq -\frac{\varepsilon}{4} \left(\|\xi_j\|_{H_0^1}^2 + \|\zeta_j\|_{L^2}^2 \right) + \frac{1}{\varepsilon} \|\xi_j\|_{L^2}^2.$$

However, the first term simplifies since $\|\xi_j\|_{H_0^1}^2 + \|\zeta_j\|_{L^2}^2 = 1$ by the orthonormality assumption on the basis. Taking the sum we have

$$\sum_{j=1}^{m} \langle D_v F(v_\mathcal{A}) \Phi_j, \Phi_j \rangle_X \leq -\frac{m}{4} \varepsilon + \frac{1}{\varepsilon} \sum_{j=1}^{m} \|\xi_j\|_{L^2}^2.$$

We claim that the last sum satisfies the estimate

$$\sum_{j=1}^{m} \|\xi_j\|_{L^2}^2 \leq \sum_{j=1}^{m} \lambda_j^{-1}, \tag{18.11}$$

where $\{\lambda_j\}_{j=1}^{m}$ are the eigenvalues of $-\Delta$ on Ω ordered from smallest to largest; cf. Chapter 6. To prove this claim, consider m vectors $w_j \in H_0^1$, which are orthogonal in L^2 to a fixed ξ_i and which are orthonormal in H_0^1. Then one can find vectors (see Exercise 18.6(b)) defined by

$$\theta_i := \xi_i + \sum_{k=1}^{i} \alpha_{ik} w_k, \qquad \langle \theta_i, \theta_j \rangle_{L^2} = (1 + \nu)\delta_{ij}, \tag{18.12}$$

where $\nu > 0$ is given and δ_{ij} is the Kronecker delta. Note that $\tilde{\theta}_i := (1 + \nu)^{-1/2}\theta_i$ are orthonormal in H_0^1 by construction. Now we can estimate

$$\sum_{j=1}^m \|\xi_j\|_{L^2}^2 \le \sum_{j=1}^m \|\theta_j\|_{L^2}^2 = (1+\nu)\sum_{j=1}^m \langle(-\Delta)^{-1}\tilde{\theta}_j, \tilde{\theta}_j\rangle_{H_0^1} \le (1+\nu)\sum_{j=1}^m \lambda_j^{-1},$$

which establishes the claim (18.11) upon taking $\nu \to 0$. $\square$

The bound in Theorem 18.4 will be crucial to get a (quite rough) upper bound on the attractor dimension for the sine-Gordon equation, as discussed further in Chapter 19. However, note that the calculation was of independent interest, as it has shown one way of dealing with variational equations for PDEs; cf. Chapters 9–10, where nonautonomous ODEs essentially played the role of a variational equation along a travelling wave. However, even if we control certain growth rates, the structure of low-dimensional attractors can be incredibly complicated; see Exercise 18.7.

Exercise 18.5. Recall the definition of the **exterior product**

$$V_1 \wedge \cdots \wedge V_m = \sum_\sigma (-1)^\sigma V_1 \otimes \cdots \otimes V_m,$$

where σ runs over all permutations of $\{1, \ldots, m\}$ and V_j are vectors in a given finite-dimensional vector space $\mathcal{V}$. (a) Use properties of the tensor product to show that the exterior product is multilinear. (b) Show that the exterior product vanishes if $V_i = V_j$ for $i \ne j$. (c) Generalize the norm considered in Example 18.3 for the two-dimensional case to $\|\cdot\|_{\wedge^m \mathcal{V}}$. Show that it gives the **volume** of the parallelepiped spanned by the vectors V_j appearing in the exterior product. $\Diamond$

Exercise 18.6. (a) Prove the inequality in (18.9). Hint: You have to use the assumption on ε and the Poincaré inequality with the sharp constant. (b) Prove that constants α_{ik} exist as claimed in (18.12). $\Diamond$

Exercise 18.7. Let $M = \mathbb{S}^1 \times \mathbb{D}^2$, where $\mathbb{D}^2$ is the unit disk in $\mathbb{R}^2$ and $\mathbb{S}^1 = \mathbb{R}/\mathbb{Z}$ is the unit circle. Consider the map

$$g(\phi, x_1, x_2) = \left(2\phi, \frac{1}{10}x_1 + \frac{1}{2}\cos\phi, \frac{1}{10}x_2 + \frac{1}{2}\sin\phi\right).$$

(a) Show that $g : M \to M$, i.e., it is a well-defined map back into M. (b) Show that g is injective. (c) Take initial conditions and iterate them forward on a computer. What do you observe? Remark: The attractor in this case is called the **Smale-Williams solenoid**. (d) Try to prove that on any Poincaré (or cross-)section $\{\phi_*\} \times \mathbb{D}^2$, the attractor has the structure of a product of an interval and a Cantor set inside two discs. $\Diamond$

Background and Further Reading: This chapter follows a few selected key points in [528] regarding the preparation for dimension bounds of attractors; see also [35, 521]. The theory of attractors for PDEs is also a large field, and we can mention only a few references here [34, 40, 108, 116, 256, 471, 496, 497]. There is also considerable overlap with constructions for inertial manifolds, so we also refer to the references in Chapter 17 as well as other previous chapters. For more on various definitions and theories regarding general attractors see [409, 478], and for the complex structure of basins of attraction we refer to [8, 314, 353]. Variational equations are a key tool in dynamical systems and appear virtually everywhere, explicitly or implicitly, if one wants to estimate dynamics near a given time-dependent solution. For the sine-Gordon equation, we also refer to Chapter 13 and references therein.

Chapter 19

Lyapunov Exponents and Fractal Dimension

In this chapter, we are going to continue our study of the (damped) sine-Gordon equation (18.1) on the square $[0, \pi]^2$ with Dirichlet boundary conditions. Theorem 18.4 provides a bound on the variational equation, and we are going to convert this into a bound on the attractor "dimension." We need two ingredients: a definition of dimension and the notion of Lyapunov exponents.

Definition 19.1. Let Z be a metric space and let $Y \subset Z$. Let $n_Y(\delta)$ the minimum number of balls of radius δ to cover Y and define

$$d_F(Y) = \limsup_{\delta \to 0} \frac{\ln n_Y(\delta)}{\ln 1/\delta}, \qquad (19.1)$$

which is called the **fractal dimension** of Y; see also Figure 19.1(a).

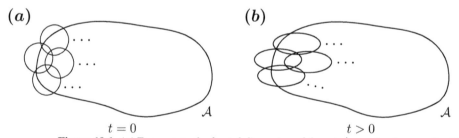

Figure 19.1. *(a) To compute the fractal dimension of the set $\mathcal{A}$, we simply cover it with balls (or solid boxes) and then count how many we need in a certain scaling limit as defined in (19.1). (b) If we think dynamically, then each ball can get deformed under the flow.*

Remark: Frequently one also refers to fractal dimension as **capacity** or **box dimension** (since one can also just take boxes instead of balls for the covering). There are many other (partially related) notions of dimension such as **Hausdorff dimension, correlation dimension**, and **pointwise dimension**.

Let $S(t) : X \to X$ be a semiflow on a Hilbert space X; we may think of the sine-Gordon semiflow $S(t)$ generated by (18.1) or $S_\varepsilon(t)$ defined via the transformation (18.2) for concreteness on the space $H_0^1(\Omega) \times L^2(\Omega)$. One may check in the sine-Gordon case that the Fréchet derivative of $u_0 \mapsto S(t)u_0$, given by

$$\mathrm{D}S(t)|_{u_0} =: L(t, u_0),$$

exists uniformly over the domain of the attractor $\mathcal{A}$, i.e., we have the limit

$$\sup_{\substack{u,\tilde{u}\in\mathcal{A} \\ \|u-\tilde{u}\|_X\le\delta}} \frac{\|S(t)u - S(t)\tilde{u} - L(t,u)(\tilde{u}-u)\|_X}{\|\tilde{u}-u\|_X} \to 0 \qquad \text{as } \delta\to 0. \qquad (19.2)$$

In fact, this limit holds also uniformly over a compact set obtained by slightly enlarging $\mathcal{A}$. For $L = L(t,u)$, define the numbers

$$\alpha_m(L) := \sup_{\substack{Y\subset X \\ \dim Y=m}} \inf_{\substack{w\in Y \\ \|w\|_X=1}} \|Lw\|_X \qquad \text{for } m\in\mathbb{N}.$$

Definition 19.2. Define the **Lyapunov exponents** $\mu_j = \mu_j(u_0)$ as the numbers

$$\mu_j(u_0) := \ln\left(\lim_{t\to\infty} \alpha_j(L(t,u_0))^{1/t}\right).$$

Roughly speaking, Lyapunov exponents are just the growth and decay coefficients associated to u_0. However, we would like to have these numbers uniformly over the attractor $\mathcal{A}$. We set

$$\omega_m(L) := \alpha_1(L)\cdots\alpha_m(L), \qquad \overline{\omega}_j(t) := \sup_{u\in\mathcal{A}} \omega_j(L(t,u)).$$

Definition 19.3. The **uniform Lyapunov exponents** over the invariant set $\mathcal{A}$ are the numbers

$$\mu_m := \ln\left(\lim_{t\to\infty}\left(\frac{\overline{\omega}_m(t)}{\overline{\omega}_{m-1}(t)}\right)^{1/t}\right).$$

The main idea is that if we have some growing/expanding directions along the attractor and some contracting ones, we should—on balance—still obtain contraction towards some set, as we have an attractor. The strength of this contraction should give us an estimate of the (fractal) dimension, as we can use it to control the growth and decay of balls covering $\mathcal{A}$; cf. Figure 19.2 and Figure 19.1(a). Recall the variational equation (18.5)

$$\frac{\mathrm{d}V}{\mathrm{d}t} = \mathrm{D}_v F(v_\mathcal{A}(t))V, \qquad V(0) = \eta\in\mathcal{A} \qquad (19.3)$$

restricted to the invariant set $\mathcal{A}$ we are interested in.

Lemma 19.4. *Under the assumptions in this chapter, and those in Chapter 18, the uniform Lyapunov exponents satisfy the following bound:*

$$\mu_1 + \cdots + \mu_m \le q_m, \qquad q_m := \limsup_{t\to\infty} q_m(t),$$

where $q_m(t)$ is a time-averaged maximal volume growth

$$q_m(t) := \sup_{u\in\mathcal{A}} \sup_{\substack{\eta_j\in X,\|\eta_j\|_X\le 1 \\ j=1,2,\ldots,m}} \left(\frac{1}{t}\int_0^t \mathrm{Tr}\, \mathrm{D}_v F(S(r)u)\circ Q_m(r)\,\mathrm{d}r\right),$$

and $S(r)$ is the given semiflow (i.e., we consider here the usual semiflow of the sine-Gordon equation or the slightly modified one $S_\varepsilon(t)$).

Remark: Of course, Lemma 19.4 is valid for a very general class of semiflows as long as one has uniform differentiability and a suitable invariant set $\mathcal{A}$ to average over.

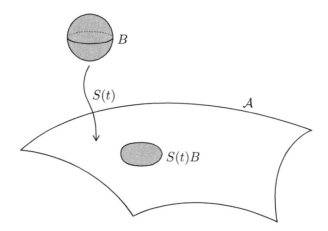

Figure 19.2. *Strong contraction of a single ball under the semiflow $S(t)$ towards the attractor $\mathcal{A}$.*

Proof. (Sketch) The main difficulties already occur for $m = 2$, so let us restrict our attention to this case to simplify the notation and present the idea. One may check (cf. Exercise 19.8) that we have the alternative characterization

$$\omega_2(L(t,u)) = \sup_{\substack{\eta_j \in X,\, \|\eta_j\|_X \leq 1 \\ j=1,2}} \|V_1(t) \wedge V_2(t)\|_{\wedge^2 X}, \qquad (19.4)$$

where V_1, V_2 are the solutions of (19.3) with initial conditions η_1, η_2. The trick is to find an evolution equation, so we differentiate, drop the subscript $\wedge^2 X$ for the inner product as well as for the norm, and write $S(t)u = v_{\mathcal{A}}(t)$, which yields

$$
\begin{aligned}
\frac{1}{2}\frac{\mathrm{d}}{\mathrm{d}t}\|V_1(t)\wedge V_2(t)\|^2 &= \left\langle \frac{\mathrm{d}}{\mathrm{d}t}\left(V_1(t)\wedge V_2(t)\right), V_1(t)\wedge V_2(t)\right\rangle \\
&= \langle V_1'(t)\wedge V_2(t), V_1(t)\wedge V_2(t)\rangle \\
&\quad + \langle V_1(t)\wedge V_2'(t), V_1(t)\wedge V_2(t)\rangle \\
&= \langle \mathrm{D}_v F(v_{\mathcal{A}}(t))V_1(t)\wedge V_2(t), V_1(t)\wedge V_2(t)\rangle \\
&\quad + \langle V_1(t)\wedge \mathrm{D}_v F(v_{\mathcal{A}}(t))V_2(t), V_1(t)\wedge V_2(t)\rangle.
\end{aligned}
$$

It is not too difficult to see using standard results about wedge products that the last two terms can be computed using the trace. We get

$$\frac{1}{2}\frac{\mathrm{d}}{\mathrm{d}t}\|V_1(t)\wedge V_2(t)\|^2 = \|V_1(t)\wedge V_2(t)\|^2 \mathrm{Tr}\left(\mathrm{D}_v F(v_{\mathcal{A}}(t))\circ Q_2\right), \qquad (19.5)$$

where Q_2, as in Chapter 18, is the time-dependent projector onto the space spanned by $V_1(t)$ and $V_2(t)$. Taking the derivative on the left-hand side in (19.5) and solving the resulting linear nonautonomous system, we get

$$\|V_1(t)\wedge V_2(t)\| = \|\eta_1 \wedge \eta_2\| \exp\left(\int_0^t \mathrm{Tr}\,\mathrm{D}_v F(S(r)u)\circ Q_2(r)\,\mathrm{d}r\right).$$

Using this result and (19.4), it follows that

$$\overline{\omega}_2(t) = \sup_{u_0 \in \mathcal{A}} \omega_2(L(t,u_0)) \leq \exp(t q_2(t)).$$

But this implies, after examining Definition 19.3 more closely, that

$$\mu_1 + \mu_2 \leq q_2 = \limsup_{t \to \infty} q_2(t).$$

The result for more general $m > 2$ just involves taking care of more summands in the same way. □

The next result shows in which way the Lyapunov exponents are most frequently used.

Theorem 19.5. *(attractor dimension bound [528, Sec. V, Thm. 3.3]) Suppose the linearization $L(u)$ is uniformly differentiable and uniformly bounded over an attractor $\mathcal{A}$. Furthermore, assume that*

$$\mu_1 + \cdots + \mu_m < 0 \qquad \text{for some } m \geq 2;$$

then the fractal dimension of $\mathcal{A}$ is bounded by

$$m\left(1 + \max_{1 \leq j \leq m-1} \frac{(\mu_1 + \cdots + \mu_j)_+}{|\mu_1 + \cdots + \mu_m|}\right), \tag{19.6}$$

where $(a)_+ = \max(a, 0)$ for $a \in \mathbb{R}$.

The full proof of Theorem 19.5 is beyond our scope. However, let us at least motivate why such a result is not completely unexpected:

- Consider a ball needed for a covering. One checks that the image of the ball under L is an ellipsoid. The length of the axes of this ellipsoid is directly related to the numbers α_j; see also Figure 19.2.

- Since the attractor $\mathcal{A}$ is compact, it can be covered by a finite number of balls. Clearly, we can just flow/iterate this covering under the semiflow, yielding a collection of ellipsoids; see Figure 19.1(b). This is again a covering using $S(t)\mathcal{A} = \mathcal{A}$.

- The sizes of the ellipsoids are related to the Lyapunov exponents via ω_j and its uniform counterpart $\overline{\omega}_j$. Furthermore, we can count how many balls we need to cover each ellipsoid.

- The last step yields a relation involving $\overline{\omega}_j$ between two different coverings. Hence, it basically shows how quickly we expect the term $\ln n_Y(\delta)$ to decay as $\delta \to 0$. In this case, we need to take limits also in the iterative covering procedure to make certain terms small.

Let us apply Theorem 19.5 to the sine-Gordon attractor more explicitly.

Theorem 19.6. *(sine-Gordon attractor dimension) Consider the damped sine-Gordon equation (18.1) on $\Omega = [0, \pi]^2$ with Dirichlet boundary conditions. Then the attractor $\mathcal{A}$ has fractal dimension bounded above by (19.6) where m is the smallest integer such that*

$$64 < \frac{m}{\ln m}.$$

Remark: The result is by no means expected to be sharp, i.e., the upper bound is just a rather abstract finite bound arising from the various estimates. To show that such a bound is sharp is, at least with current methods, for many equations not really possible.

Proof. By Theorem 19.5, we may bound the fractal dimension of $\mathcal{A}$ if we can find the smallest m such that

$$\mu_1 + \cdots + \mu_m < 0.$$

By Theorem 18.4 and Lemma 19.4, this occurs when $q_m < 0$, i.e., when

$$q_m \leq -\frac{m}{4}\varepsilon + \frac{1}{\varepsilon}\sum_{j=1}^{m}\lambda_j^{-1} = -\frac{m}{4}\varepsilon\left(1 - \frac{4}{m\varepsilon^2}\sum_{j=1}^{m}\lambda_j^{-1}\right) < 0,$$

where we can just take the boundary value $\varepsilon = \frac{1}{4}$ from (18.3) to obtain one possible (nonoptimal) condition that we should find the smallest integer m such that

$$\frac{1}{m}\sum_{j=1}^{m}\lambda_j^{-1} < \frac{1}{64}. \tag{19.7}$$

Obviously we also have to control the eigenvalues. From the explicit formulas of the eigenvalues of the (negative) Laplacian on the rectangle, as discussed in Example 6.4, we note that

$$\lambda_j \geq \frac{1}{2}\lambda_1 j = j \qquad \text{for all } j \in \mathbb{N},$$

since the first eigenvalue is just $\lambda_1 = 1^2 + 1^2$. Therefore, we find a nice upper bound

$$\sum_{j=1}^{m}\lambda_j^{-1} < \ln m.$$

In conclusion, (19.7) now implies we have to find the smallest m such that

$$64 < \frac{m}{\ln m},$$

which is the required result. $\square$

Although the result is definitely not sharp, the technique is of fundamental importance. To obtain any analytical growth bounds on the dynamics, one has to look—in one form or another—at the variational equation. Furthermore, we already remark that in Chapters 20–21, we shall see that it could even be misleading altogether in practical applications of PDEs to just look at the limit $t \to +\infty$ relevant for attractors.

Exercise 19.7. Consider Definition 19.1. (a) Prove that a compact smooth curve in $\mathbb{R}^3$ has dimension 1 while a compact smooth surface has dimension 2. (b) Recall the classical middle-third **Cantor set**: take $[0, 1]$, remove the middle third $(\frac{1}{3}, \frac{2}{3})$, then remove the middle third from $[0, \frac{1}{3}]$ and $[\frac{2}{3}, 1]$, and so on. Prove that the remaining set has fractal dimension $\ln 2 / \ln 3$. $\Diamond$

Exercise 19.8. Consider formula (19.4). (a) Prove the generalizations of this formula for any collection of solution vectors $V_1(t), \ldots, V_m(t)$ for a finite-dimensional variational equation. (b) Now generalize your proof from (a) to the Hilbert space setting. $\Diamond$

Exercise 19.9. Consider the sine-Gordon equation with a damping parameter $p > 0$, i.e.,

$$\partial_t^2 u + p\partial_t u = \Delta u - \sin(u), \qquad u = u(x, t). \tag{19.8}$$

(a) What do you expect to happen for the bound of the attractor dimension if p is increased? (b) Try to carry out the necessary modifications in the arguments given above to incorporate p. $\Diamond$

Background and Further Reading: The chapter follows parts of [528], but we also refer to [477, 497]. Lyapunov exponents, as well as their finite-time analogues, are important tools in dynamical systems [23, 171, 315, 457, 568]. Fractal dimension is a general concept relevant for many parts of analysis [237, 394], although it is by no means clear which dimension concept one should select [63, 189, 400, 440, 458, 521]. For more analysis of the dynamics and attractors for the damped sine-Gordon equation, one may consider [227, 343]. Of course, one of the main motivating examples in the theory is the Navier-Stokes equation and trying to grasp how its attractor may look [123, 352, 477, 496, 526].

Chapter 20

Metastability and Manifolds

As already indicated in Chapter 19, it might not always be useful to just consider the long-time asymptotics $t \to +\infty$ for a time-dependent PDE. This observation even holds for slight modifications of PDEs we are already familiar with. Consider the scalar **Allen-Cahn** (or **Ginzburg-Landau**, or **Nagumo**) equation for $u = u(x,t)$ given by

$$\begin{cases} \partial_t u = \varepsilon^2 \partial_x^2 u + \frac{1}{2}(u - u^3) =: \mathcal{F}(u) & \text{for } x \in \Omega := [0,1], \\ \partial_x u = 0 & \text{for } x \in \{0,1\}, \end{cases} \tag{20.1}$$

for $t \in [0,T)$ with initial condition $u(x,0) = u_0(x)$ and assume $\varepsilon > 0$ is a small parameter; the prefactor $\frac{1}{2}$ of the reaction term is inserted here for computational convenience. We set

$$f(u) = \frac{1}{2}\left(u^3 - u\right), \qquad F(u) = \frac{1}{2}\left(\frac{1}{4}u^4 - \frac{1}{2}u^2\right). \tag{20.2}$$

It can be proved that the asymptotic dynamics of (20.1) is given by convergence to steady states $\lim_{t \to +\infty} u(x,t) = u^*(x)$, where only the two homogeneous steady states $u^* = u_{\pm} := \pm 1$ are stable. The intuition for this behavior is to view (20.1) as a gradient flow; see Chapter 23. Since we are on a bounded domain, the locally stable travelling waves we encountered in Chapters 7 and 9 will eventually reach the boundary. However, if $T < +\infty$, the finite-time dynamics is very interesting as shown in Figure 20.1. In particular, separated **layers** between $u_{\pm}$ are observed to be **metastable** for a time scale $t = \mathcal{O}(e^{C/\varepsilon})$ as $\varepsilon \to 0$ for a constant $C = \mathcal{O}(1)$.

A geometric dynamics approach to study layer motion is to construct a suitable low-dimensional manifold; cf. Chapter 17. The general construction for any number of layers will not be considered here. We restrict to the case of the dynamics of a *single layer*. Exercise 20.6 provides the correct intuition that we should look at objects near standing waves. We need an auxiliary problem:

Lemma 20.1. *Given $b > 0$ such that $F''(u) > 0$ for $|u \pm 1| < b$, there exists a constant $\rho_0 > 0$ such that if $\varepsilon/l < \rho_0$ then there exists a unique solution $\phi = \phi(x,l)$ to the BVP*

$$\varepsilon^2 \frac{d^2\phi}{dx^2} = f(\phi), \quad \phi(-l/2) = 0 = \phi(l/2), \tag{20.3}$$
$$\phi(x) > 0 \quad \text{for } |x| < l/2, \quad |\phi(0) - 1| < b.$$

Proof. Exercise 20.7 shows that ϕ depends on ε, l only via the ratio $r := \varepsilon/l$. Let $v = \frac{d\phi}{dx}$; then (20.3) implies $\varepsilon^2 v \, dv = f(\phi) \, d\phi$, which yields

$$\varepsilon^2 \left(\frac{d\phi}{dx}\right)^2 = 2(F(\phi) - F(\phi(0))) \tag{20.4}$$

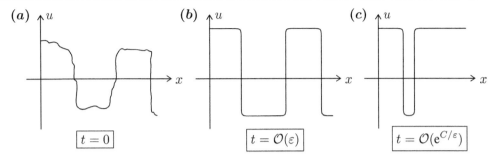

Figure 20.1. *Initial formation of layers and metastability for the Allen-Cahn equation (20.1)–(20.2). (a) Sketch of a random initial condition. (b) At the time scale $\mathcal{O}(\varepsilon)$ sharp interface layers have formed between the two locally stable states $u \equiv \pm 1$. (c) Only on exponentially long time scales do the interfaces move towards each other so one observes metastability in practice.*

for some constant $\phi(0) := z > 0$ that we may still select. Integrating (20.4), using reflection symmetry of the problem $x \mapsto -x$ and using the Dirichlet boundary conditions yields

$$\frac{1}{r} = \frac{l}{\varepsilon} = \frac{1}{\varepsilon} \int_{-l/2}^{l/2} \mathrm{d}x = \frac{2}{\sqrt{2}} \int_{z}^{\phi(l/2)} \frac{1}{(F(u) - F(z))^{1/2}} \, \mathrm{d}u$$

$$= \sqrt{2} \int_{z}^{0} \frac{1}{(F(u) - F(z))^{1/2}} \, \mathrm{d}u. \tag{20.5}$$

The integral on the right-hand side is seen to monotonically diverge as $z \nearrow 1$. Therefore, given a sufficiently small r respectively ρ_0, the integral condition (20.5) can be satisfied. □

Define the location of the single layer as $h = h(t)$. Consider a monotone function

$$\xi \in C^{\infty}(\mathbb{R}, [0,1]), \qquad \xi(x) = \begin{cases} 0 & \text{for } x \leq -1, \\ 1 & \text{for } x \geq 1. \end{cases}$$

Define the approximate metastable **layer** $u^h = u^h(x)$ by the formula

$$u^h(x) := -\left[1 + \xi\left(\frac{x-h}{\varepsilon}\right)\right] \phi(x, 2h) + \xi\left(\frac{x-h}{\varepsilon}\right) \phi(x-1, 2-2h) \tag{20.6}$$

so that $u^h(0) < 0$, $u^h(h) = 0$ and $u^h(1) > 0$; see Figure 20.2.

Definition 20.2. Fix some $\rho > 0$ and define the manifold

$$\mathcal{M} := \{u^h : 1 - \varepsilon/(2\rho) > h > \varepsilon/(2\rho)\} \subset H^1(\Omega). \tag{20.7}$$

$\mathcal{M}$ is the natural manifold containing the single layer; see Figure 20.3. Although it turns out that it is only an approximately invariant manifold, it is the most convenient object to geometrically understand the dynamics. To introduce coordinates near $\mathcal{M}$, consider the approximate tangent vector

$$\tau^h(x) := -\gamma(x)\partial_x u^h(x), \tag{20.8}$$

where $\gamma(x) := \xi((x-\varepsilon)/\varepsilon)[1 - \xi((x-1+\varepsilon)/\varepsilon)]$ just differs from 1 near the boundary of the domain. Hence, it is convenient, with a slight abuse of notation and neglecting

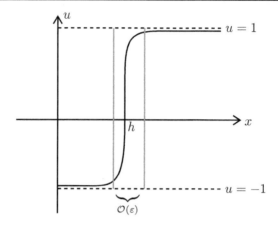

Figure 20.2. *Sketch of a single layer/interface (20.6) of thickness $\mathcal{O}(\varepsilon)$. The layer can be shifted/moved. It forms the basic building block for the analysis of metastability in the Allen-Cahn equation.*

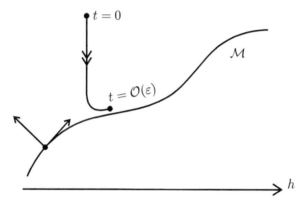

Figure 20.3. *Attracting manifold for the Allen-Cahn equation parametrized by the location of the layer. This geometric idea generalizes naturally to multiple layer locations, which is necessary to keep track of several moving interfaces as shown in Figure 20.1. The decomposition of $\mathcal{M}$ into tangential and normal components is illustrated in the lower left part.*

the boundary error, to just write $\tau^h(x) = -u_x^h$ as well as $\partial_h \tau^h = -u_{xh}^h$. One may then prove (although this is nontrivial!) that

$$u = u^h + v, \qquad 0 = \langle v, u_x^h \rangle_{L^2(\Omega)} =: \langle v, u_x^h \rangle \qquad (20.9)$$

yields a well-defined local coordinate system near $\mathcal{M}$; essentially this step is just a projection/decomposition into tangential and transversal components; see Figure 20.3. Differentiating the relations (20.9) w.r.t. t yields

$$v' = \mathcal{F}(u^h + v) - u_h^h h', \qquad (20.10)$$

$$(\langle u_h^h, u_x^h \rangle - \langle v, u_{xh}^h \rangle)h' = \langle \mathcal{F}(u^h + v), u_x^h \rangle, \qquad (20.11)$$

where $' = \partial_t$. In (20.11), the spatial component is integrated out, so we really have a one-dimensional ODE for the layer position.

Lemma 20.3. $\langle u_h^h, u_x^h \rangle \neq \langle v, u_{xh}^h \rangle$. *In particular, we may rewrite (20.11) dividing both sides by the prefactor of h'.*

Proof. (Sketch) A single layer is approximated by a combination of stationary solutions of the Dirichlet problem (20.3), which can be well approximated by the stationary problem (20.16). Therefore, we let $u^h(x) \approx \Phi(x - h)$ and obtain

$$\langle u_h^h, u_x^h \rangle \approx \int_{\mathbb{R}} \frac{1}{4\varepsilon^2} \operatorname{sech}^2 \left(\frac{x - h}{2\varepsilon} \right) \operatorname{sech}^2 \left(\frac{x - h}{2\varepsilon} \right) \, \mathrm{d}x = \frac{2}{\varepsilon},$$

as well as the second important integral

$$\langle v, u_{xh}^h \rangle \leq \|v\|_{L^\infty} \|u_{xh}^h\|_{L^1} \approx \|v\|_{L^\infty} \int_{\mathbb{R}} \frac{\tanh \left(\frac{x}{2\varepsilon} \right) \operatorname{sech}^2 \left(\frac{x}{2\varepsilon} \right)}{2\varepsilon^2} \, \mathrm{d}x = 0.$$

Even if we would not use an approximation of the layer via Φ and be more precise, it is evident that $\langle u_h^h, u_x^h \rangle$ and $\langle v, u_{xh}^h \rangle$ have different orders in ε. Hence, selecting ε sufficiently small and the neighborhood of $\mathcal{M}$ small enough so that $\|v\|_{L^\infty}$ remains bounded, the result follows. $\square$

Remark: The last proof shows that we could, in principle, just replace u_h^h by $-u_x^h$ in all arguments. However, this simplification would not work in the general proof for many layers, so we keep a slightly more complicated notation.

To rewrite (20.10)–(20.11) into a more useful form, define a projection $P^h : L^2(\Omega) \to \mathrm{T}\mathcal{M}(u^h)$ by

$$P^h w := \frac{\langle w, u_x^h \rangle}{\langle u_x^h, u_h^h \rangle} u_x^h.$$

Proposition 20.4. *Let $L^h v := -\varepsilon^2 \partial_x^2 v + f'(u^h)v$ and consider*

$$R_0 := \frac{\langle \mathcal{F}(u^h), u_x^h \rangle}{\langle u_x^h, u_h^h \rangle},$$

$$R_1 := \frac{1}{\langle v, u_{xh}^h \rangle - \langle u_x^h, u_h^h \rangle} \left(-\langle v, u_{xh}^h \rangle R_0 + \langle L^h v - R_2 v^2, u_x^h \rangle \right),$$

$$R_2 := \int_0^1 (1 - s) f''(u^h + sv) \, \mathrm{d}s.$$

Then the equations (20.10)–(20.11) can be written as

$$h' = R_0 + R_1, \tag{20.12}$$

$$v' = -L^h v + (\mathrm{Id} - P^h)\mathcal{F}(u^h) - R_1 u_h^h + R_2 v^2. \tag{20.13}$$

Proof. Basically, the main ingredient used is Lemma 20.3 so that we may divide by certain factors in (20.10)–(20.11). Classical Taylor expansion yields

$$\mathcal{F}(u^h + v) = \mathcal{F}(u^h) - L^h v + R_2 v^2. \tag{20.14}$$

Looking at (20.11) and using (20.14) gives

$$h' = \frac{\langle \mathcal{F}(u^h) - L^h v + R_2 v^2, u_x^h \rangle}{\langle u_h^h, u_x^h \rangle - \langle v, u_{xh}^h \rangle}$$

$$= \frac{\langle \mathcal{F}(u^h), u_x^h \rangle}{\langle u_h^h, u_x^h \rangle} - \frac{\langle \mathcal{F}(u^h), u_x^h \rangle}{\langle u_h^h, u_x^h \rangle} + \frac{\langle \mathcal{F}(u^h) - L^h v + R_2 v^2, u_x^h \rangle}{\langle u_h^h, u_x^h \rangle - \langle v, u_{xh}^h \rangle},$$

and combining the last two terms yields (20.12). The proof of (20.13) is now relatively easy as we just have to plug in (20.14) and (20.12) into (20.10). $\square$

The next key point is to estimate the term R_0, which turns out to be the leading-order term for the dynamics. Currently, we only know from the proof of Lemma 20.3 that $(\langle u_h^h, u_x^h \rangle)^{-1} = \mathcal{O}(\varepsilon)$. This does not explain why one observes metastability on a time scale $t = \mathcal{O}(e^{C/\varepsilon})$, so we must analyze $\langle \mathcal{F}(u^h), u_x^h \rangle$.

Proposition 20.5. *Define $\alpha(r) := F(\phi(0, l))$ for $r = \varepsilon/l$. Then in a sufficiently small neighborhood of $\mathcal{M}$, we have*

$$\langle \mathcal{F}(u^h), u_x^h \rangle = \alpha \left(\frac{\varepsilon}{2h} \right) - \alpha \left(\frac{\varepsilon}{2 - 2h} \right). \tag{20.15}$$

Proof. By construction of the layer, we have that

$$\mathcal{F}(u^h) = 0$$

if $|x - h| \geq \varepsilon$, so the following calculation simplifies. One finds

$$\langle \mathcal{F}(u^h), u_x^h \rangle = \int_{h-\varepsilon}^{h+\varepsilon} \left(u_{xx}^h - f(u^h) \right) u_x^h \, dx$$

$$= \int_{h-\varepsilon}^{h+\varepsilon} \partial_x \left(\frac{1}{2} \varepsilon^2 (u_x^h)^2 - F(u^h) \right) \, dx$$

$$= \left(\frac{1}{2} \varepsilon^2 (u_x^h)^2 - F(u^h) \right) \Bigg|_{h-\varepsilon}^{h+\varepsilon} = \alpha \left(\frac{\varepsilon}{2h} \right) - \alpha \left(\frac{\varepsilon}{2 - 2h} \right),$$

which is the claimed result. $\square$

Note that formula (20.15) also provides the interpretation that the motion of the layer/interface is driven by differences of the potential $F(\phi) - \frac{1}{2} \varepsilon^2 (\partial_x \phi)^2$. Hence, it remains to analyze $F(\phi)$ more closely, which we shall carry out in Chapter 21.

Exercise 20.6. Consider the stationary Allen-Cahn equation on $\Omega = \mathbb{R}$

$$0 = \varepsilon^2 \partial_x^2 \Phi + \frac{1}{2} \left(\Phi - \Phi^3 \right), \quad \Phi(0) = 0, \quad \lim_{x \to \pm \infty} \Phi(x) = \pm 1. \tag{20.16}$$

Prove that $\Phi(x) = \tanh(x/(2\varepsilon))$ solves (20.16). $\Diamond$

Exercise 20.7. Prove that solutions of (20.3) depend upon ε, l only via the ratio ε/l. $\Diamond$

Exercise 20.8. Consider the boundary value problem for $x \in [0, 1]$ given by

$$\frac{d^2 \phi}{dx^2} = f_0(\phi), \qquad \phi(0) = 0 = \phi(1). \tag{20.17}$$

Find a function $f_0 \in C^1(\mathbb{R}, \mathbb{R})$ such that (20.17) has no solution. Can you also find such an f_0 that (20.17) has infinitely many solutions? $\Diamond$

Background and Further Reading: Here we mainly followed [86], which provides a very detailed geometric description of interface motion; see also [74, 78, 87, 218], and for the original work of Allen and Cahn see [13]. Several generalizations, variations, and extensions of the presented results exist. For a more general approach to the eigenvalue problem consider [104], for higher dimensions see [103], and regarding relations to **mean curvature flow** we refer to [75, 187, 286, 416]. For asymptotic method approaches to interface dynamics see [198, 480] as well as Chapter 35. Similar phenomena of metastability appear in many other PDEs such as viscous conservation laws [473] and—most prominently—in the Cahn-Hilliard equation [80], where we refer to [9, 10, 450] as well as Chapter 22. For two-point boundary value problems and their numerical solution consider [32]; cf. Exercise 20.8.

Chapter 21

Exponentially Small Terms

We continue the metastability analysis for the Allen-Cahn equation (20.1) and the notation from Chapter 20. In Proposition 20.4 and Proposition 20.5, we have shown that the motion of a single layer is governed near the one-dimensional manifold $\mathcal{M}$, which contains the layer, by

$$h' = \frac{\alpha\left(\frac{\varepsilon}{2h}\right) - \alpha\left(\frac{\varepsilon}{2-2h}\right)}{\langle u_h^h, u_x^h \rangle} + R_1, \tag{21.1}$$

$$v' = -L^h v + (\mathrm{Id} - P^h)\mathcal{F}(u^h) - R_1 u_h^h + R_2 v^2, \tag{21.2}$$

where v is the dynamics transverse to $\mathcal{M}$ and h approximates the layer motion. It can be shown that R_2 is of higher order. We know already that $\langle u_h^h, u_x^h \rangle = \mathcal{O}(\varepsilon^{-1})$. It remains to analyze

$$\alpha(r) = F(\phi(0, l)), \qquad r = \varepsilon/l, \tag{21.3}$$

i.e., the solution to the Dirichlet problem (20.3). Recall from the proof of existence of ϕ in Lemma 20.1 that we reached the conclusion by considering the integral equation

$$\frac{1}{r} = \sqrt{2} \int_z^0 \frac{1}{(F(u) - F(z))^{1/2}} \, du, \tag{21.4}$$

where $z = \phi(0) = \phi(0, l)$. We want to extract a more explicit form of z from (21.4).

Lemma 21.1. *The integral equation* (21.4) *is equivalent to*

$$\frac{\sqrt{F''(z)}}{2r} = \int_0^{-z} \frac{1}{\sqrt{G(s, \delta)}} \, ds =: I(\delta) \tag{21.5}$$

for $G(s, \delta) = s^2 + 2\delta s + s^3 g(s, \delta)$ *with* g *smooth.*

Proof. Define $\delta := F'(z)/F''(z)$. Consider the equation

$$\delta - \frac{F'(z)}{F''(z)} = 0 \tag{21.6}$$

near the point $(\delta, z) = (0, 1)$. At $(0, 1)$, we use $F'(1) = 0$ and $F''(1) \neq 0$ to see that $(0, 1)$ is a solution. Furthermore, applying the Implicit Function Theorem, it follows that we may write z as a smooth function of δ near $z = 1$ with

$$z - 1 = \delta + \mathcal{O}(\delta^2) \qquad \text{as } \delta \to 0. \tag{21.7}$$

For small s, consider a Taylor expansion

$$F(z+s) - F(z) = \frac{1}{2}F''(z)\left(s^2 + 2\delta s + \mathcal{O}(s^3)\right). \tag{21.8}$$

Using this expansion and the change of variables $u = z + s$, we rewrite the integral (21.4) as follows:

$$\sqrt{2}\int_z^0 \frac{1}{(F(u) - F(z))^{1/2}}\,du = \sqrt{2}\int_0^{-z} \frac{1}{(F(z+s) - F(z))^{1/2}}\,ds$$

$$= 2\int_0^{-z} \frac{1}{\sqrt{F''(z)}\left(s^2 + 2\delta s + \mathcal{O}(s^3)\right)^{1/2}}\,ds.$$

The result now follows easily setting $G(s,\delta) := s^2 + 2\delta s + s^3 g(s,\delta)$.   □

It turns out that $I(\delta)$ has actually a very nice description.

Lemma 21.2. *There exists $\delta_0 > 0$ such that, for $\delta \in (0, \delta_0)$, we have*

$$I(\delta) = (-\ln \delta)(1 + c_1 \delta + \cdots c_k \delta^k) + I_R(\delta), \tag{21.9}$$

where c_j are constants and $I_R \in C^k([0, \delta_0], \mathbb{R})$.

Proof. The key idea of the proof is to isolate and analyze the leading-order term in G. We claim that

$$G(s,\delta) = (s^2 + 2\mu s)H(s, \mu), \quad \mu = \delta + \mathcal{O}(\delta^2), \quad H(s, \mu) = 1 + \mathcal{O}(s), \tag{21.10}$$

where H, μ are smooth. To prove the claim (21.10), consider

$$G_1(s, \mu) := \frac{1}{s}G(s, \delta) = s + 2\delta + \mathcal{O}(s^2)$$

and apply the Implicit Function Theorem to G_1 to obtain the existence of $\mu = \mu(\delta)$ such that $G_1(-2\mu(\delta), \delta) = 0$. Applying Taylor's Theorem to $G_1(s + 2\mu - 2\mu, \mu)$ yields (21.10). By smoothness of $H = 1 + \mathcal{O}(s)$, we can write the expansion

$$\frac{1}{\sqrt{H(s, \mu)}} = 1 + \sum_{j=1}^{k} a_j(\mu)s^j + R_H(s, \mu)s^{k+1}, \tag{21.11}$$

where a_j, R_H are smooth. The integral $I(\delta)$ can be split into two terms,

$$I(\delta) = \int_{s_0}^{-z} \frac{1}{\sqrt{G(s, \delta)}}\,ds + \int_0^{s_0} \frac{1}{\sqrt{G(s, \delta)}}\,ds,$$

where the first term is a C^k function of δ as G is bounded away from zero. We select $s_0, \delta_0 > 0$ sufficiently small so that (21.10) holds for $s \in [0, s_0]$ and $\delta \in (0, \delta_0)$. This implies

$$\int_0^{s_0} \frac{1}{\sqrt{G(s, \delta)}}\,ds = I_0(\delta) + \sum_{j=1}^{k} a_j(\mu)I_j(\mu) + \int_0^{s_0} \frac{R_H(s, \mu)s^{k+1}}{\sqrt{s^2 + 2\mu s}}\,ds, \tag{21.12}$$

where the integral coefficients $I_j(\mu)$ are

$$I_j(\mu) = \int_0^{s_0} \frac{s^j}{\sqrt{s^2 + 2\mu s}}\, ds, \qquad j \in \{0, 1, \dots, k\}.$$

To see that the multiplier of $I(\delta)$ in the formula (21.9) is really $(-\ln \delta)$, we evaluate

$$I_0(\delta) = \int_0^{s_0} \frac{1}{\sqrt{s^2 + 2\delta s}}\, ds = 2\ln\left(\sqrt{s} + \sqrt{s + \delta}\right)\Big|_0^{s_0}.$$

Hence, looking at the leading order near $s \to 0$, $\delta \to 0$, we find indeed the claimed logarithm. This finishes the proof of (21.9). The statement $I_R \in C^k([0, \delta_0], \mathbb{R})$ is discussed in Exercise 21.5. $\qquad\square$

The final result in the next theorem shows that the layer dynamics is very slow and exhibits metastability because the leading-order term of the ODE (21.1) is exponentially small.

Theorem 21.3. *(exponentially small terms) There exists a sufficiently small $r_0 > 0$ and a constant $K > 0$ such that for $r \in (0, r_0)$*

$$\alpha(r) - \alpha(\tilde{r}) = Ke^{-1/r}\left(1 + \mathcal{O}(r^{-1}e^{-1/(2r)})\right) \tag{21.13}$$

$$-Ke^{-1/\tilde{r}}\left(1 + \mathcal{O}(\tilde{r}^{-1}e^{-1/(2\tilde{r})})\right).$$

Proof. Lemma 21.2 entails that (21.5) is equivalent to

$$\frac{\sqrt{F''(z)}}{2r} = (-\ln \delta)(1 + c_1\delta + \cdots c_k\delta^k) + I_R(\delta). \tag{21.14}$$

Note that since we work near $z = 1$ and $\delta \to 0$, the leading-order term on the left-hand side is $\frac{\sqrt{F''(1)}}{2r} = \frac{1}{2r}$ and we absorb the remaining bounded terms from the left into $I_R(\delta)$. To actually solve for δ, and hence by (21.7) also for $z - 1$, it is natural to finally take an exponential transformation

$$\delta = \exp\left(-\frac{1}{2r} + \tau\right), \tag{21.15}$$

which defines a new parameter τ. One calculates by plugging (21.15) into (21.14), which yields

$$\frac{1}{2r} = \left(\frac{1}{2r} - \tau\right)(1 + c_1\delta + \cdots c_k\delta^k) + I_R(\delta).$$

Canceling the term $1/(2r)$, one finds an equation

$$\tau - I_R(0) + S(\tau, r) = 0, \tag{21.16}$$

where S is a C^k function by Lemma 21.2 and is small in the sense that

$$\frac{\partial^j S}{\partial r^j} = \mathcal{O}\left(r^{-(2j+1)}e^{-1/(2r)}\right) \qquad \text{for } j \in \{0, 1, \dots, k\}.$$

Therefore, we (again!) apply the Implicit Function Theorem to solve (21.16) to obtain a C^k function $\tau(r)$ for $r \in [0, r_0)$ such that $\tau(r) = I_R(0) + \mathcal{O}(r^{-1}\exp(-1/(2r)))$. Therefore, we finally get

$$\delta = e^{-\frac{1}{2r}}\left(K + \mathcal{O}(r^{-1}e^{-1/(2r)})\right) \qquad \text{as } r \to 0. \tag{21.17}$$

The result now follows by applying Taylor's Theorem to $F(z)$ using the relation between δ and z. ◻

The approach in this chapter demonstrates that, regardless whether one prefers differential-geometric dynamics or functional analysis to study the dynamics of PDEs, to provide sharp quantitative results, one has to eventually go back to the fundamentals of *classical real analysis*, which obviously underlies *both* viewpoints.

Corollary 21.4. *The first term R_0 in the vector field of the layer motion has order*

$$\mathcal{O}\left(\varepsilon\left(e^{-\frac{h}{\varepsilon}} - e^{-\frac{1-h}{\varepsilon}}\right)\right).$$

Therefore, except that we did not show here that $R_0 \gg R_1$ [86], it follows that the motion of the single layer is really extremely slow implying metastability; see also Exercise 21.7 and Figure 20.1. In some sense, the analysis presented here is sharp, and we cannot expect to get much finer details analytically. Since the proof is already quite difficult, it does make sense to also look for other techniques to just bound dynamical time scales; see Chapter 22.

Exercise 21.5. Consider the notation from the proof of Lemma 21.2. Prove that

$$\frac{\partial^j}{\partial\mu^j}R_H(s,\mu) = \mathcal{O}(1).$$

Use this result to prove that the last integral term on the right-hand side in (21.12) is C^k as a function of μ. ◊

Exercise 21.6. Consider the notation from the proof of Lemma 21.2. Prove that

$$I_j(\mu) - \mu I_{j-1}(\mu)\frac{2j-1}{2j}, \qquad j \geq 1,$$

are smooth in μ. Hint: Integration by parts. ◊

Exercise 21.7. Solve the following ODE:

$$h' = \varepsilon e^{-\frac{h}{\varepsilon}}, \qquad h(0) = h_0 \in (0,1).$$

Plot the solution for several values of $0 < \varepsilon \ll 1$ for a given initial condition, e.g., $h_0 = \frac{1}{2}$. ◊

Background and Further Reading: We followed the presentation in [86]; see also the references in Chapter 20 for classical background on metastable PDE. Exponentially small terms are an important phenomenon appearing in many contexts in differential equations, e.g., in fast-slow systems and in small splitting problems in Hamiltonian systems as well as the relation between the two fields [150, 257, 273, 303, 385, 427]. Another area in differential equations where exponentially small terms occur is noise-induced transitions and large deviations [151, 546, 561]. Large deviations directly links to elliptic/parabolic PDEs via the Kolmogorov equations (or the Fokker-Planck equation) and to Euler-Lagrange equations via minimizing the large deviation action functional, and the theory can be applied in the context of SPDEs [464]; see also Chapter 24 for more on the Fokker-Planck equation and Chapter 36 for an example of the Euler-Lagrange equation. Regarding our remark above on classical analysis, for many nice illustrations of these techniques applied to ODEs see the book [261].

Chapter 22

Coarsening Bounds and Scaling

Similar to Chapters 20–21, we are again interested in describing the correct time scale for dynamics, here in the context of so-called coarsening. We study as a benchmark PDE for this phenomenon, the **Cahn-Hilliard equation**

$$\partial_t u + \Delta(p\Delta u + 2(1 - u^2)u) = 0 \qquad \text{on } \Omega = \mathbb{T}^N, \tag{22.1}$$

i.e., we assume periodic boundary conditions for some torus of a fixed size, $u = u(x, t)$, and $(x, t) \in \Omega \times [0, T]$. In this chapter we set $p = 1$ but there are many important results for $0 < p = \varepsilon \ll 1$; cf. Chapters 20–21. The PDE (22.1) has an associated energy

$$E = E[u] := \int \frac{1}{2} \left(\|\nabla u\|^2 + (1 - u^2)^2 \right) \, dx, \tag{22.2}$$

where $\fint := \frac{1}{\text{vol}(\Omega)} \int_\Omega$ and $\text{vol}(\Omega)$ denotes the **volume** of Ω. Using the energy, (22.1) can be written as

$$\partial_t u + \nabla \cdot J = 0, \qquad J := -\nabla \mathrm{D}_u E \tag{22.3}$$

so it is related to gradient-flow PDEs, which will be discussed in Chapter 23; in fact, we use an L^2-gradient structure here. Suppose we start from an initial condition $u_0(x) = 0 + \varepsilon\xi(x)$, where $\varepsilon > 0$ is small and ξ is a bounded and suitably random; see Figure 22.1(a). Then two main stages are observed numerically:

- **Spinodal decomposition:** The solution quickly relaxes to a well-separated mixture, where u is close in most of the domain to the two stationary solutions $u = \pm 1$.

- **Coarsening:** The length scale between the two phases changes and grows in time; see Figures 22.1(a)–(c).

It helps to think intuitively of the processes in the context of the mixing process of two fluids. We restrict to **critical mixtures** with $\fint u \, dx = 0$; see also Exercise 22.5. The main results here are essentially independent of this assumption, which is made for convenience.

Definition 22.1. Given any periodic u with $\fint u(x) \, dx = 0$, its **physical length scale** is

$$L = L[u] := \sup_w \left\{ \fint uw \, dx : \text{periodic } w \text{ with } \sup_{x \in \Omega} \|\nabla w(x)\| \leq 1 \right\}. \tag{22.4}$$

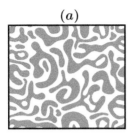 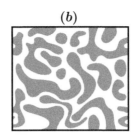 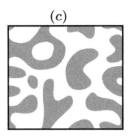

Figure 22.1. *Sketch of the dynamics of the Cahn-Hilliard equation (22.1) for two spatial dimensions. The greyscale indicates the level sets of u. (a)–(c) Sketch of the coarsening process.*

To study coarsening, one possible approach is to relate the evolution laws for energy and length scale. To state the next result, we slightly overload the notation and also write $\fint = \frac{1}{T}\int_0^T$.

Theorem 22.2 (upper coarsening bound). *Fix the initial energy E_0 and length scale L_0. Then there exists a constant $C = C(N) < +\infty$ such that*

$$\fint E^2 \, dt \geq \frac{1}{C}T^{-\frac{2}{3}} \quad provided\ T \geq CL_0^3\ and\ E_0 \leq \frac{1}{C}. \tag{22.5}$$

More concisely, the result can be stated as

$$\fint E^2 \, dt \gtrsim \int_0^T \left(t^{-\frac{1}{3}}\right)^2 \, dt \quad for\ T \gg L_0^3 \gg 1 \gg E_0, \tag{22.6}$$

where $\gtrsim$ is a shorthand notation to indicate inequality up to a positive multiplicative constant.

Given a large time in comparison to length and given small initial energy, Theorem 22.2 roughly states that *on average* the energy decays no faster than $t^{-1/3}$, hence it bounds the coarsening time scale. Formally, and from physical considerations, we would expect length scale and energy are related (e.g., a reasonable guess is $C_1 \leq EL \leq C_2$, for some $C_{1,2} > 0$) so energy bounds would yield length scale bounds, but (22.5) does not make a statement about this relation.

Lemma 22.3. $EL \gtrsim 1$ *for $E \ll 1$.*

Note that Lemma 22.3 is an abstract statement about the functionals E and L not using Cahn-Hilliard dynamics; hence we shall not prove it here and refer to [339]. The second statement is dynamically a lot more interesting as it implies **dissipation** of energy.

Lemma 22.4. $\left(\frac{dL}{dt}\right)^2 \lesssim -\left(\frac{dE}{dt}\right)$.

Proof. Using integration by parts and (22.3) we have

$$-\frac{dE}{dt} = \fint (-D_u E)(\partial_t u) \, dx = \fint |J|^2 \, dx. \tag{22.7}$$

For $t_1 < t_2$, we claim the length scale rate of change satisfies

$$|L(t_2) - L(t_1)| \leq \int_{t_1}^{t_2} \fint |J| \, dx \, dt. \tag{22.8}$$

To see this, let $w_*(x)$ be an optimal test function in the definition (22.4) of $L(t_2) = L[u](t_2)$ so that

$$L(t_2) = \fint u(x, t_2) w_*(x) \, dx,$$

where w_* is periodic with $\|\nabla w_*\| \le 1$. Now using w_* as a test function in the definition of $L(t_1)$ yields

$$L(t_2) - L(t_1) \le \fint (u(x, t_2) - u(x, t_1)) w_*(x) \, dx = \int_{t_1}^{t_2} \fint (\partial_t u) \, w_* \, dx \, dt$$

$$= \int_{t_1}^{t_2} \fint J \cdot \nabla w_* \, dx \, dt \le \int_{t_1}^{t_2} \fint |J| \, dx \, dt.$$

Interchanging the roles of t_1 and t_2, we also find

$$L(t_1) - L(t_2) \le \int_{t_1}^{t_2} \fint |J| \, dx \, dt,$$

which proves (22.8). From (22.8) and Cauchy-Schwarz it follows that

$$\left| \frac{dL}{dt} \right| \le \fint |J| \, dx \le \left(\fint |J|^2 \, dx \right)^{1/2} = \left(-\frac{dE}{dt} \right)^{1/2},$$

where the last step just uses (22.7).   □

To simplify the notation in the main proof, we set $\frac{d}{dt} = {}'$. We also may assume that E, L are differentiable with respect to t, as the argument of the following proof does not change if we mollify/smooth E and L.

Proof. (of Theorem 22.2) The inequality $(L')^2 \lesssim -E'$ from Lemma 22.4 implies that E is monotonically decreasing and L is a continuous function of E. Taking this viewpoint, we may write

$$\left(\frac{dL}{dt} \right)^2 = \left(\frac{dL}{dE} \right)^2 \left(\frac{dE}{dt} \right)^2 \lesssim |E'| \quad \Rightarrow \quad \left(\frac{dL}{dE} \right)^2 |E'| \lesssim 1, \qquad (22.9)$$

where E now also takes on a dual role as an independent variable and as $E' = E'(t)$. Multiplying (22.9) by $E(t)^2$ and integrating yields

$$\int_0^T E(t)^2 \, dt \gtrsim \int_0^T E(t)^2 \left(\frac{dL}{dE} \right)^2 |E'| \, dt = \int_{E(T)}^{E(0)} E^2 \left(\frac{dL}{dE} \right)^2 \, dE. \qquad (22.10)$$

Using a coordinate change $\hat{E} = 1/E$ in the last integral gives

$$\int_0^T E(t)^2 \, dt \gtrsim \int_{\hat{E}(0)}^{\hat{E}(T)} \left(\frac{dL}{d\hat{E}} \right)^2 \, d\hat{E}, \qquad (22.11)$$

where we have used

$$\left(\frac{dL}{dE} \right)^2 \, dE = \left(\frac{dL}{d\hat{E}} \right)^2 \left(\frac{d\hat{E}}{dE} \right) \, d\hat{E}.$$

The right-hand side of (22.11) can be bounded below by observing that it is a standard curvature minimization of functions $L(\hat{E})$ with end-point conditions $L(0)$ and $L(T)$. Hence, the minimizer is linear in $\hat{E}$ with slope $(L(T) - L(0))/(\hat{E}(T) - \hat{E}(0))$ and we find

$$h(T) := \int_0^T E(t)^2 \, dt \gtrsim \frac{(L(T) - L(0))^2}{\hat{E}(T) - \hat{E}(0)}. \qquad (22.12)$$

We split the lower bound estimation of the right-hand side of (22.12) into two cases: either $L(T) \geq 2L(0)$ or $L(T) < 2L(0)$. In the first case, we have

$$L(T) - L(0) \gtrsim L(T) \qquad \text{and} \qquad \hat{E}(T) - \hat{E}(0) \leq \hat{E}(T),$$

which implies, upon also using $EL \gtrsim 1$ from Lemma 22.3, that

$$h(T) \gtrsim \frac{(L(T) - L(0))^2}{\hat{E}(T) - \hat{E}(0)}$$

$$\gtrsim \frac{L(T)^2}{\hat{E}(T)} = L(T)^2 E(T) \gtrsim E(T)^{-1} = (h'(T))^{-1/2},$$

since $h'(T) = E(T)^2$. So we have proved

$$h'(T)h(T)^2 \gtrsim 1 \quad \text{if } L(T) \geq 2L(0). \qquad (22.13)$$

For the second case $L(T) < 2L(0)$, we have a simpler estimate

$$E(T) \gtrsim L^{-1}(T) \gtrsim L(0)^{-1} \quad \Rightarrow \quad h'(T) \gtrsim L(0)^{-2}. \qquad (22.14)$$

Combining (22.13) and (22.14) we find that

$$\frac{d}{dt}(h + L(0))^3 \simeq (h(t) + L(0))^2 h'(t) \gtrsim 1 \quad \forall t > 0.$$

Integrating the last inequality in time leads to

$$h(T) + L(0) \gtrsim T^{1/3} \qquad \forall T > 0.$$

Using the assumption $T \gg L(0)^3$, we have $h(T) \gtrsim T^{1/3}$ for all $T \gg L(0)^3 = L_0^3$, which is precisely the statement of the theorem. $\qquad \square$

The last proof shows an approach to investigate the dynamics, which heavily relies upon given *functionals* motivated by the physical structure of the problem. In Chapters 23–26, we shall encounter this view again in several variants, when we study gradient-type structures.

Exercise 22.5. Prove that the Cahn-Hilliard equation (22.1) has **mass conservation**, i.e.,

$$\fint u(x,t) \, dx = \fint u(x,0) \, dx \quad \forall t \geq 0,$$

which implies that the critical mixture assumption does not change either if we start with it. $\lozenge$

Exercise 22.6. Observe that any homogeneous constant $u \equiv u^*$ is a steady state of (22.1). (a) Write down the linearized problem near u^*. (b) Write down the linearized problem for $p = \varepsilon$, for a modified domain $\Omega = [0, L]$ (i.e., not periodic) with Neumann boundary conditions. $\lozenge$

Exercise 22.7. Continue with the linearized Cahn-Hilliard equation from (b) of Exercise 22.6. Use a Fourier ansatz

$$w(x, t) = a_0 + \sum_{k=1}^{\infty} a_k e^{\sigma_k t} e_k(x),$$

where e_k are eigenvectors for the Neumann Laplacian on $\Omega = [0, L]$, to solve the linearized Cahn-Hilliard equation. In particular, compute a_k and σ_k. Under which conditions do we have unstable directions, i.e., when is $\text{Re}(\sigma_k) > 0$? ◊

Background and Further Reading: In this chapter we follow [339], which is in the spirit of energy methods [437]; see also [2, 79, 176, 188, 199, 437] for more on the Cahn-Hilliard equation, its variants, and coarsening. For the derivation of the Cahn-Hilliard equation from physical principles, we refer to [250]. For weak solutions, semigroup properties, and abstract attractor theory of (22.1) one may look at [528]; for the existence of classical solutions see, e.g., [177]. For more on spinodal decomposition, metastability, and travelling fronts in the Cahn-Hilliard equation see [10, 450] and the references in Chapter 20.

Chapter 23

Gradient Flows and Lyapunov Functions

We have extensively studied reaction-diffusion equations with cubic nonlinearities such as the **Allen-Cahn equation**, which can be written as

$$\partial_t u = pu - u^3 + \partial_x^2 u, \tag{23.1}$$

where $(x, t) \in \Omega \times [0, +\infty)$, $p \in \mathbb{R}$ is a parameter, $\Omega \subset \mathbb{R}$, and $u = u(x, t) \in \mathbb{R}$. Suppose we forget the diffusion term in (23.1) for a moment; then we have an ODE

$$\frac{du}{dt} = pu - u^3. \tag{23.2}$$

Any *one-dimensional* ODE can be written as a **gradient system** by just integrating the right-hand side. In particular, we have for (23.2) that

$$\frac{du}{dt} = -\nabla V(u), \qquad V : \mathbb{R} \to \mathbb{R}, \; V(u) = -\frac{1}{2}pu^2 + \frac{1}{4}u^4, \tag{23.3}$$

where $\nabla V = V'$ is just a fancy way of writing the derivative. Hence, the flow of (23.2) follows the negative gradient of the **potential** V; see Figure 23.1.

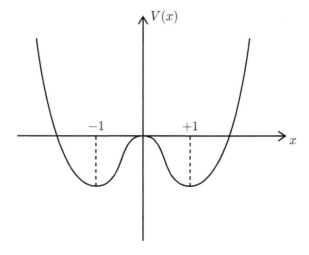

Figure 23.1. *Sketch of the classical double-well potential. The potential is balanced since both global minima have equal value $V(+1) = V(-1)$; cf. (23.3).*

121

Remark: Note very carefully that already in two dimensions, ODEs are generically not gradient flows when we consider the class of all smooth vector fields; see also Exercise 23.7.

For the Allen-Cahn equation (23.1) there is a natural analog of V if we consider the functional

$$\mathcal{V}_{AC}(u) := \int_{\Omega} -\frac{1}{2}pu^2 + \frac{1}{4}u^4 + \frac{1}{2}\left|\partial_x u\right|^2 \ dx. \tag{23.4}$$

Note that for each fixed time t, $\mathcal{V}_{AC}(u)$ is just a real number. Formally, one has to consider a Hilbert space H with inner product $\langle \cdot, \cdot \rangle$ on which the functional is defined, i.e., $u = u(\cdot, t)$, $\mathcal{V}_{AC} : H \to \mathbb{R}$, and $u \mapsto \mathcal{V}_{AC}(u)$. Note that solutions of (23.1) can be viewed as elements $u \in C^1([0, +\infty); H)$. The idea is to rewrite (23.1) as a gradient flow in H.

Definition 23.1. Given a (Fréchet) differentiable functional $\mathcal{V} : H \to \mathbb{R}$ we define the **gradient** $(\nabla_H \mathcal{V})(u^*) \in H$ at u^* as the element which satisfies $\langle (\nabla_H \mathcal{V})(u^*), h \rangle = [D\mathcal{V}(u^*)](h)$ for all $h \in H$.

Note that the previous definition is almost a tautology as it identifies the mapping $D\mathcal{V}(u^*) \in H'$ via the inner product with an element of H, which is just **Riesz' representation theorem** [207, Sec. 7.2].

Proposition 23.2. *Let $\Omega = \mathbb{R}$ and $H = L^2(\mathbb{R})$; then the Allen-Cahn equation (23.1) is a **gradient flow***

$$\partial_t u = -(\nabla_H \mathcal{V}_{AC})(u), \tag{23.5}$$

for the functional $\mathcal{V}_{AC} : H^1 \to \mathbb{R}$ defined in (23.4).

Proof. Consider $h \in C_c^{\infty}(\mathbb{R})$, suppose u is sufficiently regular, and calculate

$$D\mathcal{V}_{AC}(u)h = \int_{\mathbb{R}} -puh + u^3 h + (\partial_x u)\partial_x h \ dx$$

$$= \int_{\mathbb{R}} [-pu + u^3 - \partial_x^2 u]h \ dx,$$

where integration by parts has been used. We conclude that

$$-\nabla_H \mathcal{V}_{AC}(u) = pu - u^3 + \partial_x^2 u = \partial_t u.$$

Now one just needs a standard approximation procedure to show that the argument holds for $h \in L^2(\mathbb{R})$ and $u \in H^1(\mathbb{R})$. □

Remark: As an alternative to an approximation via density of more regular function spaces, one may also consider L^2 directly and use the subdifferential [17].

A crucial property of gradient flows is that the functional they are defined by has also direct dynamical implications. Gradient systems of ODEs are an excellent guiding principle, as the following definition and result show.

Definition 23.3. Consider the ODE

$$\frac{du}{dt} = f(u), \qquad u = u(t) \in \mathbb{R}^d, \ u(0) = u_0. \tag{23.6}$$

A continuous function $V : \mathbb{R}^d \to \mathbb{R}$ is a **Lyapunov function** for (23.6) if it is non-increasing along trajectories, i.e., for every fixed initial condition, the solution u satisfies

$$V(u(t)) \leq V(u(s)) \qquad \text{for all } t \geq s.$$

Proposition 23.4. *If (23.6) has a gradient structure $F(u) = -(\nabla V)(u)$ with $V \in C^1$, then V is a Lyapunov function.*

Proof. We just compute

$$\frac{\mathrm{d}}{\mathrm{d}t} V(u(t)) = u'(t) \cdot \nabla V(u(t)) = -\|\nabla V(u(t))\|^2 \leq 0$$

and the result follows. $\square$

Now the next generalization to the Allen-Cahn PDE is quite natural.

Proposition 23.5. *Consider the same setting as in Proposition 23.2. Then we have*

$$\frac{\mathrm{d}\mathcal{V}_{\mathrm{AC}}}{\mathrm{d}t} = -\int_{\mathbb{R}} [pu - u^3 + \partial_x^2 u]^2 \leq 0. \tag{23.7}$$

Proof. A formal calculation yields

$$\frac{\mathrm{d}\mathcal{V}_{\mathrm{AC}}}{\mathrm{d}t} = \int_{\mathbb{R}} (-pu\partial_t u + u^3 \partial_t u - \partial_x u \partial_{xt} u)\, \mathrm{d}x$$

$$= -\int_{\mathbb{R}} (\partial_t u)(pu - u^3 + \partial_x^2 u)\, \mathrm{d}x,$$

and the result follows similar to the discussion above, i.e., by approximation in regular function spaces (or by using the subdifferential). $\square$

The inequality in (23.7) can be interpreted as the result that the **(free) energy** $\mathcal{V}_{\mathrm{AC}}$ does not grow along trajectories of the system. In particular, the energy is strictly decreasing as long as u is not a stationary solution of the gradient flow since

$$\frac{\mathrm{d}\mathcal{V}_{\mathrm{AC}}}{\mathrm{d}t} = -\int_{\mathbb{R}} |\partial_t u|^2 \mathrm{d}x.$$

This argument is more general and one observes that gradient systems do not admit nontrivial time-periodic solutions, i.e., time-periodic solutions with positive minimal period.

Example 23.6. Consider the **real Ginzburg-Landau equation**

$$\partial_T A = A - A|A|^2 + \partial_X^2 A, \tag{23.8}$$

with $(X, T) \in \mathbb{R} \times [0, +\infty)$ and $A = A(X, T)$, obtained as an amplitude equation in Chapters 12 and 13. Consider the functional

$$\mathcal{V}_{\mathrm{GL}}(u) := \int_{\mathbb{R}} -|A|^2 - \frac{1}{2}|A|^4 + |\partial_x A|^2\, \mathrm{d}x. \tag{23.9}$$

Then the real Ginzburg-Landau equation is written as a gradient flow in $L^2(\mathbb{R})$ for $\mathcal{V}_{\mathrm{GL}}$. Hence, oscillatory amplitudes are not possible in this case. However, if we allow for the generalization to the **complex Ginzburg-Landau equation** given by

$$\partial_T A = A - (1 + \mathrm{i}p_1)A|A|^2 + (1 + \mathrm{i}p_2)\partial_X^2 A \tag{23.10}$$

for $p_{1,2} \in \mathbb{C}$, then (23.10) is not a gradient flow. In particular, one can show that it does support time-periodic solutions. ◆

The previous discussion about gradient flows is slightly misleading, as it partially suggests that gradient flows are purely functional-analytic objects. However, there is a more geometric description to be partially discussed in Chapters 24–26 and 28–29. Why this should be so can be motivated by a few formal calculations. Consider the **heat equation**

$$\partial_t u = \Delta u, \qquad u = u(x,t),\ x \in \mathbb{R}^d = \Omega. \tag{23.11}$$

Statistical physics suggests that a natural functional to consider is the **entropy**

$$\mathcal{E}(u) := \int_\Omega u \ln u \, \mathrm{d}x. \tag{23.12}$$

If we look for the $H = L^2$-gradient of this functional, we obtain for $h \in H$ that

$$(\mathrm{D}_u \mathcal{E})h = \int_\Omega [\ln(u) + 1]h \, \mathrm{d}x = \int_\Omega (\nabla_H \mathcal{E})(u)h \, \mathrm{d}x.$$

Hence, $\partial_t u = \Delta u \neq -(\nabla_H \mathcal{E})(u) = \ln(u) + 1$. Suppose we try to fix this to still involve the entropy as a functional on the right-hand side, which should yield the heat equation. We definitely need two derivatives and get rid of the term $1/u$ coming from the logarithm, so it is natural to look at

$$\nabla \cdot (u\nabla[(\nabla_H \mathcal{E})(u)]) = \nabla \cdot (u\nabla(\ln u + 1)) = \Delta u = \partial_t u.$$

This is precisely the desirable answer and illustrates nicely the need to distinguish between the L^2-gradient $\nabla_H = \nabla_{L^2(\mathbb{R}^d)}$ and the usual Euclidean gradient ∇. One may call gradient systems of the form

$$\partial_t u = \nabla \cdot (u\nabla[(\nabla_H \mathcal{E})(u)])$$

also **Wasserstein gradient flows**. In the next few chapters, we shall explore a few dynamical results for systems with gradient structures.

Exercise 23.7. Consider a smooth vector field $f : \mathbb{R}^d \to \mathbb{R}^d$ which is given as a gradient flow $-\nabla V = f$ for some smooth potential $V : \mathbb{R}^d \to \mathbb{R}$. Prove that, given any $\delta > 0$, there exists a smooth $\tilde{f}$ such that $f + \delta\tilde{f}$ is not a gradient flow. $\lozenge$

Exercise 23.8. Consider the **Fokker-Planck equation** given by

$$\partial u = \Delta u + \nabla \cdot (u\nabla V(u)), \quad u = u(x,t),\ x \in \mathbb{R}^d, \tag{23.13}$$

for a smooth potential $V : \mathbb{R}^d \to \mathbb{R}$ and the modified entropy functional

$$\mathcal{E}_V(u) := \int_{\mathbb{R}^d} u \ln u + u V(u) \, \mathrm{d}x.$$

Show that (23.13) is an L^2-Wasserstein gradient flow of $\mathcal{E}_V$. $\lozenge$

Exercise 23.9. Consider the **Derrida-Lebowitz-Speer-Spohn (DLSS) equation** given by

$$\partial_t u = -\partial_x^2 \left(u\partial_x^2(\ln u) \right), \quad u = u(x,t),\ x \in \mathbb{R}/\mathbb{Z} = \mathbb{T}^1, \tag{23.14}$$

where x in the torus (circle) $\mathbb{T}^1$ just means periodic boundary conditions. Consider the **Fisher information**

$$\mathcal{E}_{\text{Fisher}}(u) := \int_{\mathbb{T}^1} u[\partial_x \ln u]^2 \, \mathrm{d}x. \tag{23.15}$$

Show that (23.14) is an L^2-Wasserstein gradient flow of (23.15). $\lozenge$

Background and Further Reading: The main source for this chapter is [453]. A more abstract approach to gradient flows is given in [17]; see also [386, 517, 551]. A classical paper on Wasserstein gradient flows is [441] and we also refer to [90, 91, 228, 389, 485, 552] to illustrate a few different directions. For more details on the Fokker-Planck equation gradient flow see [21, 306], whereas the gradient flow structure of the DLSS [152] equation is discussed in [168]; for more a bit more on fourth-order equations see [399, 139]. Gradient flows also have links to various functional inequalities [83, 461, 223, 246, 364, 442]; see also Exercise 3.12. More references on gradient flows and related topics can be found in Chapters 24–26 and 28–29.

Chapter 24

Entropies and Global Decay

In Chapter 23, we saw several examples of gradient flows as well as the basic example (23.12) of an **entropy functional**, i.e., of a convex Lyapunov functional. In finite dimensions, it is relatively straightforward to use Lyapunov functionals to prove decay to a steady state. For PDEs, techniques become quite involved. Here we focus on the linear **Fokker-Planck equation**

$$\partial_t u = \nabla \cdot (\nabla u + u \nabla V) =: \mathcal{F}[u] \tag{24.1}$$

for $u = u(x,t)$, $(x,t) \in \mathbb{R}^d \times (0, +\infty)$ and we view (24.1) as an evolution equation for $u(\cdot, t) = u(t)$ using a nonnegative initial condition $u_0(x) = u_0 \in L^1(\mathbb{R}^d)$ with $\int_{\mathbb{R}^d} u_0(x) \, dx = 1$. These assumptions are reasonable as (24.1) arises as the evolution equation for a probability density of a particle in a **potential** V under the additional influence of white noise. We assume $V : \mathbb{R}^d \to \mathbb{R}$ is smooth and convex and $e^{-V} \in L^1(\mathbb{R}^d)$. The analysis idea is now as follows:

(S1) Find a steady state u^*.

(S2) Consider entropy functionals $\mathcal{E}$ such that $\mathcal{E}[u^*] = 0$.

(S3) Compute the **entropy production** $\mathcal{P}[u] := -\frac{d\mathcal{E}}{dt}[u] = \langle \mathcal{F}[u], D_u \mathcal{E}[u] \rangle$.

(S4) Prove an **entropy/entropy-production** estimate $\mathcal{E}[u] \lesssim \langle \mathcal{F}[u], D_u \mathcal{E}[u] \rangle$.

(S5) Apply Gronwall's inequality to $\frac{d\mathcal{E}}{dt} \lesssim -\mathcal{E}$ to get exponential decay.

Lemma 24.1. *The Fokker-Planck equation* (24.1) *has a positive steady state* $u^* = Ze^{-V}$ *of mass* 1, *where* Z *is a normalization constant.*

Proof. Just consider $\partial_t u = 0$. Then we check that

$$\nabla u^* + u^* \nabla V = u^* \nabla (\ln(u^*) + V) = 0$$

so u^* is indeed a steady state. $\qquad \square$

It turns out that not only one entropy functional works to show global decay to u^*, but there are many important families/classes of such functionals. Let $\phi : \mathbb{R} \to \mathbb{R}$ be a smooth, strictly convex function such that $1/\phi''$ is concave, $\phi'''(1) < 0$ and $\phi(1) = 0 = \phi'(1)$; for examples see Exercise 24.4 and Figure 24.1.

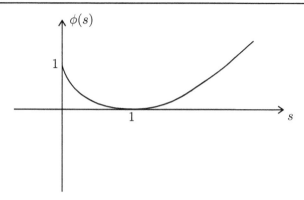

Figure 24.1. *A possible function ϕ shown only for the domain $[0, \infty)$.*

Now define the **(relative) entropy** functionals as

$$\mathcal{E}_\phi[u] := \int_{\mathbb{R}^d} \phi\left(\frac{u}{u^*}\right) u^* \, dx.$$

We want to prove the following main result of this chapter:

Theorem 24.2. *(Fokker-Planck equilibration) Suppose $\mathcal{E}[u^*] < +\infty$, and assume we may interchange the limits*

$$\lim_{t \to +\infty} \mathcal{E}_\phi[u(t)] = \mathcal{E}_\phi\left[\lim_{t \to +\infty} u\right], \qquad \lim_{t \to +\infty} \frac{d}{dt}\mathcal{E}_\phi[u(t)] = \frac{d}{dt}\mathcal{E}_\phi\left[\lim_{t \to +\infty} u\right] \quad (24.2)$$

*and that the **Bakry-Emery condition***

$$D^2 V \geq \lambda \, \text{Id} \qquad \text{for some } \lambda > 0, \qquad (24.3)$$

*holds, i.e., subtracting a positive multiple of the identity from the **Hessian matrix***

$$D^2 V = \left(\partial_{x_i x_j} V\right)_{i,j=1}^d$$

yields a positive semidefinite matrix. Then any smooth solution to (24.1) converges to the steady state with exponential rate

$$\|u(t) - u^*\|_{L^1(\mathbb{R}^d)} \leq e^{-\lambda t} \sqrt{2\mathcal{E}_\phi[u_0]/\phi''(1)}. \qquad (24.4)$$

Remark: The condition (24.2) is not necessary, as it can be proven, but it is quite technical to obtain [21]. The Bakry-Emery condition (24.3) is *crucial*.

Proof. We recall the strategy (S1)–(S5), where (S1) is done in Lemma 24.1. For (S2), let $\rho := u/u^*$ and compute

$$\frac{d}{dt}\mathcal{E}_\phi[u] = \int_{\mathbb{R}^d} \phi'(\rho)\partial_t u \, dx = -\int_{\mathbb{R}^d} \phi''(\rho)\|\nabla \rho\|^2 u^* \, dx \leq 0 \qquad (24.5)$$

using $\partial_t u = \nabla \cdot (u^* \nabla \rho)$. Since ϕ is convex, we may conclude that the **entropy production** satisfies $\mathcal{P}_\phi[u] \leq 0$. The key insight of the proof due to Bakry and Emery is that it does help to compute the second time derivative

$$\frac{d^2}{dt^2}\mathcal{E}_\phi[u] = \underbrace{-\int_{\mathbb{R}^d} \phi'''(\rho)\partial_t u \|\nabla \rho\|^2 \, dx}_{=:I_1} \underbrace{-2\int_{\mathbb{R}^d} \phi''(\rho)(\nabla \rho)^\top \partial_t \nabla \rho u^* \, dx}_{=:I_2}.$$

Using the Fokker-Planck PDE (24.1) and integrating by parts yields

$$
I_1 = \int_{\mathbb{R}^d} \left(\nabla(\phi'''(\rho)\|\nabla\rho\|^2) \right)^\top (u^*\nabla\rho)\, \mathrm{d}x,
$$

$$
= \int_{\mathbb{R}^d} \left(\phi''''(\rho)\|\nabla\rho\|^4 + 2\phi'''(\rho)(\nabla\rho)^\top \mathrm{D}^2\rho\nabla\rho \right) u^*\, \mathrm{d}x. \tag{24.6}
$$

The second integral I_2 is more difficult and we need some preparation. Observe that

$$
\nabla\partial_t\rho = \nabla\Delta\rho - \mathrm{D}^2\rho\nabla V - \mathrm{D}^2 V\nabla\rho, \tag{24.7}
$$

which follows using $u = u^*\rho$, the PDE (24.1), and the steady state condition satisfied by u^*. Furthermore, we note that

$$
(\nabla\rho)^\top\nabla\Delta\rho = \nabla\cdot(\mathrm{D}^2\rho\nabla\rho) - \|\mathrm{D}^2\rho\|_F^2, \tag{24.8}
$$

which follows from using the definition of the Hessian matrix $\mathrm{D}^2\rho$ with components $(\mathrm{D}^2\rho)_{ij} = \partial^2_{x_i x_j}\rho$ and recalling the **Frobenius (matrix) norm**

$$
\|\mathrm{D}^2\rho\|_F^2 = \sum_{i,j}^d |(\mathrm{D}^2\rho)_{ij}|^2.
$$

Using (24.7)–(24.8) for I_2 leads to

$$
I_2 = -2\int_{\mathbb{R}^d} u^*\phi''(\rho)\left((\nabla\rho)^\top\nabla\Delta\rho - (\nabla\rho)^\top\mathrm{D}^2\rho\nabla V - (\nabla\rho)^\top\mathrm{D}^2 V\nabla\rho \right)\, \mathrm{d}x
$$

$$
= -2\int_{\mathbb{R}^d} u^*\phi''(\rho)\left(\nabla\cdot(\mathrm{D}^2\rho\nabla\rho) - \|\mathrm{D}^2\rho\|_F^2 \right.
$$

$$
\left. -(\nabla\rho)^\top\mathrm{D}^2\rho\nabla V - (\nabla\rho)^\top\mathrm{D}^2 V\nabla\rho \right)\, \mathrm{d}x.
$$

The first term can be integrated by parts producing two new terms, of which one can be grouped with the third term above. In the fourth term above we apply the Bakry-Emery condition and end up with

$$
I_2 \geq 2\int_{\mathbb{R}^d} \phi'''(\rho)(\nabla\rho)^\top\mathrm{D}^2\rho\nabla\rho u^* + \phi''(\rho)\|\mathrm{D}^2\rho\|_F^2 u^*\, \mathrm{d}x
$$

$$
+2\int_{\mathbb{R}^d} \phi''(\rho)(\nabla\rho)^\top\mathrm{D}^2\rho\left(\nabla u^* + u^*\nabla V \right)\, \mathrm{d}x + \int_{\mathbb{R}^d} 2\lambda\phi''(\rho)\|\nabla\rho\|^2 u^*\, \mathrm{d}x.
$$

Using the fact that u^* is a steady state means that the second integral vanishes. By (24.5), we may replace the last integral by the negative first time derivative of the entropy. Combining these observations with (24.6) gives us an estimate for the second time derivative:

$$
\frac{\mathrm{d}^2}{\mathrm{d}t^2}\mathcal{E}_\phi[u] \geq \int_{\mathbb{R}^d} \left(\phi''''(\rho)\|\nabla\rho\|^4 + 4\phi'''(\rho)(\nabla\rho)^\top\mathrm{D}^2\rho\nabla\rho \right.
$$

$$
\left. +2\phi''(\rho)\|\mathrm{D}^2\rho\|_F^2 \right) u^*\, \mathrm{d}x - 2\lambda\frac{\mathrm{d}}{\mathrm{d}t}\mathcal{E}_\phi[u]
$$

$$
\geq 2\int_{\mathbb{R}^d} \phi''(\rho)\left| \mathrm{D}^2\rho + \frac{\phi'''(\rho)}{\phi''(\rho)}\nabla\rho\otimes\nabla\rho \right|^2 u^*\, \mathrm{d}x
$$

$$
-2\lambda\frac{\mathrm{d}}{\mathrm{d}t}\mathcal{E}_\phi[u] + \int_{\mathbb{R}^d} \left(\phi''''(\rho) - 2\frac{\phi'''(\rho)^2}{\phi''(\rho)} \right)\|\nabla\rho\|^4\, \mathrm{d}x.
$$

Although the last inequality looks too complicated, the main point is now to use the assumptions on ϕ. First, observe that $\phi'''' - 2(\phi''')^2/\phi'' = -(\phi'')^2(1/\phi'')'' \geq 0$ since $1/\phi''$ is concave, which deals with the last integral term. Second, recall that ϕ is convex, which allows one to also bound the first integral from below by zero. Hence, we end up with

$$\frac{\mathrm{d}^2}{\mathrm{d}t^2}\mathcal{E}_\phi[u] \geq -2\lambda\frac{\mathrm{d}}{\mathrm{d}t}\mathcal{E}_\phi[u]. \tag{24.9}$$

Integrating in the time t over (s, ∞) implies

$$\frac{\mathrm{d}}{\mathrm{d}t}\mathcal{E}_\phi[u(s)] - \lim_{t\to\infty}\frac{\mathrm{d}}{\mathrm{d}t}\mathcal{E}_\phi[u(t)] \leq -2\lambda\left(\mathcal{E}_\phi[u(s)] - \lim_{t\to+\infty}\mathcal{E}_\phi[u(t)]\right). \tag{24.10}$$

We claim that both $t \to \infty$ limits actually vanish. The entropy production one is dealt with in Exercise 24.5, which shows that the entropy production decays to zero. For the other limit, we use assumption (24.2) and (24.11) to get

$$0 = \lim_{t\to+\infty}\mathcal{P}_\phi[u(t)] = \mathcal{P}_\phi\left[\lim_{t\to+\infty}u(t)\right],$$

but since $\mathcal{P}_\phi$ vanishes exactly at u^*, it follows that $\lim_{t\to\infty}u(t) = u^*$. Using (24.2) again and recalling that we assumed $\phi(1) = 0$ finally gives

$$\lim_{t\to+\infty}\mathcal{E}_\phi[u(t)] = \mathcal{E}_\phi\left[\lim_{t\to+\infty}u(t)\right] = \mathcal{E}_\phi[u^*] = 0.$$

Therefore, we obtain from (24.10) the inequality

$$\frac{\mathrm{d}}{\mathrm{d}t}\mathcal{E}_\phi[u(s)] \leq -2\lambda\mathcal{E}_\phi[u(s)]$$

so that Gronwall's inequality provides exponential decay of the entropy

$$\mathcal{E}_\phi[u(s)] \leq \mathcal{E}_\phi[u_0]\mathrm{e}^{-2\lambda s}, \qquad s \geq 0.$$

To actually get the decay in L^1 as claimed in (24.4), one has to apply the **Csiszár-Kullback-Pinsker inequality**, which is stated in Theorem 24.3 below, as well as the entropy decay

$$\|u(t) - u^*\|^2_{L^1(\mathbb{R}^d)} \leq \frac{2}{\phi''(1)}\mathcal{E}_\phi[u(t)] \leq \frac{2}{\phi''(1)}\mathcal{E}_\phi[u(0)]\mathrm{e}^{-2\lambda t},$$

which finishes the proof.   □

The following inequality allows one to link a *relative entropy* to the L^1-difference.

Theorem 24.3 (Csiszár-Kullback-Pinsker inequality, [21]). *Let $\Omega \subseteq \mathbb{R}^d$ be a domain and suppose $f, g \in L^1(\Omega)$ with $f \geq 0$, $g > 0$, $\int_\Omega f(x)\,\mathrm{d}x = 1 = \int_\Omega g(x)\,\mathrm{d}x$. Let $\phi \in C^0([0,\infty)) \cap C^4(0,\infty)$ with $\phi(1) = 0$, $\phi''(1) > 0$, $\phi'''(1) < 0$, and assume that ϕ is convex and $1/\phi''$ is concave on $(0,\infty)$. Then we have the estimate*

$$\|f - g\|^2_{L^1(\Omega)} \leq \frac{2}{\phi''(1)}\int_\Omega\phi\left(\frac{f(x)}{g(x)}\right)g(x)\,\mathrm{d}x.$$

The Csiszár-Kullback-Pinsker inequality is just one example of a whole zoo of so-called **functional inequalities**.

Exercise 24.4. Consider $s \in (0, \infty)$ and prove that the functions

$$\phi(s) = s(\ln s - 1) + 1 \qquad \text{and} \qquad \phi(s) = \frac{1}{2}(s - 1)^2$$

are smooth and strictly convex and that $1/\phi''$ is concave, $\phi'''(1) < 0$, and $\phi(1) = 0 = \phi'(1)$. $\Diamond$

Exercise 24.5. In the proof of Theorem 24.2, justify why we have

$$\lim_{t \to \infty} \frac{\mathrm{d}}{\mathrm{d}t} \mathcal{E}_\phi[u(t)] = - \lim_{t \to \infty} \mathcal{P}_\phi[u(t)] = 0, \tag{24.11}$$

i.e., prove the asymptotic time-limit convergence of the entropy production to zero. $\Diamond$

Exercise 24.6. Consider the heat equation $\partial_t u = \Delta u$ for $u(x, t) = u$ with $t > 0$ on $\mathbb{R}^d$. Consider the **self-similar** variable transformation

$$y = x(2t + 1)^{-1/2}, \qquad s = \ln \sqrt{2t + 1}, \qquad \tilde{u}(y, s) = \mathrm{e}^{ds} u(\mathrm{e}^s y, (\mathrm{e}^{2s} - 1)/2)$$

and show that this converts the heat equation to a Fokker-Planck equation of the form (24.1) for $\tilde{u}$. Compute the potential V explicitly. What does Theorem 24.2 imply for the heat equation? $\Diamond$

Background and Further Reading: This chapter essentially follows the PDE viewpoint in [21] of the Bakry-Emery method [37] as presented in [312]; see also [89]. For background on Fokker-Planck (or Kolmogorov) equations in a more applied context, we refer to [209, 222, 476]. For a detailed exposition in the probabilistic context, possible references are [22, 291, 313]. Although the convergence to equilibrium for the Fokker-Planck equation [153, 264, 435] is somewhat expected, e.g., from probabilistic fundamentals under suitable assumptions, even lifting these assumptions slightly can be subtle as discussed in Chapter 29. Furthermore, for stochastic systems metastable and transient time scales are often very relevant for practical applications [54, 561]. For a prominent application of entropy ideas to geometric flows we refer to [454]. More expository approaches to problems involving geometric flows, in particular the Ricci flow, can be found in [111, 112, 336, 415, 536].

Chapter 25

Convexity and Minimizers

In Chapters 4–5, we obtained the existence of nontrivial solutions of a broad range of stationary PDEs using parameter variation and bifurcation theory. In Chapters 23–24 we assumed a gradient flow structure for PDEs and exploited it dynamically. In this chapter, we mix the two ideas. Let $V : \mathbb{R}^3 \to \mathbb{R}$ be a smooth potential with

$$\min_{x \in \mathbb{R}^3} V(x) = 0 \qquad \text{and} \qquad \lim_{R \to \infty} \inf_{\|x\| > R} V(x) = \infty,$$

where $V(x) = \|x\|^2$ would be a very simple example. Let $\psi : \mathbb{R}^3 \to \mathbb{C}$ be a function and define the **Gross-Pitaevskii energy functional** as

$$\mathcal{E}[\psi] := \int_{\mathbb{R}^3} \|\nabla \psi(x)\|^2 + V(x)|\psi(x)|^2 + p|\psi(x)|^4 \, \mathrm{d}x, \qquad (25.1)$$

where p is a parameter. Suppose we wanted to minimize $\mathcal{E}$ under the side condition

$$\int_{\mathbb{R}^3} |\psi(x)|^2 \, \mathrm{d}x = M \qquad (25.2)$$

for some $M \in \mathbb{N}$, which is actually a natural condition arising in the physical derivation of the Gross-Pitaevskii energy in a system with M particles. Define the space

$$X := \{\psi : \nabla \psi \in L^2, V|\psi^2| \in L^1, \psi \in L^4 \cap L^2\},$$

where $L^q = L^q(\mathbb{R}^3, \mathbb{C})$, and the space

$$X_M := X \cap \left\{\psi : \int_{\mathbb{R}^3} |\psi(x)|^2 \, \mathrm{d}x = M\right\}.$$

Note that a minimum for $\mathcal{E}$ in X_M may (or may not) exist but we can definitely define

$$E_*(M, p) := \inf_{\psi \in X_M} \mathcal{E}[\psi].$$

The next result is our first main goal and illustrates nicely the relation of the functional (25.1) to PDEs.

Theorem 25.1. *(Gross-Pitaevskii energy minimum) There exists $\psi^* \in X_M$ such that $E_*(M, p) = \mathcal{E}[\psi^*]$ and ψ^* solves the (stationary) Gross-Pitaevskii equation*

$$-\Delta \psi + V\psi + 4p|\psi|^2\psi = \mu\psi, \qquad (25.3)$$

where $\mu = \mu(M, p)$ is a constant. Furthermore, ψ^* is smooth, $\|\nabla \psi^*\|^2$ is unique, and one may select ψ^* such that $\psi^* > 0$.

Smoothness of ψ^* is eventually just going to follow from smoothness of V. One may actually show a certain stronger form of uniqueness for the minimizer. Both aspects, as well as positivity, will not be discussed here. The proof of Theorem 25.1 can be split up into several steps.

Lemma 25.2. *For $u \geq 0$, $\sqrt{u} \in X$, $\mathcal{E}[\sqrt{u}]$ is strictly convex in u.*

Proof. The second term in (25.1) is linear in $\sqrt{u}$ and the third one is quadratic in $\sqrt{u}$, so both terms are convex. Hence, it remains to deal with the gradient term. Consider $\psi_1 = \sqrt{u_1}$, $\psi_2 = \sqrt{u_2}$ as given functions in X_M. Observe that then

$$\psi = (\alpha u_1 + (1-\alpha)u_2)^{1/2}$$

is also in X_M for any fixed $\alpha \in (0,1)$. To obtain convexity in the gradient term we use a calculation, where ψ also acts as a "test function":

$$
\begin{aligned}
\psi \nabla \psi &= \alpha \psi_1 \nabla \psi_1 + (1-\alpha)\psi_2 \nabla \psi_2 \\
&= (\psi_1 \sqrt{\alpha})(\sqrt{\alpha} \nabla \psi_1) + \left((\psi_2 \sqrt{1-\alpha})(\sqrt{1-\alpha} \nabla \psi_2)\right) \\
&\leq \sqrt{\alpha \psi_1^2 + (1-\alpha)\psi_2^2} \sqrt{\alpha \|\nabla \psi_1\|^2 + (1-\alpha)\|\nabla \psi_2\|^2} \\
&= \psi \sqrt{\alpha \|\nabla \psi_1\|^2 + (1-\alpha)\|\nabla \psi_2\|^2}.
\end{aligned}
$$

Therefore, we have

$$\|\nabla \psi\|^2 \leq \alpha \|\nabla \psi_1\|^2 + (1-\alpha)\|\nabla \psi_2\|^2$$

and the result follows. $\square$

Lemma 25.3. *There exists $\psi^* \in X_M$ such that $\mathcal{E}[\psi^*] = E_*(M,p)$.*

Proof. Let $\{\psi_k\}_{k=1}^{\infty} \subset X_M$ be a sequence such that

$$\lim_{k \to \infty} \mathcal{E}[\psi_k] = E_*(M,p).$$

For this sequence, it follows from the structure of the functional $\mathcal{E}$ that there exists a constant $C > 0$, independent of k, such that

$$\|\psi_k\|_{H^1} < C, \qquad \|\psi_k\|_{L^4} < C, \qquad \|\psi_k\|_{L_V^2} < C, \tag{25.4}$$

where L_V^2 is the V-**weighted** L^2-space with norm

$$\|\psi\|_{L_V^2}^2 := \int_{\mathbb{R}^3} |\psi(x)|^2 V(x) \, dx.$$

Due to the bounds (25.4), there exists a weakly convergent subsequence with a limit $\psi^* \in X$, i.e.,

$$\psi_k \rightharpoonup \psi^* \text{ in } L^2 \cap L^4 \cap L_V^2, \qquad \nabla \psi_k \rightharpoonup \nabla \psi^* \text{ in } L^2.$$

Because the involved norms are lower semicontinuous, we immediately find

$$\liminf_{k \to \infty} \mathcal{E}[\psi_k] \geq \mathcal{E}[\psi^*].$$

Although we have a minimizer in X, this does not yet imply that it satisfies $\psi^* \in X_M$. However, this may be deduced from $\psi_k \in X_M$ quite directly. Indeed, we have

$$\|\psi^*\|_{L^2} \leq M \qquad \text{as } |\psi_k|^2 \to |\psi^*|^2 \text{ in } L^1_{\text{loc}}.$$

To get the reverse inequality, first note that

$$\lim_{k \to \infty} \int_\Omega |\psi_k(x)|^2 \, \mathrm{d}x = \int_\Omega |\psi^*(x)|^2 \, \mathrm{d}x \leq \|\psi^*\|_{L^2}^2 \tag{25.5}$$

for any bounded region $\Omega \subset \mathbb{R}^3$. Now argue by contradiction and suppose $\|\psi^*\|_{L^2}^2 = M - \delta$ for some positive $\delta > 0$. Hence, there exists a constant $K_\Omega > 0$ for all Ω such that

$$\int_{\mathbb{R}^3 \setminus \Omega} |\psi_k(x)|^2 \, \mathrm{d}x \geq \delta$$

for all $k \geq K_\Omega$. Since we assumed for the potential $V(x) \to \infty$ as $|x| \to \infty$, we must also have

$$\lim_{k \to \infty} \int_{\mathbb{R}^3} V(x) |\psi_k|^2 \, \mathrm{d}x = \infty.$$

This is impossible since ψ_k is a minimizing sequence for $\mathcal{E}$. $\qquad \square$

Finally we have all the required ingredients.

Proof. (of Theorem 25.1) Lemma 25.3 yields the existence of the minimizer. The uniqueness of $\|\nabla \psi^*\|^2$ follows from the strict convexity in Lemma 25.2. Hence it remains to check that it does satisfy the stationary Gross-Pitaevskii equation. Pick any test function $\psi \in C_c^\infty(\mathbb{R}^3, \mathbb{C})$ and note that stationarity of the functional $\mathcal{E}$ at ψ^* implies

$$\frac{\partial}{\partial \delta} \left(\mathcal{E}[\psi^* + \delta\phi] + \mu \|\psi^* + \delta\phi\|_{L^2}^2 \right)\big|_{\delta=0} = 0, \tag{25.6}$$

where μ is a **Lagrange multiplier** to enforce the constraint $\|\psi\|_{L^2}^2 = M$. Splitting ϕ in real and imaginary parts, the identity (25.6) for the real part reads

$$-\Delta \mathrm{Re}(\psi^*) + V\mathrm{Re}(\psi^*) + 8p\mathrm{Re}(\psi^*)|\psi^*|^2 = \mu \mathrm{Re}(\psi^*), \tag{25.7}$$

and there is an analogous equation for the imaginary part. The constant μ can actually be represented as an integral by multiplying the Gross-Pitaevskii equation with ψ^* and integrating. $\qquad \square$

In fact, the last argument of the proof can also be used to show that every solution of the (stationary) Gross-Pitaevskii equation is a minimizer and satisfies $\mathcal{E}[\psi^*] = E(M, p)$. Unfortunately, the result we have obtained may be dynamically not yet informative enough. The next result shows that one can obtain more precise information on the shape of ψ^* by using $\mathcal{E}$.

Proposition 25.4. *For all $\lambda > 0$ there exists C_λ such that*

$$\psi^*(x) \leq C_\lambda \mathrm{e}^{-\lambda|x|}.$$

In particular, $\psi^* \in L^\infty$ and it decays exponentially towards infinity.

Proof. First, we shift the potential and set $W = V + 4p|\psi^*|^2$ and observe that (25.3) yields

$$(-\Delta + \lambda^2)\psi^* = -(W - \mu - \lambda^2)\psi^*. \qquad (25.8)$$

One may check that the **Green's function** for the operator $L := (-\Delta + \lambda^2 \, \mathrm{Id})$ is given by $G_\lambda(x) = (4\pi|x|)^{-1} \exp(-\lambda|x|)$, i.e., an equation of the form $Lu = f$ is solved by $u(x) = (G_\lambda * f)(x) = \int_{\mathbb{R}^3} G_\lambda(x - y)f(y) \, \mathrm{d}y$. Using the Green's function, we may rewrite (25.8) as

$$\psi^*(x) = -\int_{\mathbb{R}^3} G_\lambda(x - y)(W(y) - \mu - \lambda^2)\psi^*(y) \, \mathrm{d}y. \qquad (25.9)$$

We know that $\psi^* > 0$, and for $\|y\| > R$ for some sufficiently large R we also have $W(y) - \mu - \lambda^2 > 0$ using the growth condition of the potential V. Therefore, (25.9) also gives

$$\psi^*(x) \leq -\int_{\|y\| < R} G_\lambda(x - y)(W(y) - \mu - \lambda^2)\psi^*(y) \, \mathrm{d}y.$$

Plugging in the expression for G and using $W\psi^* \in L^2_{\mathrm{loc}}$ now easily yields the result. $\qquad \square$

Even more information about the solution can be obtained if one knows more about the potential as shown in Exercise 25.5. The results we have developed also show that it can often be complicated to just prove the existence of a nontrivial stationary solution.

Exercise 25.5. Suppose V in (25.1) is spherically symmetric and radially monotonically increasing. Prove that then ψ^* is also spherically symmetric and radially monotonically decreasing. $\Diamond$

Exercise 25.6. It turns out that if V is convex in (25.1), then ψ^* is **log-concave**, i.e.,

$$(\psi^*)^\alpha(x)\psi^*(y)^{(1-\alpha)} \leq \psi^*(\alpha x + (1 - \alpha)y)$$

for all $x, y \in \mathbb{R}^3$ and $\alpha \in (0, 1)$. This is not easy to prove due to the nonlinear term. However, try to prove that solutions $u = u(x, t)$ of

$$\partial_t u = \Delta u \qquad \text{and} \qquad \partial_t u = -Vu$$

are log-concave if $u(x, 0)$ is log-concave. $\Diamond$

Exercise 25.7. Consider the (time-dependent) Gross-Pitaevskii equation

$$\mathrm{i}\partial_t \psi = -\partial_x^2 \psi - \psi|\psi|^2 + V(x)\psi, \qquad x \in \mathbb{R} \qquad (25.10)$$

with $V(x) = x^2$. Find conditions on the functions $f(t)$ and $l(t)$ such that the **lens transformation**

$$\psi(x, t) = l(t)\mathrm{e}^{\mathrm{i}x^2 f(t)}u(xl(t), \tau(t)), \qquad \partial_t \tau = l^2$$

implies that u satisfies the **(cubic) nonlinear Schrödinger (NLS) equation**

$$\mathrm{i}\partial_\tau u = -\partial_y^2 u - u|u|^2 + 2\mathrm{i}pu$$

for $y = xl(t)$ and $f/l^2 = p$. $\Diamond$

Background and Further Reading: The chapter follows selected parts of [377]; see also [374, 376] for more background and consider [181, 182] for derivations of the Gross-Pitaevskii (GP) equation. Of course, the GP equation can exhibit a wide range of nonlinear dynamics relevant for applications [88, 331, 332]. Essentially our viewpoint here has been **variational** as exemplified by the condition (25.6); see [141, 178, 225, 234, 309, 310, 516] for more references on the general theory of the calculus of variations and relations to PDEs. Of course, the applications and analysis of the NLS equation (see also Chapters 13 and 27) are deeply related to the GP equation, which can be directly seen from the algebraic structure; cf. Exercise 25.7 for a simple one-dimensional illustration [530].

Chapter 26

Mountain Passes and Periodic Waves

In Chapter 25, we linked minimizers of a functional to the stationary solution of a PDE. A crucial ingredient has been convexity. In this chapter, we are going to study a nonconvex functional in combination with a tool similar to the Intermediate Value Theorem to obtain nontrivial time-periodic solutions. We want to study the **cubic nonlinear wave equation**

$$\partial_t^2 u = \partial_x^2 u - u^3, \qquad u(0,t) = 0 = u(\pi, t), \qquad (26.1)$$

for $u = u(x,t)$ and $x \in (0, \pi)$. The goal is to prove the following result:

Theorem 26.1. *(time-periodic waves) There exists a nontrivial ($u \neq 0$ on a set of positive Lebesgue measure) and 2π-time-periodic solution to (26.1).*

Before we can prove Theorem 26.1, we need some preparation. Let X be a Banach space and $\mathcal{F} : X \to \mathbb{R}$ be C^1 and denote the derivative at $p \in X$ by $D\mathcal{F}(p) : X \to \mathbb{R}$; see also Definition 4.1. Suppose $\mathcal{F}$ satisfies the following conditions:

(A1) $\mathcal{F}(0) = 0$ and there exist constants $\rho, \beta > 0$ such that $\mathcal{F}(v) > 0$ for $0 < \|v\|_X =: \|v\| < \rho$ and $\mathcal{F}(v) \geq \beta$ if $\|v\| = \rho$.

(A2) There exists $v \neq 0$ such that $\mathcal{F}(v) = 0$.

(A3) Given a sequence $\{v_k\}_{k=1}^{\infty} \subset X$ with $|\mathcal{F}(v_k)|$ bounded uniformly and $\|D\mathcal{F}(v_k)\|_{X'} \to 0$ as $k \to +\infty$, then there exists a convergent subsequence.

Roughly speaking, (A1)–(A3) imply that the functional $\mathcal{F}$ has a trivial and a nontrivial zero, $\mathcal{F}$ is nonzero within a reasonable set *in between* the zeros, and $\mathcal{F}$ satisfies a certain compactness assumption.

Theorem 26.2. *(Mountain-Pass Theorem [16, 113, 186]) Suppose $\mathcal{F} \in C^1(X, \mathbb{R})$ satisfies (A1)–(A3). Then $\mathcal{F}$ has a **critical value** $c^* \in [\beta, +\infty)$, i.e., $\mathcal{F}(v^*) = c^*$, $D\mathcal{F}(v^*) = 0$ and v^* is a **critical point**.*

The intuition of Theorem 26.2 is illustrated in Figure 26.1 and is indeed reminiscent of the Intermediate Value Theorem. To construct the right Banach space X for the problem (26.1) requires preparation.

Lemma 26.3. *Let $\Omega = (0, \pi) \times (0, 2\pi)$ and assume that $f \in L^1(\Omega)$ satisfies $\langle f, \phi \rangle_{L^2(\Omega)} = 0$ for all $\phi \in L^{\infty} \cap \mathcal{N}[L]$ with $L := \partial_t^2 - \partial_x^2$. Then there exists a unique continuous*

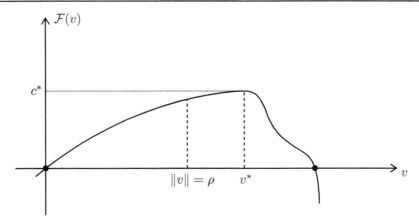

Figure 26.1. *Visualization of the setting in Theorem 26.2.*

2π-*time-periodic solution* $\mathcal{K}f$ *to*

$$\partial_t^2 u - \partial_x^2 u = f, \quad u(0,t) = 0 = u(\pi,t) \tag{26.2}$$

such that $\mathcal{K}f$ *is orthogonal to the nullspace* $\mathcal{N}[L]$ *in* $L^2(\Omega)$.

Remark: In this chapter, we shall just write $L^p = L^p(\Omega)$ and also omit the integration notation $dx\,dt$ if the integration is over the entire domain Ω. The operator L is never going to be iterated here, avoiding potential confusion of iterates with the Lebesgue space L^q.

Proof. Let $k_f \in \mathbb{R}$ be a constant to be determined and define

$$\tilde{u}(x,t) := -\frac{1}{2} \int_x^\pi \int_{t+x-y}^{t-x+y} f(y,s)\,ds\,dy + k_f \frac{\pi - x}{\pi}. \tag{26.3}$$

Then consider $w(s) := \frac{1}{2\pi} \int_0^\pi (\tilde{u}(x, s-x) - \tilde{u}(x, s+x))\,dx$ and define

$$(\mathcal{K}f)(x,t) := \tilde{u}(x,t) + w(t+x) - w(t-x). \tag{26.4}$$

Now we may check that the explicit formula (26.4) really provides the correct solution for a certain constant $k_f \in \mathbb{R}$. For example, observe that $\tilde{u}(x,t) = \tilde{u}(x, t+2\pi)$ implies $\int_0^{2\pi} w(s)\,ds = 0$ so that $w(t+x) - w(t-x)$ is in $\mathcal{N}[L]$ by Exercise 26.7(a). Then one easily calculates

$$L(\mathcal{K}f) = f$$

using the Leibniz integral rule several times. Therefore, we have that the first equation in (26.2) holds. For the boundary conditions, it is clear (see also Exercise 26.7(b)) that k_f can be adjusted uniquely to guarantee Dirichlet conditions. Checking the orthogonality conclusion essentially follows using the terms from $\mathcal{N}[L]$ in the solution. $\quad\square$

The construction of the integration region in (26.3) might seem slightly strange at first, but then one notices that the set

$$\{(y,s) \in \Omega : x \le y \le \pi, t + x - y \le s \le t - x + y\} \tag{26.5}$$

is a cone anchored at the point (x,t). Working in conical regions is very typical for wave equations, as information propagates in these regions along the characteristics for

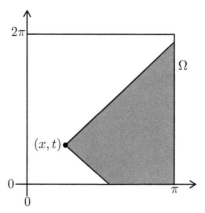

Figure 26.2. *Typical setting of information propagation from a point in the context of wave equations into a cone centered at (x, t) as defined in (26.5).*

the linear wave equation; see Figure 26.2. A good Banach space to define a suitable functional turns out to be

$$X := \left\{ v \in L^{4/3} : \langle v, \phi \rangle_{L^2} = 0 \text{ for all } \phi \in \mathcal{N}[L] \cap L^4 \right\}.$$

Lemma 26.4. *The operator $\mathcal{K} : X \to L^4$ is compact.*

Proof. We shall first show that there exists a constant $C_\mathcal{K} > 0$ such that

$$\|\mathcal{K}f\|_{C^{0,\alpha}(\Omega)} \leq C_\mathcal{K} \|f\|_{L^q} \quad \text{for } \alpha = 1 - \frac{1}{q}, \tag{26.6}$$

where $C^{0,\alpha}$ denotes α-**Hölder** functions, i.e., $g \in C^{0,\alpha}$ if and only if $|g(z) - g(\tilde{z})| \lesssim \|z - \tilde{z}\|^\alpha$ for all $z, \tilde{z} \in \Omega$. Since $k_f(1 - x/\pi)$ is smooth, it suffices to show that

$$u_f(x, t) := -\frac{1}{2} \int_x^\pi \int_{t-x+y}^{t+x-y} f(y, s) \, \mathrm{d}s \, \mathrm{d}y \tag{26.7}$$

is α-Hölder. Fix any numbers $a, b \in \mathbb{R}$, then we calculate

$$\begin{aligned} |u_f(x + a, t + b) - u_f(x, t)| &\leq \left| \int_x^{x+a} \int_{t+x-y}^{t-x+y} f(y, s) \, \mathrm{d}s \, \mathrm{d}y \right| \\ &+ \left| \int_x^\pi \int_{t+x-y}^{t+x-y+a+b} f(y, s) \, \mathrm{d}s \, \mathrm{d}y \right| \\ &+ \left| \int_x^\pi \int_{t-x+y}^{t-x+y+b-a} f(y, s) \, \mathrm{d}s \, \mathrm{d}y \right| \\ &\leq C_\mathcal{K}(|a| + |b|)^\alpha \|f\|_{L^q} \end{aligned}$$

by using Hölder's inequality in the last step; recall that **Hölder's inequality** states that for $p, q \in [1, \infty]$ with $1/p + 1/q = 1$, we have $\|fg\|_{L^1} \leq \|f\|_{L^p} \|g\|_{L^q}$. Since $C^{0,\alpha_2}(\Omega)$ is compactly embedded [232] into $C^{0,\alpha_1}(\Omega)$ for $0 < \alpha_1 < \alpha_2 < 1$, we may extract the required subsequence to get the compactness of $\mathcal{K}$ by using (26.6). □

Finally, we can define a useful functional $\mathcal{F} : X \to \mathbb{R}$ by

$$\mathcal{F}(v) := \frac{1}{2} \int_\Omega (\mathcal{K}v)v + \frac{3}{4} \int_\Omega |v|^{4/3}. \tag{26.8}$$

The derivative of the functional satisfies

$$[\mathrm{D}_u\mathcal{F}(v)]w = \int_\Omega (\mathcal{K}v)w + \int_\Omega |v|^{-2/3}vw,$$

with $[\mathrm{D}_u\mathcal{F}(v)]$ and element of the dual space X', i.e., we have a linear functional $[\mathrm{D}_u\mathcal{F}(v)] : X \to \mathbb{R}$. By the **Hahn-Banach Theorem**, one may write the derivative $\mathrm{D}_u\mathcal{F}(v)$ also as

$$\mathcal{K}v + |v|^{-2/3}v = v_\mathcal{R} + v_\mathcal{N}, \qquad v_\mathcal{N} \in \mathcal{N}[L] \cap L^4, \quad v_\mathcal{R} \in L^4, \tag{26.9}$$

and $\|v_\mathcal{R}\| = \|\mathrm{D}_u\mathcal{F}(v)\|_{X'}$. We want to check the assumptions (A1)–(A3). We start with the most difficult step.

Lemma 26.5. *The functional $\mathcal{F}$ defined in (26.8) satisfies (A3).*

Proof. Let $\{v_k\}_{k=1}^\infty \subset X$ be a sequence with $|\mathcal{F}(v_k)| \leq c$ and $\|\mathrm{D}\mathcal{F}(v_k)\|_{X'} \to 0$ as $k \to +\infty$. Using the representation (26.9), we have that $\mathcal{K}v_k + |v_k|^{-2/3}v_k = v_{\mathcal{R},k} + v_{\mathcal{N},k}$ and $v_{\mathcal{R},k} \to 0$ as $k \to +\infty$. In addition, note that using the definition of X and Hölder's inequality imply that

$$\left| \frac{1}{2} \int_\Omega (\mathcal{K}v_k)v_k + \frac{1}{2} \int_\Omega |v_k|^{4/3} \right| = \frac{1}{2} \left| \int_\Omega v_k v_{\mathcal{R},k} \right|$$

$$\leq \frac{1}{2} \|v_k\|_{L^{4/3}} \|v_{\mathcal{R},k}\|_{L^4}.$$

This estimate in combination with $|\mathcal{F}(v_k)| \leq c$ gives

$$\left(\frac{3}{4} - \frac{1}{2} \right) \int_\Omega |v_k|^{4/3} \leq c + \frac{1}{2} \|v_k\|_{L^{4/3}} \|v_{\mathcal{R},k}\|_{L^4}.$$

Therefore, the sequence $\{\|v_k\|_{L^{4/3}}\}$ is bounded. By a standard convergence result in L^p-spaces for $p \in (1, +\infty)$, bounded sequences have **weak limits** [375, Thm. 2.18]; in general, **weak convergence** of a sequence $v_k \in L^p$ to some $v_\infty \in L^p$ is denoted by $v_k \rightharpoonup v_\infty$ and just means that

$$\int v_k w \to \int v_\infty w \qquad \forall w \in L^q \text{ with } 1/p + 1/q = 1.$$

By convexity we have

$$\left| \frac{3}{4}v_\infty + \frac{1}{4}v_k \right|^{4/3} \leq \frac{3}{4}|v_\infty|^{4/3} + \frac{1}{4}|v_k|^{4/3}.$$

Some algebra and using (26.9) then leads to

$$\frac{3}{4}|v_\infty|^{4/3} - \frac{3}{4}|v_k|^{4/3} \geq |v_k|^{-2/3}v_k(v_\infty - v_k)$$

$$= (v_{\mathcal{R},k} + v_{\mathcal{N},k} - \mathcal{K}v_k)(v_\infty - v_k).$$

Upon integration and using the orthogonality condition in X, we find

$$\frac{3}{4}\int_\Omega |v_\infty|^{4/3} - \frac{3}{4}\int_\Omega |v_k|^{4/3} \geq \int_\Omega (v_{\mathcal{R},k} - \mathcal{K}v_k)(v_\infty - v_k).$$

Since $\mathcal{K}$ is compact and $\|v_{\mathcal{R},k}\|_{L^4} \to 0$, as well as $v_k \rightharpoonup v_\infty$ as $k \to +\infty$, it follows that the right-hand side of the last integral inequality tends to zero. Therefore, one obtains

$$\limsup_{k\to+\infty} \|v_k\|_{L^{4/3}} \leq \|v_\infty\|_{L^{4/3}},$$

which implies the strong convergence $v_k \to v_\infty$ in X as $k \to +\infty$ so (A3) is verified. $\qquad\square$

Lemma 26.6. *The functional $\mathcal{F}$ defined in (26.8) satisfies (A1)–(A2).*

Proof. We shall only prove (A1) and leave (A2) as Exercise 26.8. First, note that $\mathcal{F}(0) = 0$ holds trivially. Second, observe that we have the inequality $\|\mathcal{K}v\|_{L^\infty} \lesssim \|v\|_{L^1}$ directly from the construction of $\mathcal{K}$. Using this fact, and considering only v such that $\|v\|_{L^{4/3}} = r$, one finds by a direct integral estimate

$$\mathcal{F}(v) \gtrsim -\|v\|_{L^1}^2 + \frac{3}{4}\|v\|_{L^{4/3}}^{4/3}$$

$$\gtrsim -\|v\|_{L^{4/3}}^2 + \frac{3}{4}\|v\|_{L^{4/3}}^{4/3} \gtrsim \frac{3}{4}r^{4/3} - r^2 > 0$$

for all sufficiently small r, i.e., $r \in (-r_0, r_0)$ for some $r_0 > 0$. $\qquad\square$

Finally, we can put the intermediate results together and show the existence of a nontrivial solution to (26.1).

Proof. (of Theorem 26.1) Lemma 26.5 and Lemma 26.6 imply that we can apply Theorem 26.2 to obtain the existence of a critical point v^* for the functional $\mathcal{F}$ defined in (26.8). Since $D_u\mathcal{F}(v^*) = 0$, the representation (26.9) implies

$$\mathcal{K}v^* + |v^*|^{-2/3}v^* = v_\mathcal{N} \quad \text{for some } v_\mathcal{N} \in \mathcal{N}[L] \cap L^4. \tag{26.10}$$

Now let $u := v_\mathcal{N} - \mathcal{K}v^*$; then

$$(\partial_t^2 - \partial_x^2)u = L(v_\mathcal{N} - \mathcal{K}v^*) = -v^*$$

since $L(\mathcal{K}v^*) = v^*$ by construction of the operator $\mathcal{K}$ as a solution operator for (26.2). However, we also have $|v^*|^{-2/3}v^* = v_\mathcal{N} - \mathcal{K}v^* = u$ so that $u^3 = v^*$, which precisely implies

$$Lu + u^3 = \partial_t^2 u - \partial_x^2 u + u^3 = 0.$$

By using standard regularity theory, one may actually show also that the solution we found is smooth. $\qquad\square$

The structure of the proof presented in this chapter also applies to many other problems, not only to nonlinear wave equations; see Exercise 26.9. The major dynamical conclusion one obtains is that the flow has to be more complicated than just one unique attracting global steady state. However, the existence of a nontrivial solution is just one first step to really understand the dynamics completely.

Exercise 26.7. (a) Show that the nullspace $\mathcal{N}[L]$ in $L^1(\Omega)$ using Dirichlet boundary conditions and 2π time periodicity is given by functions of the form $v(t+x) - v(t-x)$ with $\int_0^{2\pi} v(s)\,\mathrm{d}s = 0$. (b) Prove that

$$k_f = \int_0^\pi \int_{t-y}^{t+y} f(y,s)\,\mathrm{d}s\,\mathrm{d}y,$$

where k_f is the constant in the proof of Lemma 26.3. $\Diamond$

Exercise 26.8. Show that property (A2) holds in the context of the proof of Lemma 26.6. $\Diamond$

Exercise 26.9. Consider the nonlinear elliptic PDE

$$\begin{aligned}
\Delta u + u^3 &= 0 \quad \text{in } \tilde{\Omega} \subset \mathbb{R}^3, \\
u &= 0 \quad \text{on } \partial\tilde{\Omega},
\end{aligned} \tag{26.11}$$

where $\tilde{\Omega}$ is a bounded connected open set with smooth boundary. Consider the functional

$$\mathcal{F}(u) := \frac{1}{2}\int_{\tilde{\Omega}} \|\nabla u\|^2\,\mathrm{d}x - \frac{1}{4}\int_{\tilde{\Omega}} |u|^4\,\mathrm{d}x \tag{26.12}$$

and show that it is C^1, when defined as a functional $\mathcal{F}: W_0^{1,2}(\tilde{\Omega}) \to \mathbb{R}$. Furthermore, prove the properties (A1)–(A2) for $\mathcal{F}$ defined in (26.12). Remark: It is possible— but more difficult—to verify that (A3) holds as well, which implies that (26.11) has a nontrivial solution. $\Diamond$

Background and Further Reading: This chapter follows a line of argument in [113]. Nonlinear wave equations are a vast field and we can only mention a few sources here to get the reader started [11, 296, 327, 372, 371, 447] which mostly focus on blow-up, a topic to be discussed in Chapter 30. For more on the Mountain-Pass Theorem initially shown in [16] and its applications see [146, 293, 468, 564]. A very nice quantitative dynamical systems application to obtain so-called **spike-layer** steady-state solutions to nonlinear elliptic problems with small diffusion can be found in [378]. We also refer to [49, 249, 379, 430, 431, 432], where a lot more on spike-layer solutions can be found, and to Chapter 35, where an asymptotic analysis approach is sketched for a related problem. Time-periodic solutions can also be obtained via bifurcation theory locally as discussed in Chapter 5, and we only refer to [307] for one example.

Chapter 27

Hamiltonian Dynamics and Normal Forms

In Chapters 23–26, we exploited gradient-type structures and energy/entropy functionals. In particular, we made use of given physical structures. **Hamiltonian dynamical systems** provide another class of problems, where additional physical structure can be exploited. Hamiltonian systems are a classical topic in the context of ODEs. In this chapter we provide a brief introduction to Hamiltonian PDEs.

Example 27.1. First, we recall an ODE example. Consider a particle at position $p = p(t) \in \mathbb{R}^d$ of mass 1 moving in a potential $V : \mathbb{R}^d \to \mathbb{R}$ according to Newton's law $p'' = -\nabla V(p)$, where $' = \mathrm{d}/\mathrm{d}t$. With $p' =: q$, we get

$$
\begin{aligned}
p' &= q &&= \partial_q H, \\
q' &= -\nabla V(p) &&= -\partial_p H,
\end{aligned}
\tag{27.1}
$$

where $H : \mathbb{R}^d \to \mathbb{R}$, $H(p, q) = \frac{1}{2}\|q\|^2 + V(p)$ is the **Hamiltonian** of the system, and (27.1) is in the canonical form of **Hamilton's equations**. One may also rewrite/reinterpret (27.1) once more, by considering the **skew-symmetric matrix**

$$
J = \begin{pmatrix} 0 & \mathrm{Id} \\ -\mathrm{Id} & 0 \end{pmatrix} \in \mathbb{R}^{d \times d},
$$

which induces a **symplectic form** ω on the **tangent space** to $\mathbb{R}^d$ via the formula $\omega(a, b) = a^\top J b$ for $a, b \in \mathbb{R}^d$, i.e., a bilinear, alternating, nondegenerate map $\omega : \mathbb{R}^d \times \mathbb{R}^d \to \mathbb{R}$; note that since we work on $\mathbb{R}^d$, we have the isomorphism for the tangent space $\mathrm{T}\mathbb{R}^d \simeq \mathbb{R}^d$ so that we may think of tangent vectors even globally as elements of $\mathbb{R}^d$ [365]. Then we immediately find

$$
\begin{pmatrix} p' \\ q' \end{pmatrix} = J\,\nabla H(p, q) := H_J(p, q),
$$

and $H_J : \mathbb{R}^d \to \mathrm{T}\mathbb{R}^d \simeq \mathbb{R}^d$ is also called a **Hamiltonian vector field**, which obeys

$$
\omega(H_J(p, q), w) = \mathrm{D}H(p, q)w \qquad \text{for all } w.
$$

In the canonical (p, q) coordinates, one may not only conveniently write H but also define the **Poisson bracket** of two functions f_1, f_2 by

$$
\{f_1, f_2\} := \sum_{j=1}^d \left(\frac{\partial f_1}{\partial q_j} \frac{\partial f_2}{\partial p_j} - \frac{\partial f_1}{\partial p_j} \frac{\partial f_2}{\partial q_j} \right).
$$

We remark that in complex notation with $z = p + iq$ one also has the more compact Hamiltonian dynamics

$$z' = -2i\frac{\partial H}{\partial \bar{z}}. \tag{27.2}$$

If $d = 1$ and $V(p) = \frac{1}{2}|p|^2$, we have the **harmonic oscillator** with phase portrait shown in Figure 27.1.

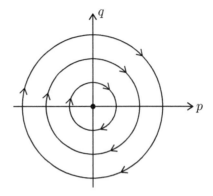

Figure 27.1. *Sketch of the phase portrait for (27.2) with $z = p + iq$, $d = 1$, and $V(p) = \frac{1}{2}|p|^2$.*

The origin is a steady state surrounded by invariant circles; steady states (or fixed points) with purely imaginary eigenvalues are also called **elliptic steady states**. A key motivating question in Hamiltonian dynamical systems is to prove which of the invariant circles (or more generally invariant tori) persist for the system

$$\begin{pmatrix} p' \\ q' \end{pmatrix} = J\,\nabla H(p, q) + \varepsilon G(p, q)$$

for some smooth map G and a sufficiently small parameter ε. ♦

Many definitions of Hamiltonian systems for ODEs, have relatively straightforward generalizations to several classes of PDEs. Let X be a Banach space and $H : X \to \mathbb{R}$ be a sufficiently smooth map (in many cases just C^2 suffices) and let ω be a symplectic form on X.

Definition 27.2. A **Hamiltonian system** on (X, ω) is given by

$$\frac{\mathrm{d}u}{\mathrm{d}t} = H_J(u), \tag{27.3}$$

where the **Hamiltonian vector field** $H_J : X \to TX \simeq X$ is defined by

$$\omega(H_J(u), v) = [DH(u)]v \qquad \text{for all } v \in X.$$

Remark: Definition 27.2 generalizes in a direct way to a Banach manifold $\mathcal{M}$ with a symplectic form $\omega = \omega(u)$ which depends upon the point $u \in \mathcal{M}$.

If X is a Hilbert space with inner product $\langle \cdot, \cdot \rangle$ we can represent the symplectic form via an operator J with $J^{-\top} = -J^{-1}$ and $\langle u, J^{-1}v \rangle = \omega(u, v)$ for all $u, v \in X$. The **gradient** is then defined by

$$DH(u)v = \langle \nabla_u H, v \rangle \qquad \text{for all } u, v \in X, \tag{27.4}$$

which then allows us to also write Hamilton's equation in a canonical form similar to (27.1); see Exercise 27.7. The PDE we are going to focus on is the **nonlinear Schrödinger equation (NLS)**

$$i\partial_t u = \partial_x^2 u - u - u|u|^2, \qquad u = u(x,t) \tag{27.5}$$

for $u : [0, \pi] \times [0, T] \to \mathbb{C}$ with Dirichlet boundary conditions $u(0, t) = 0 = u(\pi, t)$.

Lemma 27.3. *The NLS (27.5) can be written as a Hamiltonian system*

$$\frac{du}{dt} = i\nabla H(u), \qquad H(u) = \frac{1}{2}\langle -\partial_x^2 u + u, u \rangle + \frac{1}{4}\int_0^\pi |u|^4 \, dx \tag{27.6}$$

on $H_0^1([0, \pi], \mathbb{C})$ with inner product $\langle v, w \rangle = \mathrm{Re} \int_0^\pi v\bar{w} \, dx$.

Proof. The gradient is computed via the definition (27.4). So we consider the Hamiltonian in (27.6) and compute

$$DH(u)v = \mathrm{Re} \int_0^\pi -\partial_x^2 u\bar{v} + u\bar{v} + |u|^2 u\bar{v} \, dx,$$

where we have used that $\mathrm{Re}(u\bar{v}) = \mathrm{Re}(\bar{u}v)$. The result now follows upon multiplication by i to get the correct sign. $\qquad\square$

To match the result (27.6) to the canonical formulation similar to (27.1), one just looks at real and imaginary parts. One natural goal is to understand certain classes of solutions to NLS. In Chapters 7–10 we have studied travelling waves $u(x, t) = u(x - st)$. Related natural classes of solutions are periodic and **quasi-periodic** solutions

$$u(x, t) = u(\Xi), \qquad \Xi := (x, \omega_1 t, \dots, \omega_m t), \tag{27.7}$$

where $\omega := (\omega_1, \dots, \omega_m)^\top$ is a frequency vector consisting of **rationally independent** real numbers; recall that rational dependence means there exists $k \in \mathbb{Z}^m$ such that $k^\top \omega = 0$ and independence means that such a vector does not exist; see also Exercise 27.8. Time periodicity of $u(\Xi)$ is assumed in $\Xi_2, \dots, \Xi_m$, explaining the name quasi-periodicity. The **linear Schrödinger equation** (or NLS linearized around $u \equiv 0$) viewed as an evolution equation on H_0^1,

$$i\partial_t u = \partial_x^2 u - u, \qquad u(x, 0) = u_0(x), \tag{27.8}$$

provides already a guess that quasi-periodic solutions may exist.

Proposition 27.4. *The PDE (27.8) has (infinitely many) solutions on **invariant tori**.*

Proof. Consider the operator $-\partial_x^2 + \mathrm{Id}$, which has eigenvalues $\lambda_j = j^2 + 1$ and eigenfunctions $e_j(x) = \sqrt{2/\pi} \sin(jx)$ for $j \in \mathbb{N}$ in the case of Dirichlet boundary conditions; see also Example 6.3. Making the ansatz

$$u(x, t) = \sum_{j \geq 1} q_j(t) e_j(x)$$

in (27.8) easily yields the solution

$$u(x, t) = \sum_{j \geq 1} q_j(0) e^{i\lambda_j t} e_j(x), \tag{27.9}$$

where $q_j(0) = \langle e_j, u_0 \rangle_{H_0^1}$. Note that $q_j(t) = q_j(0)\mathrm{e}^{\mathrm{i}\lambda_j t}$ is periodic in t. Therefore, if we assume that the initial condition contains a certain finite or infinite number of nonzero modes, i.e., $q_j(0) \neq 0$ for

$$j \in \mathcal{J} = \{j_1 < j_2 < \cdots j_m\} \subset \mathbb{N}, \tag{27.10}$$

then there is an invariant linear subspace of complex dimension m given by

$$E_{\mathcal{J}} = \{u = q_1 e_{j_1} + \cdots + q_m e_{j_m} : q \in \mathbb{C}^m\}.$$

Each $E_{\mathcal{J}}$ is completely foliated into tori:

$$E_{\mathcal{J}} = \bigcup_{y \in (\mathbb{R}^+)^m} \mathcal{T}_{\mathcal{J}}(y),$$

where $\mathcal{T}_{\mathcal{J}}(y) = \{u = q_1 e_{j_1} + \cdots + q_m e_{j_m} : |q_j|^2 = 2y_j \text{ for } 1 \leq j \leq m\}$ is clearly isomorphic to a torus, as for each given y the argument is fixed; see also Figure 27.1. $\square$

One key idea in Hamiltonian systems is to prove when solutions *perturb*. Hence, we view the nonlinearity $u|u|^2$ as a perturbation near $u \equiv 0$ in the NLS.

Theorem 27.5. *(persistence of NLS tori [350]) For each given $\mathcal{J}$ as in (27.10) there exists a Cantor set $\mathcal{C} \subset (\mathbb{R}^+)^m$ such that the family of tori*

$$\mathcal{T}_{\mathcal{J}}[\mathcal{C}] = \bigcup_{y \in \mathcal{C}} \mathcal{T}_{\mathcal{J}}(y) \subset E_{\mathcal{J}}$$

can be mapped, via a Lipschitz embedding $\Phi : \mathcal{T}_{\mathcal{J}}[\mathcal{C}] \mapsto H_0^1$, to solutions of the NLS.

Remark: In addition, the frequencies ω of the tori are **Diophantine**, i.e., there exist constants $\alpha, \beta > 0$ such that $|k^\top \omega| \geq \alpha/|k|^\beta$ for $k \in \mathbb{Z}^m - \{0\}$.

A full proof of Theorem 27.5 is beyond our scope here. However, one strategy is to first simplify the NLS using a coordinate change and then apply a perturbation theorem. Here we illustrate part of the first step and refer to the references for the quite technical perturbation result. Let $l^{a,p}$ denote the Hilbert space of complex-valued sequences with norm

$$\|q\|_{a,p}^2 := \sum_{j \geq 1} |q_j|^2 j^{2p} \mathrm{e}^{2ja}. \tag{27.11}$$

Theorem 27.6. *(NLS normal form) There exist $a > 0$, $p > \frac{1}{2}$, and a change of coordinates so that the NLS equation (27.5) can be written in (**Birkhoff**) **normal form** (cf. (27.2)) as*

$$\frac{\mathrm{d}q_j}{\mathrm{d}t} = 2\mathrm{i}\frac{\partial \mathcal{H}}{\partial \bar{q}_j}, \qquad j \geq 1, \tag{27.12}$$

on $l^{a,p}$ where the Hamiltonian $\mathcal{H}$ in infinitely many coordinates is

$$\mathcal{H} = \Lambda + \mathcal{G} + \mathcal{O}(\|q\|_{a,p}^6), \qquad \Lambda = \frac{1}{2}\sum_{j \geq 1}\lambda_j|q_j|^2, \qquad \mathcal{G} = \frac{1}{2}\sum_{j,k \geq 1}\mathcal{G}_{ij}|q_j|^2|q_j|^2.$$

Proof. (Sketch [350]) For $q \in l^{a,p}$, using the eigenfunctions from the linearized problem, set $u = Sq = \sum_{j \geq 1} q_j e_j$. Then one verifies (Exercise 27.9(a)) that

$$\frac{\mathrm{d}q_j}{\mathrm{d}t} = 2\mathrm{i}\frac{\partial H}{\partial \bar{q}_j}, \tag{27.13}$$

where the Hamiltonian is given by

$$H = \Lambda + G, \qquad \Lambda = \frac{1}{2}\sum_{j \geq 1}\lambda_j|q_j|^2, \quad G = \frac{1}{4}\int_0^\pi |Sq|^4\,\mathrm{d}x.$$

Furthermore, one checks that solutions of (27.13) yield solutions of the NLS equation (27.5) via the formula

$$u(x, t) = \sum_{j \geq 1} q_j(t)e_j(x), \tag{27.14}$$

as discussed in Exercise 27.9(b). The key insight is that one may *simplify* the Hamiltonian with further coordinate changes. The proof idea is to use as a coordinate change M defined via the flow map of a suitable Hamiltonian vector field to cancel as many terms as possible. Let $M = X_{\mathcal{F}}^t|_{t=1}$ denote the time-1 map of the flow of the Hamiltonian vector field given by the Hamiltonian

$$\mathcal{F} = \frac{1}{4}\sum_{i,j,k,l}\mathcal{F}_{ijkl}q_iq_j\bar{q}_k\bar{q}_l$$

with coefficients

$$\mathrm{i}\mathcal{F}_{ijkl} := \begin{cases} \dfrac{G_{ijkl}}{\lambda_i + \lambda_j - \lambda_k - \lambda_l} & \text{if } i \pm j \pm k \pm l = 0 \text{ and } \{i, j\} \neq \{k, l\}, \\ 0 & \text{otherwise}, \end{cases} \tag{27.15}$$

where $i \pm j \pm k \pm l = 0$ holds for some combination of pluses and minuses for the first case, and $G_{ijkl} := \int_0^\pi e_ie_je_ke_l\,\mathrm{d}x$ are the coefficients of the "current" Hamiltonian G when expanded in the natural eigenbasis $\{e_j\}_{i \geq 1}$. It turns out that M is a real-analytic change preserving the symplectic form near the origin in $l^{a,p}$. Although we do not give the full details regarding analyticity, let us at least look carefully at the denominators $\lambda_i + \lambda_j - \lambda_k - \lambda_l$ in (27.15), which are also referred to as **divisors** in the context of normal form transformations. We claim that the conditions $i \pm j \pm k \pm l = 0$ and $\{i, j\} \neq \{k, l\}$ imply

$$\lambda_i + \lambda_j - \lambda_k - \lambda_l = i^2 + j^2 - k^2 - l^2 \neq 0. \tag{27.16}$$

We consider two cases. Suppose first that exactly one index pair is equal, say $j = l$ and $i \neq k$. This yields $i^2 + j^2 + k^2 + l^2 = i^2 - k^2 \neq 0$. Hence, suppose for the second case that no index pair is equal. We argue by contradiction and suppose

$$i^2 + j^2 = k^2 + l^2. \tag{27.17}$$

Without loss of generality we may order the indices, say $i < k \leq l \leq j$. Then one can go through all possible combinations of signs in $i \pm j \pm k \pm l = 0$ to conclude that only $i + j = k + l$ is possible. Therefore, we find

$$(i + j)^2 = i^2 + 2ij + j^2 = k^2 + 2kl + l^2 = (k + l)^2 \stackrel{(27.17)}{\Rightarrow} ij = kl.$$

This implies $i + j = k + (ij)/k$, or equivalently $(k - i)(k - j) = 0$, which is a contradiction. Therefore, (27.16) holds and this is the key step to check that M is well defined and analytic. Once this is established we start to calculate using Taylor's Theorem at $t = 0$:

$$H \circ M = H \circ X_{\mathcal{F}}^t|_{t=1} = H + \{H, \mathcal{F}\} + \int_0^1 (1 - t)\{\{H, \mathcal{F}\}, \mathcal{F}\} \circ X_{\mathcal{F}}^t \, dt$$

$$= \Lambda + G + \{\Lambda, \mathcal{F}\} + \underbrace{\{G, \mathcal{F}\} + \int_0^1 (1 - t)\{\{H, \mathcal{F}\}, \mathcal{F}\} \circ X_{\mathcal{F}}^t \, dt}_{=\mathcal{O}(\|q\|_{a,p}^6)},$$

where the Poisson bracket $\{\cdot, \cdot\}$ has been used (cf. [5, Sec. 3.3] and Example 27.1) and the last two terms are identified to have higher order. Hence, we only need to look at

$$G + \{\Lambda, \mathcal{F}\} = \frac{1}{4} \sum_{i \pm j \pm k \pm l = 0} (G_{ijkl} - \mathrm{i}(\lambda_i + \lambda_j - \lambda_k - \lambda_l)\mathcal{F}_{ijkl}) \, q_i q_j \bar{q}_k \bar{q}_l$$

$$= \frac{1}{4} \sum_{\{i,j\}=\{k,l\}} G_{ijkl} \, q_i q_j \bar{q}_k \bar{q}_l = \frac{1}{2} \sum_{i,j \geq 1} \mathcal{G}_{ij} |q_i|^2 |q_j|^2 = \mathcal{G},$$

where $2\mathcal{G}_{ii} = G_{iiii}$ and $\mathcal{G}_{ij} = G_{ijij}$. The result now follows as claimed since suitable terms have been eliminated. $\square$

It should be noted that the key idea in the last result of using an auxiliary flow $X_{\mathcal{F}}^t|_{t=1}$ to produce a coordinate change also plays a key role if applied iteratively in proving Theorem 27.5. More precisely, it helps to prove a suitable **Kolmogorov-Arnold-Moser (KAM) Theorem**, i.e., the persistence of certain tori upon perturbation. A KAM-type theorem can be proved via an iterative scheme of infinitely many time-1-map induced coordinate changes; see also the references below.

Exercise 27.7. Consider the wave equation

$$\partial_t^2 u = \partial_x^2 u \qquad \text{for } u = u(x, t) \text{ and } x \in [0, L] \tag{27.18}$$

with periodic boundary conditions. Prove that the wave equation can be rewritten as an infinite-dimensional Hamiltonian system

$$\begin{aligned} \partial_t p &= -\nabla_q H(p, q), \\ \partial_t q &= \nabla_p H(p, q), \end{aligned} \tag{27.19}$$

i.e., find a Banach space X and the Hamiltonian $H : X \to \mathbb{R}$ such that (27.19) is just a rewritten variant of (27.18) for p, q in a suitable dense subspace of X. $\Diamond$

Exercise 27.8. (a) Consider a **rotation on the circle** $\mathbb{S}^1 = [0, 1]/(0 \sim 1)$ as a discrete-time dynamical system given by

$$x_{m+1} = x_m + r \pmod 1. \tag{27.20}$$

What happens if $r \in \mathbb{Q}$? Prove that if $r \in \mathbb{R} \setminus \mathbb{Q}$, then the dynamics is **minimal**, i.e., the orbit of any initial condition x_0 is dense in $\mathbb{S}^1$. (b) Generalize the dynamics to the two-torus $\mathbb{T}^2$ (the torus obtained by identifying opposite edges of the unit square), i.e., study

$$(x_{m+1}, y_{m+1}) = (x_m + r_1, y_m + r_2) \pmod 1. \tag{27.21}$$

Show that there exist $r_1, r_2 \in \mathbb{R}$ such that (27.20) is not minimal nor filled by periodic orbits. (c) Simulate (27.20) on a computer for the following two cases: (c1) r_1, r_2 are rationally dependent and (c2) r_1, r_2 are rationally independent. What do you observe? $\Diamond$

Exercise 27.9. (a) Find a symplectic structure on $l^{a,p}$ such that (27.5) can be written in canonical Hamiltonian form given by (27.13). (b) Show that if $q = q(t)$ is an analytic solution of (27.13), then (27.14) is an analytic solution of (27.5). $\Diamond$

Background and Further Reading: The discussion of quasi-periodic solutions for the NLS equation follows [350]. For a detailed introduction to Hamiltonian dynamical systems we refer to [5, 26, 27, 398]. A very good introduction to symplectic geometry is [137] and for basics of differential geometry we refer to [365]. For chaos in infinite-dimensional Hamiltonian systems consider [257], and many concrete examples for Hamiltonian PDEs are discussed in [143]. For infinite-dimensional KAM Theorems we refer to [349, 463, 557] and references therein. Also recall Chapters 7 and 11, where we encountered the KdV equation, which is also deeply related to Hamiltonian systems. Of course, there are many other PDEs to which the same remark applies—not just KdV and NLS; to give an example, we mention the **Camassa-Holm equation** and its variants; see, e.g., [69, 81, 121, 144, 217].

Chapter 28

Empirical Measures and the Mean Field

In Chapter 27, we focused on Hamiltonian systems. A classical case is the motion of a *single* particle according to Newton's law as discussed in Example 27.1. In this chapter, we provide a strategy for general classes of *coupled particle systems*, which again links nicely questions in PDE dynamics. Consider m particles described by vectors $z_1, \ldots, z_m \in \mathbb{R}^d$ and the ODEs

$$\frac{\mathrm{d}z_j}{\mathrm{d}t} = z_j' = \frac{1}{m} \sum_{i=1}^{m} K(z_i, z_j), \qquad z_j = z_j(t), \tag{28.1}$$

with given initial conditions $Z_0 := (z_{0,1}, \ldots, z_{0,m})$ and for a kernel $K : \mathbb{R}^d \times \mathbb{R}^d \to \mathbb{R}^d$, $K \in C^1$ with symmetry $K(z, w) = -K(w, z)$ for all $z, w \in \mathbb{R}^d$ and bounded partial derivatives. Note that the symmetry implies $K(z, z) = 0$.

Example 28.1. As an example of (28.1), consider the ODEs for the motion of particles, say of mass 1, according to classical **Newtonian mechanics** with position $x \in \mathbb{R}^3$ and velocity $v \in \mathbb{R}^3$. For each individual particle we have

$$\begin{aligned} x' &= v, \\ v' &= F(x, v, t), \end{aligned} \tag{28.2}$$

where F is a (possibly time-dependent) force. If we let $z := (x, v)$ and suppose that the force F is given by collisions between particles, then we essentially have a coupled system of ODEs (28.1) with $d = 6$. ♦

The **microscopic** ODEs are exact but potentially difficult to deal with. Hence, it is of interest to derive PDEs to describe (28.1) on other temporal and spatial scales valid in the infinite particle limit $m \to \infty$.

Definition 28.2. To the state $\mathcal{Z}_m := (z_1, \ldots, z_m) \in \mathbb{R}^{md}$ we assign the **empirical measure**

$$\mu_{\mathcal{Z}_m} := \frac{1}{m} \sum_{j=1}^{m} \delta_{z_j}, \tag{28.3}$$

where δ_{z_j} is the delta- (or point-)measure at location z_j.

Observe that the kernel K can also be extended to act on Borel probability measures $\mathcal{P}(\mathbb{R}^d)$ with finite first moment

$$\mathcal{P}_1(\mathbb{R}^d) := \left\{ \nu \in \mathcal{P}(\mathbb{R}^d) : \mathbb{E}_\nu[|x|] = \int_{\mathbb{R}^d} |x| \, \mathrm{d}\nu(x) < +\infty \right\}$$

via the formula

$$Kv(z) := \int_{\mathbb{R}^d} K(z, w) \, dv(w).$$

The next result is crucial to link the particle system and the mean-field limit we eventually want to obtain.

Theorem 28.3. (*single-particle/multiparticle link*) *For given $\zeta_0 \in \mathbb{R}^d$ and $\mu_0 \in \mathcal{P}^1(\mathbb{R}^d)$, there exists a unique C^1 solution $Z = Z(\zeta_0, \mu_0, t)$ for*

$$\partial_t Z = (K\mu)Z, \qquad Z(\zeta_0, \mu_0, 0) = \zeta_0, \tag{28.4}$$

*where $\mu = \mu(t) := Z(\cdot, \mu_0, t)_* \mu_0 := \mu_0 \circ Z(\cdot, \mu_0, -t)$ is the **pushforward measure**. Furthermore, Z also satisfies*

$$z_j(t) = Z(z_{0,j}, \mu_{\mathcal{Z}_{0,m}}, t), \qquad \mathcal{Z}_{0,m} = (z_{0,1}, \ldots, z_{0,m}), \tag{28.5}$$

for any $j \in \{1, \ldots, m\}$ and all times $t \in \mathbb{R}$.

Proof. The proof [239] of the existence and uniqueness part is very similar to proving the existence of solutions for classical ODEs and we shall skip it here; cf. Theorem 2.1. Instead, we focus on the key relation (28.5), which links the space of particles z_j, the "single-particle phase space ODE" for Z, and the empirical measure μ. Let $\zeta_j(t) := Z(z_{0,j}, \mu_{\mathcal{Z}_{0,m}}, t)$ and compute

$$\mu(t) = Z(\cdot, \mu_{\mathcal{Z}_{0,m}}, t)_* \mu_{\mathcal{Z}_{0,m}} = \frac{1}{m} \sum_{j=1}^{m} \delta_{\zeta_j(t)}.$$

Therefore, applying the operator/kernel K and using (28.4) we get

$$\zeta_j' = (K\mu)\zeta_j = \frac{1}{m} \sum_{i=1}^{m} K(\zeta_i, \zeta_j).$$

Since $\zeta_j(0) = Z(z_{0,j}, \mu_{\mathcal{Z}_{0,m}}, 0) = z_{0,j}$, uniqueness of solutions to (28.1) now implies $\zeta_j(t) = z_j(t)$ for any $t \in \mathbb{R}$, so the result follows. $\square$

Let $\nu, \chi \in \mathcal{P}_1(\mathbb{R}^d)$ and denote by $\Pi(\nu, \chi)$ the set of probability measures $\pi \in \mathcal{P}(\mathbb{R}^d \times \mathbb{R}^d)$ with marginals ν and χ, i.e., integrating out the first or second set of d variables over $\mathbb{R}^d$ yields ν and χ respectively. To calculate with measures in the limit $m \to \infty$, it is helpful to consider the **Wasserstein** (or **Monge-Kantorovich**) **distance**, which we are going to use in one particular formulation. Define the Wasserstein-1 distance

$$d_1(\nu, \chi) := \inf_{\pi \in \Pi(\nu, \chi)} \int_{\mathbb{R}^d \times \mathbb{R}^d} \|x - y\| \, d\pi(x). \tag{28.6}$$

The intuition is that one transports with minimal effort the mass of ν to χ; see Figure 28.1. The next estimate is crucial, as it gives bounds which interact nicely with the limit $m \to \infty$.

Theorem 28.4. (*Dobrushin bound*) *Consider $\mu_0, \tilde{\mu}_0 \in \mathcal{P}_1(\mathbb{R}^d)$ and let*

$$\mu(t) := Z(\cdot, \mu_0, t)_* \mu_0 \qquad and \qquad \tilde{\mu}(t) := Z(\cdot, \tilde{\mu}_0, t)_* \tilde{\mu}_0. \tag{28.7}$$

Let κ be the Lipschitz constant of the operator K. Then we have

$$d_1(\mu(t), \tilde{\mu}(t)) \leq e^{2\kappa|t|} d_1(\mu_0, \tilde{\mu}_0) \qquad \forall t \in \mathbb{R}. \tag{28.8}$$

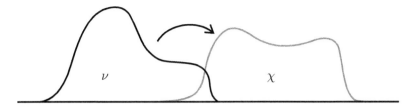

Figure 28.1. *Transport of a measure ν (in black) to a measure χ (in grey).*

Proof. The result (28.8) indicates that we should try to set up a Gronwall-type inequality. Just writing solutions of (28.4) as $Z = Z(\zeta, \mu, t)$ and $\tilde{Z} = Z(\tilde{\zeta}, \tilde{\mu}, t)$ in integral form and subtracting yields

$$Z - \tilde{Z} = \zeta - \tilde{\zeta} + \int_0^t \int_{\mathbb{R}^d} K(Z(\zeta, \mu, s), w) \, d\mu(w) \, ds$$

$$- \int_0^t \int_{\mathbb{R}^d} K(Z(\tilde{\zeta}, \tilde{\mu}, s), w) \, d\tilde{\mu}(w) \, ds$$

$$= \zeta - \tilde{\zeta} + \int_0^t \int_{\mathbb{R}^d} K(Z(\zeta, \mu_0, s), Z(w, \mu_0, s)) \, d\mu_0(w) \, ds$$

$$- \int_0^t \int_{\mathbb{R}^d} K(Z(\tilde{\zeta}, \tilde{\mu}_0, s), Z(\tilde{w}, \tilde{\mu}_0, s)) \, d\tilde{\mu}_0(\tilde{w}) \, ds,$$

where (28.7) has been used in the last step. From the last formula, one understands why the Wasserstein-1 distance should be useful here. For each coupling $\pi_0 \in \Pi(\mu_0, \tilde{\mu}_0)$ we have

$$Z - \tilde{Z} = \zeta - \tilde{\zeta} + \int_0^t \int_{\mathbb{R}^d \times \mathbb{R}^d} K(Z(\zeta, \mu_0, s), Z(w, \mu_0, s))$$

$$- K(Z(\tilde{\zeta}, \tilde{\mu}_0, s), Z(\tilde{w}, \tilde{\mu}_0, s)) \, d\pi_0(w, \tilde{w}) \, ds,$$

which is the crucial step in the Dobrushin bound. Recall that the kernel K is Lipschitz by assumption so that

$$\|K(w, \tilde{w}) - K(z, \tilde{z})\| \le \|K(w, \tilde{w}) - K(z, \tilde{w})\| + \|K(z, \tilde{w}) - K(z, \tilde{z})\|$$

$$\le \kappa \|w - z\| + \kappa \|\tilde{w} - \tilde{z}\|$$

for all $w, \tilde{w}, z, \tilde{z} \in \mathbb{R}^d$. Therefore, we can use the Lipschitz estimate and that π_0 is a probability measure to obtain

$$\|Z - \tilde{Z}\| \le \|\zeta - \tilde{\zeta}\| + \kappa \int_0^t \|Z(\zeta, \mu_0, s) - Z(\tilde{\zeta}, \tilde{\mu}_0, s)\| \, ds$$

$$+ \kappa \underbrace{\int_0^t \int_{\mathbb{R}^d \times \mathbb{R}^d} \|Z(w, \mu_0, s) - Z(\tilde{w}, \tilde{\mu}_0, s)\| \, d\pi_0(w, \tilde{w})}_{=:D[\pi_0](s)} \, ds,$$

and integrating the last inequality with respect to $d\pi_0(\zeta, \tilde{\zeta})$ yields

$$D[\pi_0](t) \le \int_{\mathbb{R}^d \times \mathbb{R}^d} \|\zeta - \tilde{\zeta}\| \, d\pi_0(\zeta, \tilde{\zeta}) + 2\kappa \int_0^t D[\pi_0](s) \, ds$$

$$= D[\pi_0](0) + 2\kappa \int_0^t D[\pi_0](s) \, ds.$$

Gronwall's inequality now implies

$$D[\pi_0](t) \leq D[\pi_0](0)e^{2\kappa|t|}. \tag{28.9}$$

Now we can just calculate

$$
\begin{aligned}
d_1(\mu(t), \tilde{\mu}(t)) &= \inf_{\pi \in \Pi(\mu(t), \tilde{\mu}(t))} \int_{\mathbb{R}^d \times \mathbb{R}^d} \|\zeta - \tilde{\zeta}\| \, d\pi(\zeta, \tilde{\zeta}) \\
&\leq \inf_{\pi_0 \in \Pi(\mu_0, \tilde{\mu}_0)} \int_{\mathbb{R}^d \times \mathbb{R}^d} \|Z(\zeta, \mu_0, s) - Z(\tilde{\zeta}, \tilde{\mu}_0, s)\| \, d\pi_0(\zeta, \tilde{\zeta}) \\
&= \inf_{\pi_0 \in \Pi(\mu_0, \tilde{\mu}_0)} D[\pi_0](t) \\
&\overset{(28.9)}{\leq} e^{2\kappa|t|} \inf_{\pi_0 \in \Pi(\mu_0, \tilde{\mu}_0)} D[\pi_0](0) = e^{2\kappa|t|} d_1(\mu_0, \tilde{\mu}_0)
\end{aligned}
$$

and the result (28.8) follows. $\qquad\square$

The next result shows that the empirical measure in the limit $m \to \infty$ can be viewed as a solution to a transport-type mean-field PDE.

Theorem 28.5. *(mean-field limit) Suppose $u_0 = u_0(z)$ is a probability density on $\mathbb{R}^d$ with finite first moment. Then the **mean-field** PDE*

$$\partial_t u + \nabla_z \cdot (u(Ku)) = 0, \qquad u(z, 0) = u_0(z) \tag{28.10}$$

has a unique weak solution $u \in C(\mathbb{R}, L^1(\mathbb{R}^d))$. Let $\mathcal{Z}_{0,m}$ denote the initial condition of the m-particle ODE (28.1) and denote the associated solution at time t by $\mathcal{Z}_{t,m}$. Assume that the empirical measure converges at time zero,

$$d_1(\mu_{\mathcal{Z}_{0,m}}, u_0) \to 0 \qquad as\ m \to \infty,$$

i.e., convergence in the infinite particle limit occurs. Then we have

$$d_1(\mu_{\mathcal{Z}_{t,m}}, u(\cdot, t)\text{Leb}) \to 0 \qquad as\ m \to \infty, \tag{28.11}$$

where Leb *is the Lebesgue measure.*

Proof. The existence of a weak solution to the mean-field PDE is quite standard, as it is essentially a general proof for transport PDEs. The convergence (28.11) follows directly from the Dobrushin bound since

$$d_1(\mu_{\mathcal{Z}_{t,m}}, u(\cdot, t)\text{Leb}) \leq e^{2\kappa|t|} d_1(\mu_{\mathcal{Z}_{0,m}}, u_0) \to 0$$

as $m \to \infty$. $\qquad\square$

In addition to Theorem 28.5, one may actually also prove weak convergence of the empirical measure in the space of probability measures $\mathcal{P}(\mathbb{R}^d)$. Furthermore, several extensions to other classes of particle systems exist, which can lead to a mean-field limit PDE. However, if the particle interactions become very complicated, it is unlikely that one can link a *dynamical system* defined by ODEs rigorously to a simple mean-field limit *PDE*.

Exercise 28.6. Let $\nu, \chi \in \mathcal{P}(\mathbb{R})$, $c : \mathbb{R}^2 \to \mathbb{R}$ be continuous, $\phi_1 \in L^1(\mathrm{d}\nu)$, $\phi_2 \in L^1(\mathrm{d}\chi)$, $\pi \in \Pi(\nu, \chi)$, and define

$$I[\pi] := \int_{\mathbb{R}^2} c(x, y) \, \mathrm{d}\pi(x, y), \qquad J(\phi_1, \phi_2) := \int_{\mathbb{R}} \phi_1(x)\mathrm{d}\nu(x) + \int_{\mathbb{R}} \phi_2(y)\mathrm{d}\chi(y).$$

Let Φ_c be the set to be all measurable functions $(\phi_1, \phi_2) \in L^1(\mathrm{d}\nu) \times L^1(\mathrm{d}\chi)$ such that

$$\phi_1(x) + \phi_2(y) \le c(x, y).$$

Prove that

$$\sup_{\Phi_c \cap L^1} J(\phi_1, \phi_2) \le \inf_{\pi \in \Pi(\nu, \chi)} I[\pi]. \tag{28.12}$$

Remark: The statement (28.12) is the first step towards proving **Kantorovich duality**, which basically states that (28.12) also holds with equality. $\lozenge$

Exercise 28.7. Consider the **incompressible Euler equation**

$$\partial_t u + (u \cdot \nabla)u = -\nabla p, \qquad \nabla \cdot u = 0,$$

for the density $u = u(x, t)$ and pressure $p = p(x, t)$ for $x \in \mathbb{R}^3$. Define the **vorticity** as

$$\omega := \nabla \times u \in \mathbb{R}^3.$$

(a) Find the PDE ω satisfies. (b) Now take $x \in \mathbb{R}^2$ and define the *scalar* vorticity as $\omega := \nabla \times u = \partial_{x_1} u_2 - \partial_{x_2} u_1$. Show that it satisfies the transport-type PDE

$$\partial_t \omega + u \cdot \nabla \omega = 0. \tag{28.13}$$

Remark: The vorticity equation (28.13) can be viewed as the mean-field limit of the motion and interaction of a large number of fluid vortices. $\lozenge$

Exercise 28.8. Consider (28.10) and use the **method of characteristics** to derive ODEs, which have to be satisfied; see Chapter 11. Hint: Compare your results to the ODEs in Theorem 28.3. $\lozenge$

Background and Further Reading: The presentation in this chapter follows several key parts of the notes [239]. *Kinetic theory*, which deals broadly speaking with equations describing the distribution function of particles on a mesoscopic level, has evolved into quite a mature field [97, 369, 509, 550]. There are a wide variety of kinetic equations which can be derived as mean-field limits as above, such as the **Vlasov-Poisson equation** [381, 460] or the time-dependent **Hartree-Fock equation** [100]. Although not strictly speaking a mean-field limit, probably the most famous kinetic equation is the **Boltzmann equation** [98, 244] for a single-particle density $u = u(x, y, t)$ for $(x, y) \in \mathbb{R}^3 \times \mathbb{R}^3$, in gas dynamics:

$$\partial_t u + y \cdot \nabla_x u = \mathcal{Q}(u, u), \tag{28.14}$$

where $\mathcal{Q}$ is a (quadratic) integral operator describing the collisions between particles. From kinetic equations one can then often pass to *macroscopic* PDEs describing coarse-grained variables [43, 66, 102]. Many details on **optimal transportation** in the Monge-Kantorovich/Wasserstein distance can be found in the books [551, 552]. The Dobrushin bound was shown in [156]. For a detailed discussion regarding particle limits in the context of fluid dynamics see [396] and for introduction to vortex dynamics see [392]. For gas dynamics applications one detailed book is [99].

Chapter 29

Two Effects: Hypocoercivity and Turing

One fundamental dynamics challenge is to understand the effect of different terms in a PDE. In this chapter, we describe two cases which look a bit counterintuitive at first sight. We start with a scenario which naturally connects to the gradient-structure and energy-functional techniques from Chapters 23–26 as well as to kinetic theory as discussed in Chapter 28. Consider the linear evolution equation

$$\partial_t u + Tu = Lu, \qquad u = u(x,y,t), \quad u_0(x,y) = u(x,y,0), \qquad (29.1)$$

with $(x,y,t) \in \mathbb{R}^d \times \mathbb{R}^d \times (0,+\infty)$ and for linear operators T and L. Typical examples would be to consider for T a **transport operator**

$$Tu = y \cdot \nabla_x u - \nabla_x V(x) \cdot \nabla_y u, \quad V : \mathbb{R}^d \to \mathbb{R} \qquad (29.2)$$

with a given, say sufficiently smooth, potential V and for L a **Fokker-Planck operator**

$$Lu = \nabla_y \cdot (\nabla_y u + yu) \qquad (29.3)$$

or an integral **collision operator**

$$(Lu)(x,y) = \int_{\mathbb{R}^d} k_1(v,y)u(x,y) - k_2(v,y)u(x,v) \, \mathrm{d}v \qquad (29.4)$$

for given kernels $k_{1,2}$; cf. equation (28.14). The combination of (29.2) and (29.4) is a typical example of a **kinetic equation** as derived from microscopic principles in Chapter 28. If (29.1) has a global steady state u^*, one may hope to prove conditions under which $u(x,y,t) \to u^*$ as $t \to +\infty$. The next key ODE example illustrates the problems one may face by just looking at the two linear operators separately.

Example 29.1. As a model problem, consider (29.1) in the setting of planar linear ODEs

$$w' = \frac{\mathrm{d}w}{\mathrm{d}t} = (L_2 - T_2)w, \qquad w \in \mathbb{R}^2, \qquad (29.5)$$

where $L_2, T_2 \in \mathbb{R}^{2 \times 2}$ and the unique steady state is $w^* = 0$. Suppose

$$T_2 = \begin{pmatrix} 0 & 1 \\ -1 & 0 \end{pmatrix}, \qquad L_2 = \begin{pmatrix} -2 & 0 \\ 0 & 0 \end{pmatrix}. \qquad (29.6)$$

For the ODE $w' = L_2 w$ we cannot prove global convergence to w^* as the nullspace $\mathcal{N}[L_2] = \{w_1 = 0\}$ is an invariant subspace of steady states; see Figure 29.1(a).

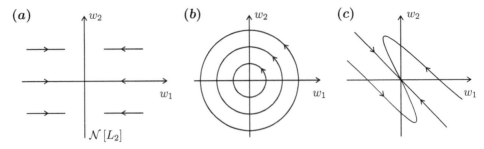

Figure 29.1. *Phase portraits for the different linear systems based upon* (29.5). *(a)* $w' = L_2 w$. *(b)* $w' = -T_2 w$. *(c)* $w' = (L_2 - T_2)w$.

Also for $w' = -T_2 w$ convergence to w^* fails as the flow is purely rotational; see Figure 29.1(b). However, considering (29.5)–(29.6) then we easily find that the eigenvalues of $(L_2 - T_2)$ are $\lambda_1 = -1$ and $\lambda_2 = -1$ so global convergence to $w^* = 0$ holds; see Figure 29.1(c). ◆

The innocent-looking observation from Example 29.1 can be generalized to (29.1). We suppose that $L - T$ generates a strongly continuous semigroup e^{L-T} on a given Hilbert space H with inner product $\langle \cdot, \cdot \rangle_H = \langle \cdot, \cdot \rangle$; see also Chapter 14. Let Q denote the orthogonal projection onto $\mathcal{N}[L]$. We make the following assumptions:

(A1) $-L$ is symmetric and **microscopically coercive**, i.e., there exists a constant $C_1 > 0$ such that

$$-\langle Lu, u \rangle \geq C_1 \|(\mathrm{Id} - Q)u\|^2 \qquad \text{for all } u \in \mathcal{D}(L). \qquad (29.7)$$

(A2) T is skew-symmetric and **macroscopically coercive**, i.e., there exists a constant $C_2 > 0$ such that

$$\|TQu\|^2 \geq C_2 \|Qu\|^2 \qquad \text{for all } u \in H \text{ with } Qu \in \mathcal{D}(T). \qquad (29.8)$$

Example 29.2. (Example 29.1 continued) Let us check that (A1)–(A2) are indeed satisfied in the motivating example for the Hilbert space $H = \mathbb{R}^2$ with the standard Euclidean inner product

$$\langle L_2 w, w \rangle = -2(w_1)^2 = -2\|(\mathrm{Id} - Q)w\|^2,$$
$$\|T_2 Qw\| = (w_2)^2 = \|Qw\|^2,$$

so (29.7) and (29.8) hold with $C_1 = 2$ and $C_2 = 1$, respectively. ◆

A strategy to show global decay for (29.1) to a steady state is to consider the **modified entropy**

$$\mathcal{E}[u] := \frac{1}{2}\|u\|^2 + \varepsilon \langle Bu, u \rangle, \qquad B := (\mathrm{Id} + (TQ)^*(TQ))^{-1}(TQ)^*, \qquad (29.9)$$

where $(\cdot)^*$ denotes the **adjoint operator** and $\varepsilon > 0$ will be chosen later; see also Chapter 24 for more on entropy methods. We need additional assumptions to guarantee global convergence:

(A3) Assume that $BT(\mathrm{Id} - Q)$ and BL are bounded and there exists a constant $C_3 > 0$ such that for all $u \in H$

$$\|BT(\mathrm{Id} - Q)u\| + \|BLu\| \leq C_3 \|(\mathrm{Id} - Q)u\|. \qquad (29.10)$$

(A4) We assume the following projection-transport relation $QTQ = 0$.

Example 29.3. (Example 29.1 continued) We want to interpret the modified entropy $\mathcal{E}[w]$ for the ODE case. Note that we have

$$B := (\mathrm{Id} + (TQ)^*(TQ))^{-1}(TQ)^* = \begin{pmatrix} 1 & 0 \\ 0 & 2 \end{pmatrix}^{-1} \begin{pmatrix} 0 & 0 \\ 1 & 0 \end{pmatrix} = \begin{pmatrix} 0 & 0 \\ \frac{1}{2} & 0 \end{pmatrix}.$$

Therefore, $\mathcal{E}[w] = \frac{1}{2}\left((w_1)^2 + (w_2)^2 + \varepsilon w_1 w_2\right)$ so that the modified entropy actually takes into account a mixed term between w_1 and w_2. Furthermore, one easily checks that (A3)–(A4) hold; see Exercise 29.10(a). ♦

We may now state the global decay result for the class (29.1).

Theorem 29.4. (*hypocoercive decay*) *Suppose (A1)–(A4) hold. Then there exist constants $C > 0$ and $\lambda > 0$, explicitly computable in terms of $C_{1,2,3}$ such that for any $u_0 \in H$ we have*

$$\|e^{t(L-T)}u_0\| \leq Ce^{-\lambda t}\|u_0\| \qquad \forall t \geq 0.$$

To prove Theorem 29.4, one would like to follow a similar strategy as discussed in Chapter 24. However, quite a number of new/different calculation tricks are required. We need two preliminary results.

Lemma 29.5. *There is an **entropy/entropy-production** relation*

$$\frac{\mathrm{d}}{\mathrm{d}t}\mathcal{E}[u] = -\mathcal{P}[u] \tag{29.11}$$

*where the **entropy-production functional** is given by*

$$\mathcal{P}[u] := -\langle Lu, u \rangle + \varepsilon\langle BTQu, u \rangle + \varepsilon\langle BT(\mathrm{Id} - Q)u, u \rangle - \varepsilon\langle TBu, u \rangle - \varepsilon\langle BLu, u \rangle.$$

Furthermore, the sum of the first two terms in $\mathcal{P}[u]$ is coercive.

Proof. A direct differentiation of the modified entropy with respect to time yields

$$\frac{\mathrm{d}}{\mathrm{d}t}\mathcal{E}[u] = \langle u, Lu \rangle + \varepsilon\langle BLu, u \rangle - \varepsilon\langle BTu, u \rangle - \varepsilon\langle Bu, Tu \rangle + \varepsilon\langle Bu, Lu \rangle.$$

We claim that $LB = 0$ so that the last term vanishes. Indeed, note that $B = QB$ since

$$(TQ)^*u = v + (TQ)^*(TQ)v \quad \Leftrightarrow \quad v = -QTu + QT^2Qv \in QH,$$

which is a very similar resolvent argument as in the proof of Theorem 5.7. Therefore, we do find $LB = LQB = 0 \cdot B = 0$. Using symmetry of L and skew-symmetry of T, the result (29.11) now follows easily. For the second claim, we compute

$$BTQ = (1 + (TQ)^*(TQ))^{-1}(TQ)^*(TQ)$$

so that (A1)–(A2) imply

$$-\langle Lu, u \rangle + \varepsilon\langle BTQu, u \rangle \geq C_1\|(\mathrm{Id} - Q)u\|^2 + \frac{\varepsilon C_2}{1 + C_2}\|Qu\|^2$$

$$\geq \min\left(C_1, \frac{\varepsilon C_2}{1 + C_2}\right)\|u\|^2$$

and coercivity is proved.   □

Lemma 29.6. *Suppose (A3) holds. Then B and TB are bounded. In addition, for all $u \in H$ we have the estimates*

$$\|Bu\| \leq \frac{1}{2}\|(\mathrm{Id} - Q)u\| \qquad \text{and} \qquad \|TBu\| \leq \|(\mathrm{Id} - Q)u\|.$$

The proof of Lemma 29.6 is Exercise 29.10(b). With the entropy/entropy-production relation and the free parameter ε, the last step is to show that $\mathcal{E}[u]$ is really a Lyapunov (-type) functional.

Proof. (of Theorem 29.4) First, note that by the first inequality in Lemma 29.6 we have

$$\frac{1}{2}(1 - \varepsilon)\|u\|^2 \leq \mathcal{E}[u] \leq \frac{1}{2}(1 + \varepsilon)\|u\|^2 \tag{29.12}$$

so $\mathcal{E}[u]$ is equivalent to $\|u\|^2$ if $\varepsilon \in (0, 1)$. Hence, if we can show the decay of $\mathcal{E}[u]$, the decay of $\|u\|$ follows immediately. By the second inequality from Lemma 29.6 and (A1)–(A4), we can now estimate

$$\mathcal{P}[u] \geq C_1\|(\mathrm{Id} - Q)u\|^2 + \frac{\varepsilon C_2}{1 + C_2}\|Qu\|^2 - \varepsilon(1 + C_3)\|(\mathrm{Id} - Q)u\|\|u\|$$

$$\geq \left(C_1 - \frac{\varepsilon(1 + C_3)}{2\delta}\right)\|(\mathrm{Id} - Q)u\|^2 + \varepsilon\left(\frac{C_2}{1 + C_2} - \frac{(1 + C_3)\delta}{2}\right)\|Qu\|^2$$

for an arbitrary positive constant $\delta > 0$. Selecting δ sufficiently small and then selecting ε sufficiently small, there exists a positive constant $C_4 > 0$ such that $\mathcal{P}[u] \geq C_4\|u\|^2$. This result together with (29.12) gives

$$\frac{\mathrm{d}}{\mathrm{d}t}\mathcal{E}[u] \leq -\frac{2C_4}{1 + \varepsilon}\mathcal{E}[u]. \tag{29.13}$$

Now the main result follows by taking $u = e^{t(L-T)}u_0$ and defining $\lambda := C_4/(1 + \varepsilon)$ and $C := \sqrt{1 + \varepsilon}/\sqrt{1 - \varepsilon}$. $\qquad\square$

In summary, the ODE Example 29.1 guided us correctly to define the right functional for decay to equilibrium. However, the next classical effect demonstrates that being just guided by a (badly chosen!) ODE example can be very misleading.

One typical term occurring in many PDEs is a classical diffusion term given by a Laplacian. Frequently, diffusion is viewed as a stabilizing effect; see also Example 6.3. The insight regarding the **Turing mechanism** is that it can also be destabilizing. We shall present this effect in the context of the one-dimensional **Schnakenberg equation**

$$\begin{aligned}
\partial_t u &= d_u \partial_x^2 u + p_u - u + u^2 v, \\
\partial_t v &= d_v \partial_x^2 v + p_v - u^2 v,
\end{aligned} \tag{29.14}$$

where $d_u, d_v, p_u, p_v, p > 0$ are parameters, $u = u(x,t)$, $v = v(x,t)$, $(x,t) \in (0, p) \times [0, T]$, suitable initial conditions are fixed, and we consider Neumann boundary conditions

$$\partial_x u(0, t) = 0 = \partial_x u(p, t) \quad \text{and} \quad \partial_x v(0, t) = 0 = \partial_x v(p, t). \tag{29.15}$$

We shall mostly study the case $p_u = 0, p_v = 2$ to simplify the notation but remark that results are going to hold also for open sets of the two parameters.

Lemma 29.7. *For $p_u = 0, p_v = 2$, the PDE (29.14)–(29.15) has a unique homogeneous positive steady state $(u^*, v^*) = (2, \frac{1}{2})$. If $d_u = 0 = d_v$ then (u^*, v^*) is locally asymptotically stable for the ODE associated to (29.14).*

Proof. First one just solves $-u + u^2 v = 0 = 2 - u^2 v$ to find $(u^*, v^*) = (2, \frac{1}{2})$, as $u = 0$ is excluded by the positivity assumption. Regarding stability of the ODE, we consider $(u, v)^\top = (u^*, v^*)^\top + \varepsilon w$ for $w \in \mathbb{R}^2$, which to leading order yields the linearized system

$$\frac{dw}{dt} = \begin{pmatrix} -1 + 2u^* v^* & (u^*)^2 \\ -2u^* v^* & -(u^*)^2 \end{pmatrix} w = \underbrace{\begin{pmatrix} 1 & 4 \\ -2 & -4 \end{pmatrix}}_{=:A} w, \tag{29.16}$$

and the eigenvalues $\lambda_\pm = \frac{1}{2}\left(-3 \pm i\sqrt{7}\right)$ of A satisfy $\mathrm{Re}(\lambda_\pm) = -\frac{3}{2} < 0$. Hence, the result follows from the Hartman-Grobman Theorem [529] by locally conjugating the linearized ODE to the full nonlinear ODE near the equilibrium. $\square$

Theorem 29.8. *Suppose $p_u = 0, p_v = 2, d_u = 1, p = 2\pi$ and let $d = d_v$. Then there exists $d_c > 0$ such that for all $d > d_c$ the steady state $(u^*, v^*) = (2, \frac{1}{2})$ is linearly unstable for the PDE (29.14)–(29.15), i.e., increasing a diffusion coefficient can destabilize.*

Proof. To analyze linear stability of (u^*, v^*), we have to study the PDE

$$\partial_t w = D\Delta w + Aw, \qquad w = w(x, t), \quad D := \begin{pmatrix} 1 & 0 \\ 0 & d \end{pmatrix} \tag{29.17}$$

for $x \in (0, 2\pi)$ with Neumann boundary conditions, and A is defined in (29.16). Let $(\lambda_k, e_k(x)) = (-\frac{1}{4}k^2, \cos(\frac{1}{2}kx))$ for $k \geq 0$ be the eigenvalues and eigenfunctions of the Neumann Laplacian on $(0, 2\pi)$, i.e., $\Delta e_k = \lambda_k e_k$. To show instability, we just have to find a solution of (29.17) satisfying our boundary conditions. We make the **Fourier series** ansatz

$$w(x, t) = \sum_{k=0}^{\infty} e^{\sigma t} e_k(x) c_k, \tag{29.18}$$

where $c_k \in \mathbb{R}^2$ are vector-valued coefficients, which could be calculated from the initial data, and $\sigma = \sigma(k)$ depends upon the wave number k. For each mode k, we get the equation

$$\sigma e_k = [\lambda_k D + A] e_k \qquad \Leftrightarrow \qquad [\sigma\,\mathrm{Id} - \lambda_k D - A] e_k = 0. \tag{29.19}$$

Nontrivial solutions of (29.19) occur if and only if $\det(\sigma\,\mathrm{Id} - \lambda_k D - A) = 0$. So we calculate and solve for $\sigma = \sigma(k)$, which yields

$$\sigma_\pm(k) = \frac{\pm\sqrt{d^2\lambda_k^2 - 2d\lambda_k^2 - 10d\lambda_k + \lambda_k^2 + 10\lambda_k - 7} + d\lambda_k + \lambda_k - 3}{2}.$$

To have instability, we must require a nontrivial spatial perturbation with $k \geq 1$ to have a growing mode, i.e., $\mathrm{Re}(\sigma(k)) > 0$. One checks from the explicit formula that $\mathrm{Re}(\sigma_+(k)) > 0$ for $k = 1$ if $d > \frac{80}{3}$. $\square$

Remark: The formula for $\sigma = \sigma(k)$ in the last proof can be recognized as another instance of a **dispersion relation**; see Chapter 8. Furthermore, the Turing mechanism (or diffusion-driven instability) is often mixed terminologically with a **Turing instability**, or even **Turing bifurcation**, which is also used in the context of bifurcations not driven by diffusivity constants.

Theorem 29.8 has far-reaching generalizations to several classes of reaction-diffusion equations as shown in [421, Sec. 2.3]; see also Exercise 29.9.

A common lesson from both, quite different, effects discussed in this chapter is that PDE dynamics is often governed by the *interplay* between different terms/operators in the equation. One has to develop a general feel for particular classes of PDEs before one can be certain which roles different terms should *really play*.

Exercise 29.9. Consider the **Gierer-Meinhardt equation** given by

$$\partial_t u = d_u \partial_x^2 u + p_1 - p_2 u + \frac{u^2}{v},$$
$$\partial_t v = d_v \partial_x^2 v + u^2 - v, \tag{29.20}$$

with boundary conditions (29.15). Fix $d_u = 1$. Prove that there are parameters $p_{1,2} > 0$, $d_v = d$, and $p > 0$ so that (29.20) also exhibits a diffusion-driven instability (or Turing mechanism) upon increasing d for a positive homogeneous steady, i.e., prove analogous statements to Lemma 29.7 and Theorem 29.8. ◊

Exercise 29.10. Consider the ODE case from Example 29.1. Show that assumptions (A3)–(A4) hold in this context. (b) Prove Lemma 29.6. ◊

Exercise 29.11. Consider the transport operator T given by (29.2). (a) Check that the **Maxwellian**

$$u^*(x, y) = e^{-\frac{1}{2}\|y\|_2^2 - V(x)} \tag{29.21}$$

satisfies $Tu^* = 0$. (b) Consider now also L given by (29.4). Find conditions on nonzero kernels $k_{1,2}$ such that $Lu^* = 0$. ◊

Background and Further Reading: The first part of this chapter on the interplay between transport and contraction operators for kinetic equations follows [163]. The approach is one instance, with a particularly nice geometric ODE dynamics interpretation, of the concept of **hypocoercivity** [553]. In fact, designing good Lyapunov-type functionals such as (29.9) and/or norms can be crucial to prove global decay to equilibrium for kinetic equations [417]. The concept of entropy/entropy-production from Lemma 29.5 is a theme which occurs for kinetic and Fokker-Planck type equations in the PDE context [21] as well as from a probabilistic viewpoint of these equations [38]; see also references to Chapter 24. The second part of this chapter on a diffusion-driven instability (or a Turing [539] mechanism) follows the presentation in [421]. Turing's general work was certainly groundbreaking and helped to establish the field of pattern formation [134, 135, 283].

Chapter 30

Blow-up in Cross-Diffusion Systems

In the analysis of the Turing mechanism in Chapter 29, we have learned that already constant diffusion terms in two-component reaction-diffusion systems can have quite unexpected effects upon parameter variation. In this chapter, we expand this theme to nonconstant diffusion terms. A classical PDE to describe collective movement via a population density $u = u(x,t)$ under the influence of a (chemical) concentration $v = v(x,t)$ is the **Keller-Segel equation**. It is given by

$$\begin{pmatrix} \partial_t u \\ \partial_t v \end{pmatrix} = \begin{pmatrix} \nabla \cdot (\nabla u - u \nabla v) \\ p \Delta v - v + pu \end{pmatrix} \tag{30.1}$$

on a smooth bounded domain $\Omega \subset \mathbb{R}^N$ with Neumann boundary conditions

$$\vec{n} \cdot \nabla u = 0 = \vec{n} \cdot \nabla v \quad \text{on } \partial\Omega \tag{30.2}$$

and initial data $u(x,0) = u_0(x)$ and $v(x,0) = v_0(x)$; $p > 0$ is a parameter. Note that (30.1) is a **cross-diffusion** system as the concentration v influences the second-order diffusion term for u.

Remark: Steady-state bifurcation analysis similar to Chapters 4–6, and travelling wave existence and stability results similar to Chapters 7–10 have been studied for (30.1)–(30.2); see also the references and exercises to this chapter.

One key ingredient for the Keller-Segel equation is to observe mass conservation properties. It is convenient to use the notation

$$\overline{w} := \int_\Omega w(x) \, \mathrm{d}x.$$

Lemma 30.1. *Suppose the solution u, v of the Keller-Segel equation (30.1)–(30.2) is smooth for $t \in [0,T]$. Then the model has **mass conservation** for u, i.e., $\overline{u} = \overline{u_0}$ for all $t \in [0,T]$. Furthermore, we have $\frac{1}{p}(\partial_t + 1)\overline{v} = \overline{u_0}$.*

Proof. Using smoothness, the equation for u, homogeneous Neumann boundary conditions, and the Divergence/Stokes' Theorem we have

$$\partial_t \int_\Omega u(x,t) \, \mathrm{d}x = \int_\Omega \nabla \cdot (\nabla u - u \nabla v) \, \mathrm{d}x = \int_{\partial\Omega} \vec{n} \cdot (\nabla u - u \nabla v) \, \mathrm{d}x = 0,$$

and the first result easily follows imposing the initial condition u_0. The second result is Exercise 30.7. $\qquad \square$

Unfortunately, the assumption of smoothness of the solution is sometimes too strong. In particular, solutions to (30.1)–(30.2) may have a **blow-up** in finite time.

Example 30.2. Indeed, the quadratic nonlinearity in the u equation of the Keller-Segel equations is not harmless. As a reminder, recall that the ODE

$$\frac{du}{dt} = u^2 \qquad \text{for } u = u(t),\, u(0) = u_0, \tag{30.3}$$

from Exercise 2.12 can have solutions, which become unbounded in finite time. ♦

Definition 30.3. The solution (u, v) of the Keller-Segel equation (30.1)–(30.2) exhibits **(finite-time) blow-up** if there exists a time $T_{\max}$ with $0 < T_{\max} < +\infty$ such that

$$\limsup_{t \to T_{\max}} \|u(x, t)\|_{L^\infty(\Omega)} = \infty \quad \text{or} \quad \limsup_{t \to T_{\max}} \|v(x, t)\|_{L^\infty(\Omega)} = \infty. \tag{30.4}$$

To explain why we expect blow-up for the Keller-Segel equation under certain conditions, we are going to study a simplified system. Let $\tilde{v} := \overline{v} - v$ so that, using the second part of Lemma 30.1, it follows that

$$\frac{1}{p}(\partial_t + 1)\tilde{v} = \Delta\tilde{v} - u + \overline{u_0}.$$

Consider the *formal* limit $p \to +\infty$ and assume $\overline{u_0} > 0$. Then a rescaling $\tilde{\tilde{v}} := \tilde{v}/\overline{u_0}$, $\tilde{u} := u/\overline{u_0}$ and dropping the tildes leads to

$$\begin{pmatrix} \partial_t u \\ 0 \end{pmatrix} = \begin{pmatrix} \Delta u - \overline{u_0}\nabla \cdot (u\nabla v) \\ \Delta v - u + 1 \end{pmatrix}. \tag{30.5}$$

Theorem 30.4. *(Keller-Segel blow-up) Consider* (30.5) *with boundary conditions* (30.2). *Suppose $N = 2$ and $\Omega = \{x \in \mathbb{R}^2 : \|x\| \leq 1\} =: \mathcal{B}(0, 1)$.*

(a) If $\overline{u_0} < 4\pi$, a bounded classical solution exists for all $t \geq 0$.

(b) There exists (u_0, v_0) such that finite-time blow-up occurs.

Remark: Many generalizations and variations of these results exist; see the references below for more details and for the proof of (a). Here we restrict ourselves proving (b).

The proof of (b) is constructive and works only with radial solutions and radial initial conditions. In fact, it is known that radial initial conditions imply local-in-time radial smooth solutions.

Lemma 30.5. *Let $r := \|x\|$ and define $U(r, t) := \int_0^{\sqrt{r}}(u(s, t) - 1)s\, ds$ for $0 \leq r \leq 1$. Then $U = U(r, t)$ satisfies the PDE*

$$\partial_t U = 4r\partial_r^2 U + \overline{u_0}\partial_r U^2 + \overline{u_0}U, \tag{30.6}$$

with initial and boundary conditions given by

$$U(r, 0) = \int_0^{\sqrt{r}}(u_0 - 1)s\, ds, \qquad U(0, t) = 0 = U(1, t). \tag{30.7}$$

Proof. We integrate the first equation in (30.5) over $\mathcal{B}(0, \sqrt{r})$. Considering each term involves calculating derivatives and suitable transformations. As an example, we start with the Laplacian term using a change to polar coordinates:

$$\int_{\mathcal{B}(0,\sqrt{r})} \Delta u \, dx = \int_0^{\sqrt{r}} \int_0^{2\pi} \left(\partial_s^2 u + \frac{1}{s} \partial_s u + \frac{1}{s^2} \partial_\theta^2 u \right) s \, d\theta \, ds.$$

The ∂_θ^2 term vanishes, as solutions are radial, so that we have

$$\int_{\mathcal{B}(0,\sqrt{r})} \Delta u \, dx = 2\pi \int_0^{\sqrt{r}} \partial_s(s\partial_s u) \, ds$$

$$= 2\pi \sqrt{r} \partial_{\sqrt{r}} u(\sqrt{r}, t) = 8\pi r \partial_r^2 U(r, t),$$

where it is convenient to compute the right-hand side of the last equation to realize the last equality. Similarly we get

$$\int_{\mathcal{B}(0,\sqrt{r})} \partial_t u \, dx = 2\pi \partial_t U(r, t), \tag{30.8}$$

$$-\int_{\mathcal{B}(0,\sqrt{r})} \nabla \cdot (u \nabla v) \, dx = 2\pi (\partial_r U^2 + U) \tag{30.9}$$

as discussed in Exercise 30.7. The last three results prove (30.6). Equation (30.7) and the initial condition are clear, and we leave checking the boundary conditions as an exercise. $\square$

Proof. (of Theorem 30.4) Considering the PDEs that U and W satisfy, the idea is to construct a (weak) **subsolution** $W(r, t)$ to $U(r, t)$, i.e., $W(r, t) \leq U(r, t)$ for all t, r such that

$$\partial_t W < 4r\partial_r^2 W + \overline{u_0} \partial_r W^2 + \overline{u_0} W, \tag{30.10}$$

and W satisfies the same boundary conditions as given for U in (30.7) and $W(r, 0) \leq U(r, 0)$. For the subsolution, we want to show that there exists $\bar{t}$ such that for any $K > 0$ we have

$$\limsup_{t \to \bar{t}} W(r, t) \geq K > 0 \qquad \text{for each } \epsilon > 0.$$

This construction would then provide a blow-up of U, and hence also u, at the center of the disc $\mathcal{B}(0, 1)$ at some time $t^* \leq \bar{t}$. The geometric idea of how to define W is illustrated in Figure 30.1.

We fix r_1, r_2 such that $0 < r_1 < r_2 < 1$ and let $\tau := a_0 - a_1 t$

$$W(r, t) = \begin{cases} \frac{a_2 r}{r + \tau^3} & \text{for } r < r_1, \\ b(t) \left(1 - r - \frac{(r_2 - r)_+^2}{r_2} \right) & \text{for } r \geq r_1, \end{cases} \tag{30.11}$$

where $(\cdot)_+$ denotes the positive part, a_0, a_1, a_2 are positive parameters, and

$$b(t) = \frac{1}{\left(1 - r_1 - \frac{(r_2 - r_1)^2}{r_2} \right)} \frac{a_2 r_1}{r_2 + \tau^3}.$$

Note that (30.11) does exhibit blow-up at the origin for $a_2 \neq 0$, as the derivative becomes unbounded in any neighborhood of the origin as $(\partial_r W)(0, t)$ becomes unbounded upon suitable choices of the parameters. Indeed, we calculate for $r < r_1$

$$\partial_t W = \frac{3a_1 \tau^2}{r + \tau^3} W, \quad \partial_r W = \frac{a_2 \tau^3}{(r + \tau^3)^2}, \quad r\partial_r^2 W = -\frac{2\tau^3}{(r + \tau^3)^2} W.$$

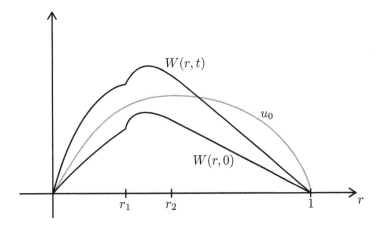

Figure 30.1. *Construction and time evolution of the subsolution $W = W(r,t)$ (piecewise smooth black curves), which lies below the actual initial condition u_0 at $t = 0$. However, W develops a singularity in finite time at the origin as t increases.*

This allows us to check (30.10):

$$\partial_t W - 4r\partial_r^2 W - \overline{u_0}\partial_r W^2 - \overline{u_0}W = \left(\frac{3a_1\tau^2}{r+\tau^3} + (8-2a_2)\frac{\tau^3}{(r+\tau^3)^2} - \overline{u_0} \right)W,$$

which is indeed negative upon fixing the a_j parameters suitably, taking $\overline{u_0}$ sufficiently large, constructing a positive initial condition, and using the maximum principle to ensure that W remains positive. Similarly, considering $r \geq r_1$ we find

$$\partial_t W = \frac{b'(t)}{b(t)}W = \frac{2a_1\tau^2}{r_1+\tau^3}W, \quad \partial_r W = -b(t)\left(1 - \frac{2(r_2-r)_+}{r_2} \right),$$

and $r\partial_r^2 W = -2b(t)/r_2$ for $r < r_2$ and $r\partial_r^2 W = 0$ for $r \geq r_2$. Hence, we get

$$\partial_t W - 4r\partial_r^2 W - \overline{u_0}\partial_r W^2 - \overline{u_0}W \leq \left(\frac{2a_1a_0^2}{r_1} + \frac{8}{r_2(1-r_2)} - \left(\overline{u_0} - \frac{2a_2\overline{u_0}}{1-r_2} \right) \right)W,$$

which can again be made smaller than zero by making a_2 sufficiently small and then taking $\overline{u_0}$ sufficiently large. For the case $r = r_1$, the derivative with respect to r has a positive jump and hence the correct sign. In particular, we do have a subsolution as long as we can show that $W(r,0) \leq U(r,0)$. Since we only want to exhibit blow-up for some (weak) solution, we may prepare the initial condition such that $U(r,0) > 0$ for $r \in (0,1)$. Note that we can also select $u_0(0)$ and $u_0(1)$ so that

$$\partial_r U(0,0) = \frac{1}{2}(u_0(0) - 1) > 0 \quad \text{and} \quad \partial_r U(1,0) = \frac{1}{2}(u_0(1) - 1) < 0.$$

Furthermore, we have

$$\partial_r W(0,0) = \frac{a_2}{a_0^3}, \qquad \partial_r W(1,0) = -b(0).$$

We can now select a_0 large enough so that $0 < \partial_r W(0,0) < \partial_r U(0,0)$ and we can make a_2 sufficiently small such that $0 > \partial_r W(1,0) > \partial_r U(1,0)$. Therefore, we can assure the initial comparison $W(r,0) \leq U(r,0)$, which concludes the proof. $\square$

Of course, Theorem 30.4 is only a step towards understanding the dynamics of the Keller-Segel model. Roughly speaking, it says that there are different regimes for initial conditions, which is already a topic easily visualized in the context of ODE theory; see Exercise 30.8.

Exercise 30.6. For the stationary version ($\partial_t u = 0 = \partial_t v$) Keller-Segel equations, show that each stationary solution corresponds to a solution of

$$p\Delta v - v + \kappa p e^v = 0 \qquad \text{in } \Omega \tag{30.12}$$

with Neumann boundary condition $\vec{n} \cdot \nabla v = 0$ for some $\kappa \in \mathbb{R}$. Can you compute κ? $\Diamond$

Exercise 30.7. Prove the second statement in Lemma 30.1. Prove equations (30.8)–(30.9). $\Diamond$

Exercise 30.8. Design an ODE with a smooth vector field, which has the property that one open set $\mathcal{U}_1$ of initial conditions exhibits blow-up, while another open set $\mathcal{U}_2$ of initial conditions does not. Geometrically sketch some cases, by which types of sets $\mathcal{U}_1$ and $\mathcal{U}_2$ could be separated. $\Diamond$

Background and Further Reading: A survey of mathematical results for Keller-Segel [322] containing a number of historical references is [280, 281]. The blow-up result [294] we followed in this chapter spawned several lines of research [60, 265, 282, 422, 565]; see also [267, 425, 511] for more on chemotaxis modeling and blow-up in biology. Cross-diffusion systems appear in several branches of modeling, and we can just refer to a few references here [311, 500]. We remark that the one-component problem (30.12) is particularly useful for bifurcation analysis [492] as well as to analyze **spike-layer solutions** for small diffusion [430], which we mentioned already in the references in Chapter 26. Of course, there are many types of PDEs [122, 212] where blow-up problems have been studied, e.g., nonlinear hyperbolic/wave equations [360, 513], as mentioned in the references to Chapter 26.

Chapter 31

Self-Similarity and Free Boundaries

In Chapter 30, we focused on blow-up in cross-diffusion systems. In this chapter, we focus on another important class of nonlinear diffusion processes occurring already for scalar PDEs. The main equation we consider is the **porous medium equation**

$$\partial_t u = \Delta(u^m), \qquad m > 1, \tag{31.1}$$

for $u = u(x, t)$ and $(x, t) \in \Omega \times [0, T) \subseteq \mathbb{R}^d \times [0, \infty)$ with suitable initial conditions understood. The porous medium equation can be formally viewed as a nonlinear diffusion equation (the classical diffusion/heat equation occurs for $m = 1$); the case of the **fast diffusion equation** for $m < 1$ will not be discussed here.

Definition 31.1. Let $\eta = xt^{-\beta}$ and suppose

$$u(x, t) = t^{-\alpha} f(\eta) \tag{31.2}$$

solves (31.1); then (31.2) is called a **self-similar solution** and the exponents $\alpha, \beta \in \mathbb{R}$ are called **similarity exponents**.

Remark: The concept of similarity pertains to essentially all types of solutions (strong, weak, mild, etc), which we want to consider here.

Self-similar solutions play a key role in the analysis of the porous medium equation. They link nicely to dynamics of ODEs and provide explicit guidelines in many applications. We start deriving the most important family of self-similar solutions. Inserting (31.2) into (31.1) yields

$$\partial_t u = -\alpha t^{\alpha-1} f(\eta) + t^{-\alpha}(-\beta t^{-\beta-1}x) \cdot \nabla f(\eta)$$
$$= -t^{-\alpha-1}(-\alpha f(\eta) + \beta \eta \cdot \nabla f(\eta))$$

as well as

$$\Delta(u^m) = t^{-\alpha m}\Delta_x(f^m(xt^{-\beta})) = t^{-\alpha m - 2\beta}\Delta_\eta f^m(\eta).$$

Therefore, we obtain the (time-dependent) profile equation

$$-t^{-\alpha-1}(-\alpha f(\eta) + \beta \eta \cdot \nabla f(\eta)) = t^{-\alpha m - 2\beta}\Delta_\eta f^m(\eta). \tag{31.3}$$

It is evident from (31.3) that we can obtain a stationary profile equation if we choose the relation

$$\alpha(m - 1) + 2\beta = 1 \tag{31.4}$$

between the similarity exponents. This leads to the **profile equation**

$$\Delta f^m + \beta \eta \cdot \nabla f + \alpha f = 0. \tag{31.5}$$

Of course, (31.5) may need conditions such as boundary or growth conditions at infinity to be abstractly solvable, which then yields restrictions on the parameter β. Here we circumvent this abstract step and impose **mass conservation**:

$$\partial_t \int_{\mathbb{R}^d} u(x,t)\, \mathrm{d}x = 0. \tag{31.6}$$

Since we have assumed the self-similar structure of the solution, mass conservation gives

$$\partial_t \int_{\mathbb{R}^d} t^{-\alpha} f(xt^{-\beta})\, \mathrm{d}x = \partial_t \left(t^{-\alpha+\beta d} \int_{\mathbb{R}^d} f(\eta)\mathrm{d}\eta \right) = 0. \tag{31.7}$$

This implies the restriction $\alpha = d\beta$, and together with (31.4) it follows that

$$\beta = \frac{1}{d(m-1)+2} \quad \text{and} \quad \alpha = \frac{d}{d(m-1)+2}. \tag{31.8}$$

For the next step, ODE techniques are very helpful.

Proposition 31.2. *Suppose $f > 0$; then the profile equation (31.5) with the condition (31.8) has a radially symmetric solution*

$$f = \left(C - \frac{(m-1)}{2m(d(m-1)+2)} r^2 \right)^{1/(m-1)}, \tag{31.9}$$

where C is a constant depending upon the mass; see also Figure 31.1.

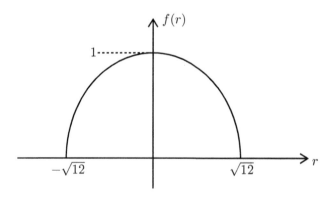

Figure 31.1. *Sketch of the radially symmetric solution (31.9).*

Proof. (of Proposition 31.2) Suppose $f = f(r)$ with $r = \|\eta\|$. In radial coordinates the profile equation is given by

$$\frac{1}{r^{d-1}} (r^{d-1}(f^m)')' + \beta r f' + d\beta f = 0, \qquad ' = \frac{\mathrm{d}}{\mathrm{d}r}.$$

One can combine the derivatives nicely (cf. Example 7.1) to obtain

$$(r^{d-1}(f^m)' + \beta r^d f)' = 0.$$

Integrating once and setting the integration constant equal to zero (as we have as yet free choice of it) yields

$$(f^m)' = -\beta r f.$$

Just differentiating the left-hand side, making the assumption that $f > 0$, and integrating the resulting first-order problem

$$m f^{m-2} f' = -\beta r$$

yields (31.9) upon inserting β from (31.8). The integration constant C can clearly be fixed once the mass of u is fixed.   □

Unfortunately, the last result has a severe drawback as it requires, a priori, that f is positive. Hence, it is often more natural to look at self-similar solutions of the form

$$f = \left(K_1(m, d) - K_2(m, d) r^2\right)_+^{1/(m-1)}, \tag{31.10}$$

where $(\cdot)_+$ denotes the positive part so that f is zero outside a certain ball and $K_{1,2} > 0$ are constants depending upon the problem parameters and K_1 also depends upon the chosen mass. Solutions of the form (31.10) are referred to as **Zel'dovich-Kompaneets-Barenblatt** or just **Barenblatt source-type solutions**; see also Exercise 31.5. The structure of (31.10) shows that we expect a **free boundary** Γ to appear for the porous medium equation (31.1) given by

$$\Gamma := (\partial \mathcal{P}_u) \cap (\Omega \times [0, T)),$$

where $\mathcal{P}_u := \{(x, t) \in \Omega \times [0, T) : u(x, t) > 0\}$ is the positivity set; see Figure 31.2.

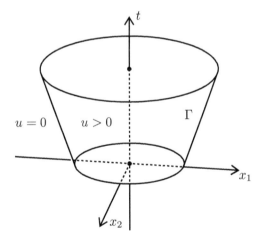

Figure 31.2. *Illustration of the positivity set $\mathcal{P}_u$ and its intersection Γ with the space-time domain; here we just show the case of two spatial dimensions.*

The next estimate will be helpful to characterize the positivity set.

Lemma 31.3. *Let u be a weak solution of (31.1) satisfying*

$$u(x, 0) = u_0(x) > 0 \qquad and \qquad u(x, t) \geq 0 \, for \, (x, t) \in \Omega \times [0, T);$$

then it holds, in the sense of distributions on $\Omega \times [0, T)$, that

$$\partial_t u \geq -\frac{u}{(m-1)t} \qquad for \, t > 0. \tag{31.11}$$

Proof. (Sketch) let u_k be an approximation to the solution u, which satisfies $u_k > 0$; such an approximation can be constructed by adding on a factor $1/k$ locally near the parts where u degenerates to zero. Consider the function

$$v := (m - 1)t\partial_t u_k + u_k.$$

Differentiation shows that v satisfies the PDE

$$\partial_t v = \Delta(m u_k^{m-1} v)$$

and $v(x, t) = u_k(x, t) \geq 0$ on the boundary as well as $v(x, 0) \geq 0$. Applying the standard parabolic maximum principle (i.e., the parabolic version of Theorem 3.8) gives that

$$v(x, t) \geq 0$$

in the entire space-time domain. This last result yields, upon passing to the limit $k \to \infty$, in a suitable weak formulation, the inequality (31.11). □

Theorem 31.4. (*free boundary growth*) *The set* $\mathcal{P}_u(t) := \{(x, t) \in \Omega \times [0, T) : u(x, t) > 0\}$ *cannot shrink for solutions as in Lemma 31.3, i.e.,*

$$\mathcal{P}(t_1) \subseteq \mathcal{P}(t_2) \qquad \text{for } 0 < t_1 < t_2 < T.$$

Proof. Fix any $x \in \Omega$ and consider the function

$$g(t) := u(x, t) t^{1/(m-1)}.$$

Then we find

$$\frac{dg}{dt}(t) = t^{1/(m-1)-1} \frac{1}{m-1} u + (\partial_t u) t^{1/(m-1)}$$

$$= t^{1/(m-1)} \left(\frac{u}{(m-1)t} + \partial_t u \right) \geq 0,$$

where the last inequality follows from Lemma 31.3. If u is positive at some point in time, say at time t_1, then $g(t_1) > 0$. Since g is nondecreasing in time we must have $g(t_2) > 0$ for $t_2 > t_1$, which implies u is also positive at this later time at the fixed location x. □

The last two results indicate already that solutions of the porous medium equation satisfy a maximum principle [548, Sec. 3]. However, this is basically due to the algebraic structure of the porous medium equation but does not hold for many other free boundary problems, which may complicate the analysis considerably; see Exercises 31.6–31.7.

Exercise 31.5. Prove that source-type solutions $u(x, t) = t^{-\alpha} f(x t^{-\beta})$ for the porous medium equation of the form (31.10) satisfy

$$\lim_{t \to 0^+} u(x, t) = K \delta_0,$$

in the sense of measures for some constant $K > 0$ and δ_0 being the point measure at zero. Can you compute K? ◊

Exercise 31.6. Consider the **thin-film equation** in the form

$$\partial_t u + \partial_x(u^p \partial_x^3 u) = 0, \qquad (x,t) \in \{u > 0\}, \tag{31.12}$$

where $u = u(x,t)$ and $p \in (0,3)$ is a parameter. Use the self-similar ansatz

$$h(x,t) = t^{-1/(p+4)} f(\eta), \qquad \eta = t^{-1/(p+4)} x \tag{31.13}$$

to compute a third-order ODE for f. $\Diamond$

Exercise 31.7. Consider the degenerated-linear thin film equation (31.12) for $p = 0$. Show that solutions of this linear PDE can change sign in time and compare this to the second-order diffusion/heat equation. Remark: This provides a first hint that the maximum principle does not hold for the free-boundary problem for the thin film equation. $\Diamond$

Background and Further Reading: The chapter follows a few selected parts of the very extensive monograph [548] on the porous medium equation [28], where techniques frequently also pertain to other classes of degenerate parabolic PDEs [155, 547]. There are many additional works on the porous medium equation, frequently focusing on decay to equilibrium [18, 29, 92, 219]; see also Chapters 23–24. For more on self-similarity in the context of PDE dynamics we refer to [44, 230, 231]. For basic background on the thin film free-boundary problem and self-similarity in this problem see [55, 56, 57, 93].

Chapter 32

Spirals and Symmetry

In Chapter 31, we encountered special solutions, which are singled-out by certain purely algebraic scalings. This suggests taking an even more abstract algebraic perspective and looking at symmetries in PDEs. Consider a system of reaction-diffusion equations given by

$$\partial_t u = D\Delta u + f(u; p), \qquad u(x, 0) = u_0(x), \tag{32.1}$$

where $u = u(x, t) \in \mathbb{R}^N$, $p \in \mathbb{R}$ is a parameter, D is a positive definite diagonal diffusion matrix, $(x, t) \in \Omega \times [0, T] \subset \mathbb{R}^d \times [0, \infty)$, and f is a smooth nonlinearity chosen such that solutions of (32.1) exist globally in time and $f(0; p) \equiv 0$. We assume, as considered in Chapter 15, that (32.1) generates a well-defined semiflow $S(t)u_0 = u(\cdot, t)$.

We have already studied in Chapters 7–10 for $x \in \mathbb{R}$ the case of travelling waves, which have a translation symmetry. The translate of a travelling wave is still a travelling wave so the involved symmetry group in this context is just $\mathbb{R}$. Here we give an introduction to a topic which can only occur for dimension $d > 1$, namely **spiral waves**; see Figure 32.1. Spiral waves are linked to more complicated symmetries.

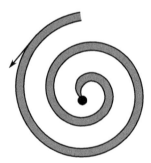

Figure 32.1. *Suppose $x \in \mathbb{R}^2$ and consider a point (black dot), which we interpret as the center of the spiral wave. The wave rotates around this center as indicated by the arrow. The shaded area indicates an approximation to the level set of u. In many situations, there is a rather steep gradient between different regions so that a spiral structure emerges.*

Example 32.1. Consider the two-component system

$$\begin{aligned} \partial_t u_1 &= d_1 \Delta u_1 + u_1 - u_2 + u_1^3, \\ \partial_t u_2 &= d_2 \Delta u_2 + 2u_1 - pu_2. \end{aligned} \tag{32.2}$$

Just on a formal level, let us look at the pure reaction system first ($d_1 = 0 = d_2$) despite the obvious caveats as discussed in Chapter 29. We have

$$\frac{\mathrm{d}u}{\mathrm{d}t} = \underbrace{\begin{pmatrix} 1 & -1 \\ 2 & -p \end{pmatrix}}_{=:A(p)} u + \begin{pmatrix} u_1^3 \\ 0 \end{pmatrix},$$

where $u = 0$ is a steady state for all parameters $p \in \mathbb{R}$. One checks that $A(p) = D_u f(0; p)$ has two eigenvalues $\lambda_\pm = \alpha(p) \pm i\beta(p)$ such that the real part $\alpha(p)$ satisfies $\alpha(1) = 0$. We also have $\beta(1) = 1$ and $\alpha'(1) \neq 0$. Therefore, the system undergoes a **Hopf bifurcation**, where a pair of complex conjugate eigenvalues crosses the imaginary axis. Generically, there are two possibilities: **supercritical Hopf bifurcation** as shown in Figure 32.2(a), or **subcritical Hopf bifurcation** shown in Figure 32.2(b).

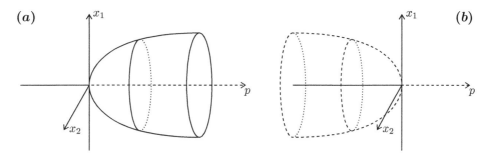

Figure 32.2. *(a) Supercritical Hopf bifurcation; a family of attracting periodic orbits emerges for $p > 0$. (b) Subcritical Hopf bifurcation; a family of repelling periodic orbits exists for $p < 0$.*

Which type of Hopf bifurcation occurs can be calculated from the first Lyapunov coefficient as discussed in Exercise 32.9. In both cases, small periodic orbits are generated. ◆

The last example leads one to the conjecture that for (32.1) nontrivial time-periodic structures might be possible. In fact, reactions and mixing via diffusion can even generate spiral structures as shown in Figure 32.1.

Definition 32.2. The **special Euclidean Group** SE(d) for $d \in \mathbb{N}$ is the semidirect product of the **special orthogonal group** SO(d) and translations, i.e.,

$$\mathrm{SE}(d) = \mathrm{SO}(d) \dotplus \mathbb{R}^d$$

with composition on the product

$$(R, S)(\tilde{R}, \tilde{S}) = (R\tilde{R}, S + R\tilde{S}).$$

In particular, SE(d) consists of rotations and translations of Euclidean space; see also Exercise 32.10.

Lemma 32.3. *Solutions of (32.1) are **equivariant** with respect to SE(d), i.e., they are invariant under the action*

$$((R, S)u)(x, t) = u(R^{-1}(x - S), t).$$

Proof. Recall that the Laplacian is rotation invariant. Indeed, let $R = (r_{ij})$ be a rotation matrix and set $y = Rx$; then we have

$$\Delta_x = \sum_{i=1}^{d} \partial_{x_i x_i} = \sum_{i=1}^{d} \sum_{j,k=1}^{d} r_{ji} r_{ki} \partial_{y_j y_k}$$

$$= \sum_{j,k=1}^{d} \left(\sum_{i=1}^{d} r_{ji} r_{ki} \right) \partial_{y_j y_k} = \sum_{j,k=1}^{d} \delta_{jk} \partial_{y_j y_k} = \Delta_y,$$

where δ_{jk} is the Kronecker delta and we used $RR^\top = \mathrm{Id}$ in the second to last equality. The result now follows easily. □

For an abstract evolution equation of the form $u' = F(u)$, we also say it is **equivariant** under the action by a group G if for all $g \in G$ the relation $g \cdot F(u) = F(g \cdot u)$ holds. In our context, we work with $G = \mathrm{SE}(d)$.

Definition 32.4. A solution u^* is called a **relative equilibrium** for a group G and a semiflow $S(t)$ if its time orbit $\{S(t)u^* : t \geq 0\}$ is contained in its **group orbit** $\{gu^* : g \in G\}$.

Definition 32.5. Fix $d = 2$. **(Rigidly rotating) spiral waves** are relative equilibria of the form

$$u(x,t) = u(R_{st}x, 0), \qquad s \neq 0, \tag{32.3}$$

where R_θ is a **rotation** around the origin by an angle $\theta \in \mathbb{R}/(2\pi\mathbb{Z})$.

In particular, spiral waves are steady in a rotating frame. Other typical classes of spiral waves are **meandering spiral waves**, which are periodic in a rotating frame, and **drifting spiral waves**, which are periodic in a moving frame; see also Figure 32.3(b).

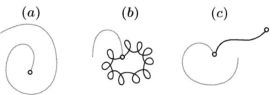

(a) (b) (c)

Figure 32.3. *Centers (black circle) of a spiral wave (indicated in grey) can have various behaviors. (a) Rigidly rotating spiral wave with a fixed center. (b) Meandering spiral wave: the spiral center (black curve) undergoes a periodic path. (c) Drifting spiral wave: the spiral center moves/drifts (black curve).*

Here we restrict to the rigidly rotating case for

$$d = 2$$

from now on. One approach to obtain the existence of spiral wave solutions is to use a reduction to a **center manifold** near Hopf bifurcation, i.e., to find a finite-dimensional invariant manifold which tracks only the imaginary pair of eigenvalues. In this context, a variant of (symmetric) Lyapunov-Schmidt reduction as discussed in Chapter 4 is not directly applicable as we are on an unbounded domain $\mathbb{R}^2$ so that the linearization has essential spectrum; see also Chapter 9. Interestingly, in $\mathbb{R}^2$ one may proceed directly and still obtain a center manifold. We need preliminary transformations.

Lemma 32.6. *The system* (32.1) *(for* $d = 2$*) restricted to spiral wave solutions can be rewritten as*

$$\partial_r^2 u + \frac{1}{r}\partial_r u + \frac{1}{r^2}\partial_\theta^2 u = -D^{-1}f(u; p) + sD^{-1}\partial_\theta u \qquad (32.4)$$

for $r \in (0, \infty)$ *and* $\theta \in \mathbb{R}/(2\pi\mathbb{Z}) = \mathbb{S}^1$.

Proof. Consider polar coordinates $x = (r\cos\theta, r\sin\theta)^\top$ and recall (or quickly check) that the Laplacian in polar coordinates is given by

$$\Delta_{r,\theta} = \partial_r^2 + \frac{1}{r}\partial_r + \frac{1}{r^2}\partial_\theta^2.$$

Furthermore, the ansatz (32.3) yields $\partial_t u(R_{st}x, 0) = s\partial_\theta u(R_{st}x, 0)$ so that the result follows upon some algebra and using that D is positive definite and diagonal. $\qquad\square$

Unfortunately, (32.4) is still problematic near $r \approx 0$. So fix $r_* > 0$ and introduce a (radial) **blow-up transformation**

$$\tau(r) := \begin{cases} \ln r & \text{if } r \leq r_*, \\ r & \text{if } r \geq 2r_*, \end{cases} \qquad (32.5)$$

with a smooth interpolation on the interval $(r_*, 2r_*)$; see also Exercise 32.11 for the usefulness of singular transformations and polar coordinates in a slightly different context. One may then check that for $r < r_*$ we get from (32.4)

$$\frac{d^2 u}{d\tau^2} = A_\theta u - e^{2\tau}(D^{-1}f(u; p) - B_\theta u). \qquad (32.6)$$

with $A_\theta = -\partial_\theta^2$ and $B_\theta = sD^{-1}\partial_\theta$, while the equation remains unchanged for $r > 2r_*$. The intermediate smooth part will not matter for the following results. We already suggestively wrote (32.6) as an evolution equation viewing τ as a time variable. We work in the Sobolev space

$$H^k = H^k(\mathbb{S}^1, \mathbb{R}^N) \qquad \text{for some } k \geq 0,$$

on which A_θ is self-adjoint with spectrum $\{l^2 : l \in \mathbb{Z}\}$. Since the spectrum of A_θ is well understood, the key spectral information to obtain Hopf bifurcation from the homogeneous solution $u \equiv 0$ upon parameter variation of p is obviously contained in the operator

$$B_{s,p} := -D^{-1}D_u f(0; p) + B_\theta, \quad B_{s,p} : H^{k+1} \to H^k. \qquad (32.7)$$

For later use, we let $P^c = P^c(s, p)$ denote the **spectral projection** for $B_{s,p}$ on $(-\infty, 0] \subset \mathbb{C}$, i.e., we use the **Dunford integral** and set

$$P^c(s, p) := \frac{1}{2\pi i}\int_\Lambda \frac{dz}{z - B_{s,p}},$$

where Λ is a contour in $\mathbb{C}$ enclosing $(-\infty, 0]$ but no other parts of the spectrum of $B_{s,p}$; see also Chapter 14. Next, rewrite (32.6) as a first-order system

$$\frac{d}{d\tau}\begin{pmatrix} u \\ v \end{pmatrix} = L(\tau)\begin{pmatrix} u \\ v \end{pmatrix} + \mathcal{F}(u, v), \quad v := \frac{du}{d\tau} \qquad (32.8)$$

on $X := H^{k+1} \times H^k$ with k sufficiently large, and where L is a linear operator collecting all the linear terms from the linearization around $u \equiv 0$ and $\mathcal{F}$ collects the remaining terms. In particular, the linear problem is

$$\frac{d}{d\tau}\begin{pmatrix} U \\ V \end{pmatrix} = L(\tau)\begin{pmatrix} U \\ V \end{pmatrix}, \qquad (U(0), V(0)) = (U_0, V_0). \tag{32.9}$$

To construct a center manifold, one has to *restrict* the function spaces according to the dynamics. Motivated by the strategy of exponential dichotomies from Chapter 10, we set

$$Y_\delta := \{(U, V) \in C^0(\mathbb{R}, X) : \|U\|_{Y_\delta} < \infty\},$$

with an exponentially weighted in-time supremum norm

$$\|U\|_{Y_\delta} := \sup_{\tau \in \mathbb{R}} e^{-\delta|\tau|} \|(U(\tau), V(\tau))\|_{X_\tau},$$

and where the norm on X_τ is defined as

$$\left\|\begin{pmatrix} U(\tau) \\ V(\tau) \end{pmatrix}\right\|_{X_\tau} := \begin{cases} \tau^{-1}\|U(\tau)\|_{H^{k+1}} + \|U(\tau)\|_{H^{k+1/2}} + \|V(\tau)\|_{H^k} & \text{if } \tau > 2r_*, \\ \|U(\tau)\|_{H^{k+1}} + \|V(\tau)\|_{H^k} & \text{if } \tau < 2r_*. \end{cases}$$

Of course, these choices are far from obvious, but essentially the space Y_δ just measures Sobolev regularity in space and distinguishes solutions with different growth or decay rates via the parameter δ. Then we denote the (eventual) center eigenspace by

$$E^c(\tau) := \{(U_0, V_0) : \|U\|_{Y_\delta} < \infty\},$$

i.e., the linear space of all initial conditions at time τ, whose associated solutions to the linear problem (32.9) remain bounded in Y_δ; cf. this construction to Chapters 16–17.

Theorem 32.7. *(spiral wave center manifold [493]) Suppose f is sufficiently smooth and satisfies growth conditions implying global existence for (32.1). Suppose that for $s = s_0$ and $p = p_0$ the spectral projection is nonempty $P^c(s, p) \neq 0$; then there exists a local (in parameter space), unique, finitely differentiable, invariant, and finite-dimensional center manifold*

$$\mathcal{M}^c \subset X \times \mathbb{R},$$

which contains all bounded solutions of (32.8). Furthermore, it holds that

- *$\mathcal{M}^c$ is given as a graph over $\{E^c(\tau) : \tau \in \mathbb{R}\}$,*

- *for fixed $\tau = \tau_0$, $\mathcal{M}^c$ is tangent to $E^c(\tau_0)$,*

- *$\mathcal{M}^c$ is invariant under the diagonal action of $SO(2)$ on X.*

The proof of Theorem 32.7 is beyond our scope. Very roughly speaking, it employs exponential dichotomies to obtain the required hyperbolicity to perturb the linear space to a nonlinear manifold. However, it illustrates very nicely that concepts we previously encountered can be combined and that symmetry is *inherited* in this case on the finite-dimensional ODE dynamics on the center manifold as shown by the last conclusion of the theorem.

Example 32.8. The key application to (32.1) of Theorem 32.7 is the planar two-component situation with a homogeneous state:

$$d = 2, \qquad N = 2, \qquad f(0; p) \equiv 0.$$

Suppose we **complexify** and identify $\mathbb{R}^2 \simeq \mathbb{C}$ and assume $u \equiv 0$ undergoes a nondegenerate Hopf bifurcation

$$\left(\frac{\mathrm{d}}{\mathrm{d}u} D^{-1} F\right)(0; p) = \alpha + \mathrm{i}\beta, \qquad \beta \neq 0, \tag{32.10}$$

and α passes through zero upon variation of p. Then there exists a finite-dimensional center manifold. Furthermore, using a (complex) Fourier series

$$u = \sum_{k \in \mathbb{Z}} u_k(r) \mathrm{e}^{\mathrm{i}k\theta},$$

one may show—after quite a few nontrivial computations—that m-armed spiral waves bifurcate, i.e., for any m there is a rotation speed s such that the subspace

$$\mathrm{span}\{(0, \mathrm{e}^{\mathrm{i}m\theta})^{\top}, (\mathrm{e}^{\mathrm{i}m\theta}, 0)^{\top}\}$$

governs the linear dynamics. ◆

Travelling waves and spiral waves are just two classes of PDE solutions where symmetry is important; see the references below. A center manifold $\mathcal{M}$ can also be viewed as containing the effective "slow" dynamics governed by eigenvalues on the imaginary axis; see Figure 32.4.

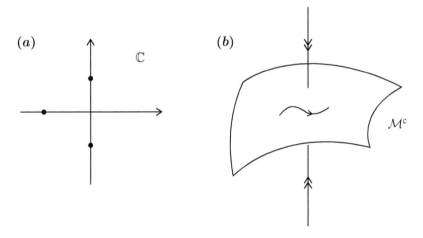

Figure 32.4. *(a) Spectrum of the linearized problem for a differential equation containing eigenvalues on the imaginary axis as well as some eigenvalues with negative real part. (b) Phase space dynamics showing the strong/fast contraction (double arrows) towards a center manifold $\mathcal{M}^c$. On $\mathcal{M}^c$ the dynamics (single arrow) is governed by slow/center components.*

This is another case of *scale separation*, a concept we encountered already in Chapters 12–13. In Chapters 33–36, we are going to expand this theme in considerable detail.

Exercise 32.9. Consider the following **Hopf bifurcation normal form** for ODEs:

$$\frac{du}{dt} = \underbrace{\begin{pmatrix} p_1 & -1 \\ 1 & p_1 \end{pmatrix}}_{=:\tilde{A}(p)} u + \begin{pmatrix} p_2 u_1 (u_1^2 + u_2^2) \\ p_2 u_2 (u_1^2 + u_2^2) \end{pmatrix} \tag{32.11}$$

for parameters p_1, p_2. (a) Show that $u = 0$ undergoes a Hopf bifurcation under variation of p_1. (b1) Convert (32.11) to polar coordinates. (b2) Convert (32.11) to a single complex coordinate. (c) Use a result from (b) to check under which conditions on p_2 the Hopf bifurcation is sub- or supercritical. $\Diamond$

Exercise 32.10. (a) Find a matrix representation of the group $SE(d)$. (b) Use this representation to prove that $SE(2)$ is homeomorphic to $\mathbb{R}/\mathbb{Z} \times \mathbb{R}^2$. (c) Check that $SE(d)$ is a **Lie group** (i.e., a group and a differentiable manifold). Can you find its **Lie algebra** (i.e., the tangent space to the identity)? $\Diamond$

Exercise 32.11. Consider the ODE

$$\begin{aligned} u_1' &= u_1^2 - 2u_1 u_2, \\ u_2' &= u_2^2 - 2u_1 u_2. \end{aligned} \tag{32.12}$$

(a) Show that $u = 0$ is a nonhyperbolic steady state. (b) Convert (32.12) to polar coordinates obtaining a vector field $F = F(\theta, r)$ with $(\theta, r) \in \mathbb{S}^1 \times [0, \infty)$. (c) Divide the resulting vector field by r, i.e., consider $\frac{1}{r} F$ and check that all six of its steady states on the set $\mathbb{S}^1 \times \{r = 0\}$ are hyperbolic. $\Diamond$

Background and Further Reading: In this chapter we have followed a combination of [488, 493] but see also [240, 469]. Symmetry in differential equations, and exploiting it for reduction and computation, is a rather diverse field [110, 241, 242, 283, 490]. The concept of center-manifold reduction to eventually contain all non-exponentially growing (or "critical") modes is key to reduce many bifurcation problems geometrically [85, 247, 258, 351, 545], and it provides a comparable tool to Lyapunov-Schmidt reduction. Working with polar coordinates [165] and complexification [24, 288] is often very helpful in computations in low-dimensional phase or spatial spaces; see also the introduction in [346] following up Exercise 32.11.

Chapter 33

Averaging and Ergodicity

Having considered oscillation phenomena already several times, e.g., rotating spiral waves in Chapter 32, we pick up this idea in the context of finding a suitable method to average out oscillations in a PDE. **Averaging** (or **homogenization**) encompasses a relatively broad spectrum of topics. The key theme is to approximate "fast (oscillating) dynamics" governed by a small parameter $0 < \varepsilon \ll 1$ over a certain time scale by a limiting equation as $\varepsilon \to 0$. Here we shall focus on **transport equations** for $u^\varepsilon = u^\varepsilon(x,t)$ given by

$$\begin{cases} \partial_t u^\varepsilon - f^\varepsilon \cdot \nabla u^\varepsilon = 0 & \text{in } \mathbb{R}^d \times (0,\infty), \\ u^\varepsilon(\cdot,0) = g & \text{for } x \in \mathbb{R}^d, \end{cases} \tag{33.1}$$

where $f^\varepsilon(x) = f(x/\varepsilon)$ is a smooth vector field $f^\varepsilon : \mathbb{R}^d \to \mathbb{R}^d$, which we shall assume to be one-periodic:

$$f\left(\frac{x}{\varepsilon} + \vec{1}\right) = f\left(\frac{x}{\varepsilon}\right) \quad \forall x \in \mathbb{R}^d, \varepsilon > 0, \tag{33.2}$$

where $\vec{1} = (1,1,\ldots,1)^\top \in \mathbb{R}^d$. Studying (33.1) is also motivated directly by a link to ODEs. Indeed, consider a smooth vector field $h : \mathbb{R}^d \to \mathbb{R}^d$ and the operators

$$Lv := h \cdot \nabla v \quad \text{and} \quad L^*v := -\nabla \cdot (hv). \tag{33.3}$$

If $z = z(t) \in \mathbb{R}^d$ solves

$$\frac{\mathrm{d}z}{\mathrm{d}t} = z' = h(z), \qquad z(0) = z_0, \tag{33.4}$$

then we find

$$\frac{\mathrm{d}}{\mathrm{d}t} v(z(t)) = \nabla v(z(t)) \cdot z'(t) = Lv(z(t)).$$

Using the last observation, and/or recalling the method of characteristics from Chapter 11, it is easy to see that classical solutions of the transport equation $\partial_t v = Lv$ with (infinitesimal) generator L and initial condition $v_0(z) = v(z,0)$ are linked to solutions of ODEs via the formula

$$v(z,t) = v_0(\varphi(z,t)), \tag{33.5}$$

where φ is the flow associated to $z' = h(z)$; see Chapter 2. Similarly, one may use the (formal) adjoint L^* to consider the **Liouville equation**

$$\partial_t \rho = L^*\rho, \qquad \rho(z,0) = \rho_0(z). \tag{33.6}$$

The Liouville equation has to be viewed as the forward equation. In particular, if solutions ρ are regular enough for an initial probability distribution ρ_0, then (33.6) tells us how likely it is, via $\rho(z, t)$, to find a particle at position $z = z(t)$ at time t.

Remark: One also refers to L as the generator of the **Koopman operator** or **transfer operator**, while L^* is the generator of the **Perron-Frobenius operator**.

To analyze (33.1), the formal starting point is essentially the method of multiple scales as in Chapter 12. Assume that f is periodic, introduce a second variable, and decompose the differential operator

$$y := \frac{x}{\varepsilon}, \qquad L = \frac{1}{\varepsilon} f(y) \cdot \nabla_y + f(y) \cdot \nabla_x =: \frac{1}{\varepsilon} L_0 + L_1$$

to study the resulting PDE

$$\partial_t u^\varepsilon = \left(\frac{1}{\varepsilon} L_0 + L_1 \right) u^\varepsilon. \tag{33.7}$$

Inserting a multiple scales ansatz for u^ε, we easily find the two leading-order equations

$$L_0 u_0 = 0, \tag{33.8}$$
$$L_0 u_1 = \partial_t u_0 - L_1 u_0, \tag{33.9}$$

as discussed in Exercise 33.4, where $u_j = u_j(x, y, t)$ depends on two types of spatial variables. It is evident that without information on the nullspace $\mathcal{N}[L_0]$, it will be impossible to solve for u_0 in (33.8). Exercise 33.5 discusses $\mathcal{D}(L_0)$ and $\mathcal{D}(L_0^*)$. Let us assume for now that

$$\mathcal{N}[L_0] = \operatorname{span}(1) \qquad \text{and} \qquad \mathcal{N}[L_0^*] = \operatorname{span}(\rho^*), \tag{33.10}$$

where $\rho^* = \rho^*(y)$ is the unique stationary solution to

$$L_0^* \rho^* = 0, \qquad \int_{\mathbb{T}^d} \rho^*(y) \, dy = 1. \tag{33.11}$$

Using the first assumption in (33.10) to solve (33.8) we may conclude that u_0 does not depend upon y, so we may set

$$u_0(x, y, t) =: u(x, t).$$

Inserting this result in (33.9), integrating against ρ^* over $\mathbb{T}^d$, and using the second part of (33.10) gives the desired averaged equation

$$\partial_t u - f_* \cdot \nabla_x u = 0, \qquad f_* := \int_{\mathbb{T}^d} f(y) \rho_*(y) \, dy. \tag{33.12}$$

The PDE (33.12) has been derived on purely formal grounds but it shows that we may expect that only if the Liouville equation has a unique invariant density/solution, then we may average the transport coefficient against it to obtain an averaged PDE. This raises two questions: (I) Which dynamics for vector fields f can we allow? (II) How can we prove a result such as (33.12) rigorously? Question (II) is answered in Chapter 34, while we discuss (I) now.

Example 33.1. Consider the one-dimensional case $y \in \mathbb{T}^1$ and periodic vector fields $f(y + 1) = f(y)$. Then we (formally) have

$$L_0^* \rho = -\nabla_y(f\rho) = -\rho \partial_y f - f \partial_y \rho = 0$$

if $\rho \in \mathcal{N}[L_0^*]$. Therefore, $\partial_y \rho = -\frac{\rho}{f} \partial_y f$ holds as long as $f \neq 0$, say without loss of generality $f > 0$. This already shows that it will be troublesome if f has a steady state in this context. Furthermore, one easily solves the ODE for ρ, uses the mass-one normalization, and obtains

$$\rho^*(y) = \frac{1}{f(y)} \left(\int_0^1 \frac{1}{f(y)} \, \mathrm{d}y \right)^{-1},$$

where the normalization constant is also known as the **harmonic average** of f. This result already hints at the more general requirement that ρ should sample the entire phase space well enough in a suitable sense. ♦

Definition 33.2. Consider the flow $\varphi(z_0, t) = z(t)$ for (33.4) on $\mathbb{T}^d$. A measure μ on $\mathbb{T}^d$ is called **invariant** if for any μ-measurable set A we have

$$\mu(\varphi(A, t)) = \mu(A) \qquad \text{for } t \geq 0. \tag{33.13}$$

An invariant measure μ is called **ergodic measure** if for every **invariant set** A (i.e., $\varphi(A, t) = A$ for all $t \in \mathbb{R}$), we have either $\mu(A) = 0$ or $\mu(A) = 1$; we also refer to φ as an **ergodic flow** in this case.

The next result summarizes several, yet definitely not all, important properties for ergodic dynamical systems associated to ODEs on $\mathbb{T}^d$.

Theorem 33.3. *(invariant measures and ergodicity [356, 446]) The following statements hold for the flow φ:*

(T1) A measure μ with density $\rho^* \in C^1(\mathbb{T}^d)$ with respect to Lebesgue measure is invariant if and only if $L^* \rho^* = 0$.

(T2) Let μ be an invariant probability measure. If μ is ergodic then for any $\psi \in L^1(\mathbb{T}^d; \mu)$ we have

$$\lim_{t \to \infty} \frac{1}{t} \int_0^t \psi(\varphi(z, s)) \, \mathrm{d}s = \int_{\mathbb{T}^d} \psi(w) \, \mathrm{d}w \quad \text{for } \mu\text{-a.e. } z \in \mathbb{R}^d. \tag{33.14}$$

(T3) Let μ be an invariant probability measure. μ is ergodic if and only if the equation $Lv = 0$ for $v \in \mathcal{D}(L)$ has only constant solutions (w.r.t. μ-a.e.) in L^∞.

Remark: The statement (T2) of Theorem 33.3 is the classical **Ergodic Theorem** stating that time average equals space average. This reinforces the idea that we may need some variant of ergodicity to apply averaging. We shall not prove (T2) here but refer to standard books on ergodic theory. However, (T1) and (T3) provide direct links to PDEs, so we shall provide the basic ideas for the proof.

Proof. (Sketch of (T1) and (T3); see [356]) We need some preliminary considerations. First, note that since L and L^* are adjoints, so are the semigroups e^{tL} and e^{tL^*}. This yields for $v \in \mathcal{D}(L^*)$, $w \in \mathcal{D}(L)$, and any (suitably nonsingular) measure μ that

$$\int_{\mathbb{T}^d} \left(\mathrm{e}^{tL^*} v \right) w \, \mathrm{d}\mu = \int_{\mathbb{T}^d} \left(\mathrm{e}^{tL} w \right) v \, \mathrm{d}\mu = \int_{\mathbb{T}^d} w(\varphi(x, t)) v(x) \, \mathrm{d}\mu(x).$$

Now take any μ-measurable set A and the indicator function $\mathbf{1}_A = w$ to conclude

$$\int_A \left(e^{tL^*} v \right)(x) \, \mathrm{d}\mu(x) = \int_{\varphi(A,-t)} v(x) \, \mathrm{d}\mu(x), \qquad (33.15)$$

which can actually be viewed as an equivalent definition of the Perron-Frobenius operator $e^{tL^*} : L^1 \to L^1$. From (33.15) is easily follows that μ is an invariant measure for φ with density ρ^* if and only if

$$e^{tL^*} \rho^* = \rho^* \qquad \text{for all } t \geq 0. \qquad (33.16)$$

Now we can prove one implication in (T1). Indeed, suppose μ (with density ρ^*) is invariant; then the characterization of L^* as an infinitesimal generator gives

$$L^* \rho^* = \lim_{t \searrow 0} \frac{e^{tL^*} \rho^* - \rho^*}{t} \overset{(33.16)}{=} 0.$$

The converse in (T1) is less obvious and requires the **Hille-Yosida Theorem** [179, 448]. To give the main idea, suppose $L^* \rho^* = 0$ and consider

$$L_\lambda^* = \lambda R_\lambda^* L^*, \qquad R_\lambda := (\lambda \, \mathrm{Id} - L^*)^{-1} \qquad (33.17)$$

for $\lambda > 0$ so that R_λ^* is just the resolvent of L^*, and L_λ^* is the **Hille-Yosida approximation** of L^*. From (33.17), we now get $L_\lambda^* \rho^* = 0$ and from the series representation

$$e^{tL_\lambda^*} \rho^* = \sum_{k=0}^{\infty} \frac{t^k}{k!} (L_\lambda^*)^k \rho^* = \rho^*.$$

Due to the convergence of the resolvent approximation in the Hille-Yoshida Theorem, we obtain

$$\lim_{\lambda \to \infty} e^{tL_\lambda^*} \rho^* = L^* \rho^* = \rho^* \qquad \forall t \geq 0.$$

This implies by (33.16) that ρ^* is the density of an invariant measure μ, concluding the proof of (T1). For (T3), assume first that there are only constant solutions to $Lv = 0$ and we want to show that then μ is ergodic. Note that

$$0 = Lv = \lim_{t \searrow 0} \frac{e^{tL} v - v}{t}$$

implies that the only fixed points of e^{tL} are constants. Now we argue by contraposition. If μ is not ergodic, then there is a nontrivial invariant subset $\mathcal{M} \subset \mathbb{T}^d$, i.e.,

$$\varphi(\mathcal{M}, -t) = \mathcal{M} \qquad \text{for } t \geq 0.$$

Now let $v = \mathbf{1}_\mathcal{M}$, which implies

$$e^{tL} v = \mathbf{1}_\mathcal{M} \circ \varphi(\cdot, t) = \mathbf{1}_{\varphi(\mathcal{M}, -t)} = \mathbf{1}_\mathcal{M} = v,$$

and since $\mathcal{M}$ is not a trivial set (i.e., neither measure zero nor full measure) by construction, it follows that v is not a constant function showing the required contraposition. For the reverse implication in (T3), we refer to [356, Thm. 7.7.2./7.8.3], which is just another limiting argument playing with the Koopman and Perron-Frobenius operators. $\square$

In Chapter 34, we are actually going to use the existence of an invariant measure and ergodicity as key assumptions for averaging. However, one should be careful that ergodicity is not to be confused with the stronger notion of **mixing** as discussed in Exercise 33.6.

Exercise 33.4. Use a multiple scales ansatz $y = x/\varepsilon$ given by

$$u^\varepsilon(x,t) = u_0(x,y,t) + \varepsilon u_1(x,y,t) + \cdots, \tag{33.18}$$

where u_j is one-periodic in y for each j, in (33.7) to formally derive the equations (33.8)–(33.9) at orders $\mathcal{O}(\varepsilon^{-1})$ and $\mathcal{O}(1)$. $\lozenge$

Exercise 33.5. Let f be smooth. Consider the operator $L_0 := f(y) \cdot \nabla_y$ defined classically on continuously differentiable periodic functions in $C^1_{\mathrm{per}}(\mathbb{T}^d)$. Use (33.5) (and recall Definition 14.2) to extend L_0 to $L^\infty_{\mathrm{per}}(\mathbb{T}^d)$, respectively, $\mathcal{D}(L_0) \subset L^\infty_{\mathrm{per}}(\mathbb{T}^d)$. Now try a similar strategy for L_0^* on $L^1_{\mathrm{per}}(\mathbb{T}^d)$. $\lozenge$

Exercise 33.6. A flow $\varphi(x,t)$ with invariant measure μ is called **mixing** if

$$\lim_{t \to \infty} \mu(A \cap \varphi(B,-t)) = \mu(A)\mu(B)$$

for any two μ-measurable sets A, B. Consider the circle rotation flow

$$\varphi(x,t) = x + \omega t \pmod 1$$

for $x \in \mathbb{T}^1 = [0,1]/(0 \sim 1)$ for $\omega \neq 0$. Show that this flow is ergodic but not mixing. $\lozenge$

Background and Further Reading: The first part of this chapter follows [446], and the exposition to ergodic theory and the operator-theoretic viewpoint for it are summarized from [356]. There are many specialized textbooks for classical ergodic theory and its extensions [1, 129, 174, 323, 334, 395, 459, 555]; for introductions to the topic in the context of general dynamical systems we refer to the books [71, 320]. The Ergodic Theorem is often also attributed to Birkhoff [58]. The operator-theoretic approach to the topic is not only interesting from an abstract PDE viewpoint [175] but it has also found a number of applications in global ODE dynamics and spatial data analysis [149, 216, 402]. In some sense, this is not surprising as ergodic properties are well known to be important in nonlinear time series analysis [316]. For references regarding averaging and homogenization methods, please see the end of Chapter 34.

Chapter 34

Two-Scale Convergence

We continue with our study of the **transport PDE** (33.1) under the same assumptions as in Chapter 33, i.e., f is a smooth 1-periodic vector field and ε is a small positive parameter. Recall from the ansatz (33.18) that the key issue was to deal with functions of two spatial arguments simultaneously:

$$x \in \Omega \quad \text{and} \quad y \in \mathbb{T}^d := \mathbb{R}^d / \mathbb{Z}^d. \tag{34.1}$$

Since we need time dependence for the transport equation, we look at points $(x, y, t) \in \Omega \times \mathbb{T}^d \times (0, \infty)$.

Definition 34.1. Fix a time $T > 0$. The sequence of functions $u^\varepsilon \in L^2(\Omega \times (0, T))$ **two-scale converges** to $u \in L^2(\Omega \times \mathbb{T}^d \times (0, T))$ if for every test function $\phi \in L^2(\Omega \times (0, T); C_{\text{per}}(\mathbb{T}^d))$ we have

$$\lim_{\varepsilon \to 0} \int_0^T \int_\Omega u^\varepsilon(x, t) \phi\left(x, \frac{x}{\varepsilon}, t\right) \, \mathrm{d}x \, \mathrm{d}t = \int_0^T \int_\Omega \int_{\mathbb{T}^d} u(x, y, t) \phi(x, y, t) \, \mathrm{d}y \, \mathrm{d}x \, \mathrm{d}t.$$

Two-scale convergence, also denoted by $u^\varepsilon \xrightarrow{2} u$, turns out to be a good notion to capture the limit of our transport problem. In particular, an abstract analysis of the concept yields the following toolbox-type result:

Theorem 34.2. *(existence of a two-scale limit; see, e.g., [12]) Suppose u^ε is a bounded sequence in $L^2(\Omega \times (0, T))$. Then there exists a subsequence, say still denoted by u^ε, and a function $u \in L^2(\Omega \times \mathbb{T}^d \times (0, T))$ such that $u^\varepsilon \xrightarrow{2} u$. Furthermore, we may average over the unit cell in the weak limit, i.e.,*

$$u^\varepsilon \rightharpoonup \int_{\mathbb{T}^d} u(\cdot, y, \cdot) \, \mathrm{d}y \quad \text{weakly in } L^2(\Omega \times (0, T)).$$

Roughly speaking, we need some boundedness/compactness statement about the sequence u^ε; then we get a suitable form of *weak* convergence. The main goal in this chapter is the next result.

Theorem 34.3. *(transport averaging) Consider the PDE*

$$\begin{cases} \partial_t u^\varepsilon - f^\varepsilon \cdot \nabla u^\varepsilon = 0 & \text{in } \mathbb{R}^d \times (0, \infty), \\ u^\varepsilon(\cdot, 0) = g & \text{for } x \in \mathbb{R}^d, \end{cases} \tag{34.2}$$

for $u^\varepsilon = u^\varepsilon(x, t)$ with $g \in C_b^\infty(\mathbb{R}^d)$. Assume that $f^\varepsilon(x) = f(x/\varepsilon)$, where

182

- f is a smooth 1-periodic vector field on $\mathbb{R}^d$,

- f generates an ergodic flow (see Definition 33.2), and

- f is divergence-free ($\nabla \cdot f = 0$).

Then $u^\varepsilon \xrightarrow{2} u \in L^2(\mathbb{R}^d \times (0,\infty); L^2_{\mathrm{per}}(\mathbb{T}^d))$ and $u = u(x,t)$ is the weak solution of the transport PDE

$$\begin{cases} \partial_t u - f_* \cdot \nabla u = 0 & in \ \mathbb{R}^d \times (0,\infty), \\ u(\cdot, 0) = g & for \ x \in \mathbb{R}^d, \end{cases} \tag{34.3}$$

where $f_* = \int_{\mathbb{T}^d} f(y) \, \mathrm{d}y$.

Before sketching the proof of Theorem 34.3, we briefly discuss the resulting PDE (34.3) in comparison to the formal limit (33.12).

Lemma 34.4. *Let f be smooth, 1-periodic, divergence-free, and ergodic; then there exists a unique classical periodic solution to*

$$L_0^* \rho := f(y) \cdot \nabla_y \rho = 0, \qquad \int_{\mathbb{T}^d} \rho(y) \, \mathrm{d}y = 1 \tag{34.4}$$

given by $\rho(y) \equiv 1$.

Proof. By ergodicity, we know from Theorem 33.3(T3) that $\mathcal{N}[L_0] = \mathrm{span}(1)$. However, if f is also divergence-free then L_0^* is **skew-symmetric** since

$$\int_{\mathbb{T}^d} a(y)(f(y) \cdot \nabla_y b(y)) \, \mathrm{d}y = \int_{\mathbb{T}^d} a(y) \nabla_y \cdot (f(y)b(y)) \, \mathrm{d}y$$
$$= -\int_{\mathbb{T}^d} b(y)(f(y) \cdot \nabla_y a(y)) \, \mathrm{d}y$$

using $\nabla \cdot f = 0$ and integration by parts under the assumption that the functions a, b, f involved are all periodic. Therefore, $L_0^* = -f(y) \cdot \nabla_y$, and we know already the nullspace of L_0 consists only of constants. The normalization condition in (34.4) finishes the proof. $\qquad \square$

Lemma 34.4 shows that our formal result (33.12) is consistent with Theorem 34.3 as for divergence-free vector fields we may replace the stationary distribution $\rho^*(y)$ just by 1.

Remark: The divergence-free constraint naturally appears in fluid dynamics and is explicit in the **incompressible Navier-Stokes equation** (15.10).

Proof. (of Theorem 34.3) The first part of the proof will only be sketched as it follows from the standard weak solution theory for general transport PDEs. However, it is relevant to recall the main ideas to see, where the boundedness comes in. The idea is to use the method of **vanishing viscosity** and to regularize (34.2) via a diffusive/viscous second-order term

$$\partial_t u^{\varepsilon,\delta} - f^\varepsilon \cdot \nabla u^{\varepsilon,\delta} + \delta \Delta u^{\varepsilon,\delta} = 0. \tag{34.5}$$

Furthermore, one may approximate the initial condition g by a smooth function with compact support also using the same approximation parameter $0 < \delta \ll 1$. Then classical parabolic regularity theory (cf. Chapter 3 and references therein) provides a sufficiently regular unique solution, which actually converges to the desired weak solution of (34.2) in the limit $\delta \to 0$. In addition, we claim that this weak solution u^ε satisfies

$$\|u^\varepsilon\|_{L^\infty((0,T);L^2(\mathbb{R}^d))} \le \|g\|_{L^2(\mathbb{R}^d)}. \tag{34.6}$$

Indeed, integration by parts and the fact that f is divergence-free show, very similarly to the proof in Lemma 34.4, that

$$\int_{\mathbb{R}^d} (f^\varepsilon \cdot \nabla u^\varepsilon) u^\varepsilon \, \mathrm{d}x = 0. \tag{34.7}$$

Multiplying the transport equation by u^ε and integrating over $\mathbb{R}^d$ yields

$$\frac{1}{2}\|u^\varepsilon\|_{L^2(\mathbb{R}^d)}^2 - \int_{\mathbb{R}^d} (f^\varepsilon \cdot \nabla u^\varepsilon) u^\varepsilon \, \mathrm{d}x = 0. \tag{34.8}$$

Therefore, employing (34.7) in (34.8) gives the a priori bound (34.6) as claimed. Since (34.6) holds, we may apply Theorem 34.2 to obtain a subsequence, still denoted by u^ε, which two-scale converges to some $u \in L^2((0,\infty) \times \mathbb{R}^d; L^2_{\text{per}}(\mathbb{T}^d))$. To derive the limiting equation, we first consider the weak formulation of (34.2), which can be derived from integrating by parts in x and t, and using the divergence-free condition so that

$$\int_0^\infty \int_{\mathbb{R}^d} \left(\partial_t \phi^\varepsilon - f^\varepsilon \cdot \nabla \phi^\varepsilon\right) u^\varepsilon \, \mathrm{d}x \, \mathrm{d}t + \int_{\mathbb{R}^d} g(x)\phi^\varepsilon(x,0) \, \mathrm{d}x = 0 \tag{34.9}$$

for every compactly supported test function $\phi^\varepsilon(x,t) \in C_c^\infty$. The trick is now to consider a two-scale test function of the form

$$\phi^\varepsilon(x,t) = \varepsilon \phi\left(x, \frac{x}{\varepsilon}, t\right) = \varepsilon \phi(x,y,t),$$

where $\phi \in C_c^\infty((0,\infty) \times \mathbb{R}^d; C_{\text{per}}^\infty(\mathbb{T}^d))$. Inserting this function into (34.9), using the chain rule for the gradient, as ϕ depends upon two spatial variables, and then taking the limit $\varepsilon \to 0$ gives

$$\int_0^\infty \int_{\mathbb{R}^d} \int_{\mathbb{T}^d} f(y) \cdot \nabla_y \phi(x,y,t) \, u(x,y,t) \, \mathrm{d}y \, \mathrm{d}x \, \mathrm{d}t = 0. \tag{34.10}$$

Integrating (34.10) by parts shows that it is the weak formulation of $L_0 u = 0$. Since f is ergodic we may conclude, by Theorem (33.3)(T3), that (34.10) holds if and only if u is independent of y. Therefore, we can use a y-independent test function $\phi^\varepsilon = \phi(x,t)$ and pass to the limit $\varepsilon \to 0$ to obtain the limiting weak formulation

$$\int_0^\infty \int_{\mathbb{R}^d} \int_{\mathbb{T}^d} \left(\partial_t \phi - f \cdot \nabla_x \phi\right) u \, \mathrm{d}y \, \mathrm{d}x \, \mathrm{d}t + \int_{\mathbb{R}^d} g(x)\phi(x,0) \, \mathrm{d}x = 0.$$

Evaluating the y integrals above gives precisely 1 for the first integral, as we work with the unit torus, and we get the average f_* in the second integral. Therefore, we obtain the weak formulation (34.3) as claimed. $\square$

Many generalizations of Theorem 34.3 exist. However, ergodicity and at least some relatively strong control over the dynamics of f is necessary to prove averaging results; see also references below. It is important to keep in mind that averaging requires the small parameter ε to appear in a particular way, i.e., to induce fast/small-scale oscillations. In Chapter 35, we shall illustrate that there are many other natural occurrences of small parameters.

Exercise 34.5. Let a be a smooth 1-periodic function with $0 < a_0 \leq a(y)$ for some constant $a_0 > 0$. Consider the boundary value problem

$$-\frac{\mathrm{d}}{\mathrm{d}x}\left(a\left(\frac{x}{\varepsilon}\right)\frac{\mathrm{d}u^\varepsilon}{\mathrm{d}x}\right) = h, \qquad u^\varepsilon(0) = 0 = u^\varepsilon(L)$$

for $u^\varepsilon(x)$ with $x \in [0, L]$ and with smooth h. Then one may show that a limiting **homogenized** (i.e., basically second-order averaged) problem is given by

$$-\frac{\mathrm{d}}{\mathrm{d}x}\left(a_*\frac{\mathrm{d}u}{\mathrm{d}x}\right) = h, \qquad u(0) = 0 = u^\varepsilon(L),$$

where $a_* = \int_0^1 a(y) + a(y)\chi'(y)\,\mathrm{d}y$, where χ satisfies the **cell problem**

$$-\frac{\mathrm{d}}{\mathrm{d}y}\left(a\left(y\right)\frac{\mathrm{d}\chi}{\mathrm{d}y}\right) = \frac{\mathrm{d}a}{\mathrm{d}y}, \quad \chi(0) = \chi(1), \quad \int_0^1 \chi(y)\,\mathrm{d}y = 0. \qquad (34.11)$$

Use the cell problem to show that

$$a_* = \left(\int \frac{1}{a(y)}\,\mathrm{d}y\right)^{-1},$$

i.e., a_* is the harmonic average. $\Diamond$

Exercise 34.6. Prove that the vector field $f(y) = (\sin(2\pi y_2), \sin(2\pi y_1))$ is divergence-free. Furthermore, show that it is not ergodic by considering the nullspace $\mathcal{N}[L_0]$ of the generator L_0 of the Koopman operator for f. $\Diamond$

Exercise 34.7. Write out a generalization of ergodicity with respect to maps. Then consider the iterated mapping on $\mathbb{T}^2$ defined via applying the matrix

$$A := \begin{pmatrix} 2 & 1 \\ 1 & 1 \end{pmatrix} : \mathbb{T}^2 \to \mathbb{T}^2$$

repeatedly. Show that this discrete-time dynamical system is ergodic. Remark: This map is an example of a **hyperbolic toral automorphism**. $\Diamond$

Background and Further Reading: This chapter is based upon a condensed summary of the transport averaging theory from [446]. Averaging classically appears in time-dependent ODEs with fast oscillations [247, 346, 484]. **Homogenization** is often used as a synonym for averaging and vice versa in the PDE community; in [446] a reasonable distinction is made between first-order perturbation methods (averaging) and higher-order methods (homogenization); see also Exercise 34.5. The literature on averaging/homogenization is quite large [36, 62, 279, 295, 439, 522], and many results are now quite classical. For two-scale convergence we refer to [12, 429]; other methods for homogenization proofs are also available [118]. Two-scale convergence is only one instance where a convergence notion and/or limiting object are tailor-made for a problem class. Other classical cases are Γ-**convergence** [64, 213, 214, 485] as well as the theory of **Young measures** [39, 419, 449, 479], both used frequently in certain variational problems [33]; see also Chapters 25–26.

Chapter 35

Asymptotics and Layers

We have already seen in several instances (cf. Chapters 12–13, 20–21, and 33–34) the important role played by techniques using small parameters. Here we shall develop yet another, actually very classical, facet of this theme. More precisely, we are going to discuss basic principles of (formal) asymptotic matching in the context of the **Gierer-Meinhardt equation**

$$\begin{cases} \partial_t u = \varepsilon^2 \partial_x^2 u - u + \frac{u^2}{v}, \\ \tau \partial_t v = \partial_x^2 v - v + u^2, \end{cases} \tag{35.1}$$

on the one-dimensional spatial domain $\Omega = [-1, 1]$ with Dirichlet boundary conditions

$$u(-1, t) = 0 = u(1, t). \tag{35.2}$$

The parameter τ is going to be considered in the singular limit $\tau = 0$ so that we have a coupled parabolic-elliptic system; cf. Chapter 30. We are interested in so-called **spike (or layer) solutions**, which actually tend to infinity as $\varepsilon \to 0$. The correct scaling to renormalize/scale this divergence to ensure $\mathcal{O}(1)$ variables u, v as $\varepsilon \to 0$ turns out to be

$$u = \varepsilon^{-1} \tilde{u} \qquad \text{and} \qquad v = \varepsilon^{-1} \tilde{v}. \tag{35.3}$$

Inserting (35.3) into (35.1), dropping the tildes, and using $\tau = 0$ yields

$$\begin{cases} \partial_t u = \varepsilon^2 \partial_x^2 u - u + \frac{u^2}{v}, \\ 0 = \partial_x^2 v - v + \varepsilon^{-1} u^2. \end{cases} \tag{35.4}$$

To construct stationary solutions $(u^*, v^*) = (u^*(x), v^*(x))$ to (35.4) having M spikes, as shown in Figure (35.1) for the case $M = 3$, we assume that all the spikes have equal height and are located symmetrically with equal maxima of u as well as equal values of v at the locations

$$x_m = -1 + \frac{1 + 2m}{M}, \qquad m = 0, 1, \ldots, M - 1. \tag{35.5}$$

In particular, we have for all j that

$$\frac{\mathrm{d}u^*}{\mathrm{d}x}(x_j) = 0 \qquad \text{and} \qquad v^*(x_j) = v_H,$$

where $v_H \in (0, \infty)$ is fixed and independent of j.

The main idea of **matched asymptotic expansions** is to decompose the solution into two regions:

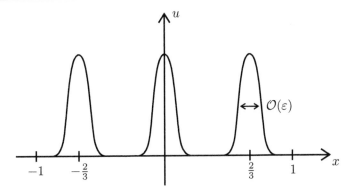

Figure 35.1. *Three-spike solution for the Gierer-Meinhardt equation* (35.1).

- **inner layers** (or inner regions) near the locations, where the maxima of u are reached and a high/fast variation of the solution is observed;

- **outer layers** (or outer regions) in the parts, where u is expected to be exponentially small.

We start with the inner layer. Define the variables near the mth spike as

$$y_m = \varepsilon^{-1}(x - x_m), \qquad U(y_m) = u(x_m + \varepsilon y_m), \qquad V(y_m) = v(x_m + \varepsilon y_m).$$

Then we have

$$\frac{\mathrm{d}u}{\mathrm{d}x} = \frac{\mathrm{d}u}{\mathrm{d}y_j}\frac{\mathrm{d}y_j}{\mathrm{d}x} = \frac{1}{\varepsilon}\frac{\mathrm{d}u}{\mathrm{d}y_j} \tag{35.6}$$

and the same for v. Inserting the new variables into (35.4) and imposing stationarity ($\partial_t u = 0$), we get

$$\begin{cases} 0 = U'' - U + \frac{U^2}{V}, \\ 0 = \frac{1}{\varepsilon^2}V'' - V + \frac{1}{\varepsilon}U^2, \end{cases} \tag{35.7}$$

where differentiation is with respect to y_j. The next step is to consider an **asymptotic series**

$$\{g_k(\varepsilon)\}_{k=0}^{\infty}, \qquad \lim_{\varepsilon \to 0}\frac{g_{k+1}(\varepsilon)}{g_k(\varepsilon)} = 0 \;\forall k \in \mathbb{N} \cup \{0\}.$$

The standard choice is to consider a power series, i.e., $g_k(\varepsilon) = \varepsilon^k$; see also Chapter 12. Each power has an unknown coefficient, and we make the ansatz

$$\begin{aligned} U(y_j) &= U_0(y_j) + \varepsilon U_1(y_j) + \mathcal{O}(\varepsilon^2), \\ V(y_j) &= V_0(y_j) + \varepsilon V_1(y_j) + \mathcal{O}(\varepsilon^2). \end{aligned} \tag{35.8}$$

Plugging (35.8) into (35.7), collecting problems with different orders of ε, and looking at the first three ODEs gives

$$\begin{cases} 0 = U_0'' - U_0 + \frac{U_0^2}{V_0}, \\ 0 = V_0'', \\ 0 = V_1'' + U_0^2, \end{cases} \tag{35.9}$$

posed on a domain with $y_j \in \mathbb{R}$. At the spike location, which has been moved to $y_j = 0$, we require $U_0'(0) = 0$ and $V_0(0) = v_H$. Since we also need to **match** the inner solution

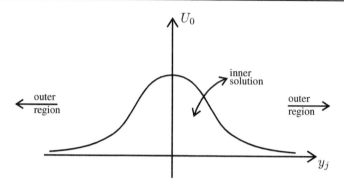

Figure 35.2. *Illustration of the broader ($\mathcal{O}(1)$-scale) inner solution, which has to be matched at $\pm\infty$ to the outer solution.*

to the outer solution, we also require that V_0 is bounded as $|y_j| \to \infty$ and $U_0 \to 0$ as $|y_j| \to \infty$; see Figure 35.2.

From (35.9), we consider first

$$0 = V_0'', \qquad V_0(0) = v_H, \qquad V_0 \text{ is bounded as } |y_j| \to \infty,$$

which has the solution $V_0 \equiv v_H$. Since V_0 is constant, it is helpful to scale U_0 to simplify the remaining two equations from (35.9). Let us consider

$$U_0 v_H^{-1} =: U_c.$$

A quick calculation shows

$$U_0'' - U_0 + \frac{U_0^2}{V_0} = v_H U_c'' - v_H U_c + \frac{v_H^2}{v_H} U_c^2 = v_H U_c'' - v_H U_c + v_H U_c^2,$$

so it remains to study the problem

$$\begin{cases} 0 = U_c'' - U_c + U_c^2, \\ 0 = V_1'' + U_c^2 v_H^2, \end{cases} \tag{35.10}$$

with $U_c \to 0$ as $|y_j| \to \infty$ and $U_c'(0) = 0$. One checks that the equation for U_c together with the required conditions is satisfied by the function

$$U_c(y_j) = \frac{3}{2} \operatorname{sech}^2\left(\frac{y_j}{2}\right). \tag{35.11}$$

Of course, this solution is not easy to guess or find if one does not know it, but see Exercise 35.3 as well as Example 7.1. The next step is to integrate $V_1'' = -U_c^2 v_H^2$ once over $\mathbb{R}$ to obtain

$$\lim_{y_j \to +\infty} V_1'(y_j) - \lim_{y_j \to -\infty} V_1'(y_j) = -v_H^2 \int_{-\infty}^{+\infty} U_c(w)\, \mathrm{d}w = -6v_H^2. \tag{35.12}$$

Essentially, (35.12) provides a jump condition of the derivative, which makes sense as it does change from one side of the spike to the other. We have to employ this jump condition for the outer solution. Outside of each x_j, we assume that u is exponentially small and v can be expanded as

$$v(x) = v_0(x) + \mathcal{O}(\varepsilon) \qquad \text{as } \varepsilon \to 0. \tag{35.13}$$

Inserting (35.13) into (35.4) gives at the lowest order

$$\frac{d^2 v_0}{dx^2} = v_0 \qquad \text{for } x \in (-1, 1).$$

Since v_0 should also be continuous across each inner region and satisfy the jump condition (35.12) by **matching** the inner and outer solution, we must solve the modified problem

$$\frac{d^2 v_0}{dx^2} - v_0 = -6 v_H^2 \sum_{m=0}^{M-1} \delta(x - x_m), \quad x \in (-1, 1), v_0'(\pm 1) = 0, \qquad (35.14)$$

where δ is the Dirac delta distribution. One may solve (35.14) using a classical **Green's function** approach and consider the basic boundary value problem

$$\frac{d^2 G}{dx^2} - G = -\delta(x - x_m) \qquad \text{for } x \in (-1, 1) \text{ and } \frac{dG}{dx}(\pm 1; x_m) = 0. \qquad (35.15)$$

Standard theory [32, 186, 232] shows that one may construct G and study its properties. However, here (35.15) is simple enough that even an explicit formula is possible. One checks that

$$G(x; x_m) = \begin{cases} \frac{A_k \cosh(1+x)}{\cosh(1+x_k)} & \text{for } x \in (-1, x_k), \\ \frac{A_k \cosh(1-x)}{\cosh(1-x_k)} & \text{for } x \in (x_k, 1), \end{cases} \qquad (35.16)$$

where $A_k = [\tanh(1 - x_k) + \tanh(1 + x_k)]^{-1}$. By linearity, we now find the solution to (35.14) as

$$v_0(x) = 6 v_H^2 \sum_{m=0}^{M-1} G(x; x_m).$$

One may now also fix the constant v_H by setting $v_H = v_0(x_j)$ using the assumption of equal heights of the spikes. One may even check that this assumption is not strictly necessary, as the sum

$$\sum_{m=0}^{M-1} G(x_j; x_m) \qquad (35.17)$$

is independent of j if the spikes are equally spaced; see Exercise 35.4. Therefore, we must have

$$v_H = \frac{1}{6 u_G}, \qquad u_G := \sum_{m=0}^{M-1} G(x_j; x_m).$$

Although one has to mention that the next result has been obtained via formal asymptotic calculations, it is evident that it provides an excellent step to anchor rigorous proofs and further analysis.

Theorem 35.1. (*Gierer-Meinhardt stationary spikes*) *There exists a formal asymptotic M-spike stationary solution as $\varepsilon \to 0$ to (35.4) given by*

$$u^*(x) \sim v_H \sum_{m=0}^{M-1} U_c(\varepsilon^{-1}(x - x_m)),$$

$$v^*(x) \sim \frac{v_H}{u_G} \sum_{m=0}^{M-1} G(x; x_m),$$

which is spatially symmetric with maxima located according to (35.5).

The high degree of symmetry in the result may be initially quite intriguing. The next example links this to a classical perturbation technique in dynamical systems.

Example 35.2. We look at the first ODE in (35.10). In Exercise 35.3(b) the phase plane has been sketched for the system

$$u_1' = u_2, \tag{35.18}$$
$$u_2' = u_1 - u_1^2, \tag{35.19}$$

which is easily checked to be **Hamiltonian** with Hamiltonian function

$$H(u_1, u_2) = \frac{1}{2}u_2^2 - \frac{1}{2}u_1^2 + \frac{1}{3}u_1^3,$$

or its negative depending upon the convention; see Example 27.1. Note from the phase plane or by considering the Hamiltonian that (35.18) has a hyperbolic saddle equilibrium point at $(u_1, u_2) = (0,0)$ to which a homoclinic orbit $\Gamma(t)$, given actually by formula (35.11), is attached. $\Gamma(t)$ itself encloses a continuous family of periodic orbits as shown in Figure 35.3; also cf. Chapter 7 and Chapter 27.

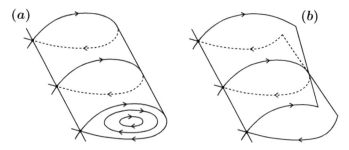

Figure 35.3. *(a) Singular situation corresponding to (35.18), where there is no perturbation and we just have a family of homoclinic orbits with periodic orbits inside. (b) Perturbed system (35.20), where for each ϕ we expect different dynamics for the u coordinates. The generic breaking of the family of homoclinic orbits results in a transversal intersection of stable and unstable manifolds of the family of saddle equilibria.*

Suppose we would have perturbed the original Gierer-Meinhardt system by a suitable space-periodic perturbation such that (35.18) becomes

$$
\begin{aligned}
u_1' &= u_2 + \zeta h_1(u, \phi) =: f_1(u) + \zeta h_1(u, \phi), \\
u_2' &= u_1 - u_1^2 + \zeta h_2(u, \phi) =: f_2(u) + \zeta h_2(u, \phi), \\
\phi' &= \omega,
\end{aligned}
\tag{35.20}
$$

where $0 < \zeta \ll 1$, $\omega > 0$ is a parameter representing the frequency, and $h = (h_1, h_2)$ is periodic in ϕ so that (35.20) is posed on $\mathbb{R}^2 \times \mathbb{S}^1 = \mathbb{R}^2 \times [0,1]/(0 \sim 1)$. Depending upon the type of perturbation, it is geometrically evident by looking at the stable and unstable manifolds of the line of equilibria

$$\{(u, \phi) \in \mathbb{R}^2 \times \mathbb{S}^1 : u = 0, \phi \in [0,1)\}$$

that the single torus of homoclinics as well as the tori of periodic orbits shown in Figure 35.3(a) may break upon applying the perturbation, e.g., as shown in Figure 35.3(b).

Similar to the situation in Chapter 27 for tori, a natural question to ask is, whether certain homoclinic orbits persist. Although we cannot give a full answer here, let us at least indicate that a key role in deciding persistence is played by the **Melnikov function**

$$M(t_0) := \int_{-\infty}^{\infty} f(\Gamma(t - t_0)) \wedge h(\Gamma(t - t_0), \omega t) \, \mathrm{d}t, \qquad (35.21)$$

where $f \wedge h = f_1 h_2 - f_2 h_1$, which helps to measure when there are actually intersections of certain stable and unstable manifolds in a perturbed homoclinic orbit; see Exercise 35.5 and compare the geometric setup to Lin's method in Chapter 7. ♦

The last example strongly indicates that we may understand many phenomena in spatially extended systems using suitable ODE systems. We have already encountered this interplay throughout the entire book. In the next chapter, we shall see that the theory of multiple time scale dynamics is a very flexible approach to capture problems with small parameters.

Exercise 35.3. (a) Verify that $U_c(y) = \frac{3}{2}\operatorname{sech}^2\left(\frac{y}{2}\right)$ solves the ODE

$$\begin{aligned} 0 &= \frac{\mathrm{d}^2 U_c}{\mathrm{d}y^2} - U_c + U_c^2, \qquad y \in \mathbb{R}, \\ U_c(y) &\to 0 \text{ as } |y| \to \infty, \qquad U_c'(0) = 0. \end{aligned} \qquad (35.22)$$

(b) Consider (35.22) as a planar ODE and sketch the phase plane. ◊

Exercise 35.4. (a) Check that the Green's function (35.16) is indeed the **fundamental solution** to (35.15). (b) Use the assumption (35.5) and the explicit formula (35.16) to prove that the sum (35.17) is independent of j. ◊

Exercise 35.5. Consider Example 35.2. (a) Try to calculate the Melnikov function M from (35.21) for several simple perturbations h, e.g., using sin and cos terms or other simple periodic functions. (b) Find one example in (a) where M has a simple zero at some t_0. (c) Try to geometrically interpret the zero of the Melnikov function on a Poincaré section chosen transversally to the families of homoclinic and periodic orbits. ◊

Background and Further Reading: The main argument for the existence of spikes using matched asymptotics follows the derivation in [290]. There are many introductory texts on asymptotic methods and matching for ODEs as well as PDEs [198, 268, 328, 346, 426]. The Gierer-Meinhardt equation was derived/suggested in [229] and has been considered mathematically particularly in the context of locally concentrating solutions [159, 290, 433, 558]. Of course, the stationary patterns we have derived can then be analyzed from a bifurcation-theoretic point of view as studied in Chapters 4–6. A closely related model to the Gierer-Meinhardt equations where many asymptotic techniques also apply is the **Gray-Scott equation**

$$\begin{cases} \partial_t u = \Delta u - uv^2 + p_1(1 - u), \\ \partial_t v = \varepsilon \Delta v + uv^2 - p_2 v, \end{cases}$$

where $p_{1,2}$ are parameters and $\varepsilon > 0$ is usually assumed to be small. We refer to [158, 160, 341, 436] for more details regarding the Gray-Scott equation. The Melnikov method has evolved from a tool initially designed for periodic perturbations of Hamiltonian systems to a very general technique in dynamics [226, 247, 272, 563].

Chapter 36

Fast-Slow Systems: Periodicity and Chaos

In Chapter 35 asymptotic matching was used to study PDEs with small parameters. In this chapter, we briefly sketch two quite different paths to obtain fast-slow ODEs with a small parameter from PDEs. Furthermore, we briefly highlight results for these two ODEs obtainable from dynamical systems methods.

As a first example, consider the minimization problem

$$\min_{u \in X} \mathcal{E}_\varepsilon[u], \quad \mathcal{E}_\varepsilon[u] := \int_0^1 \varepsilon^2 (\partial_x^2 u)^2 + \frac{1}{4}((\partial_x u)^2 - 1)^2 + u^2 \, dx, \qquad (36.1)$$

where X is a given function space and $\varepsilon > 0$ is a small parameter. One possible setting is to consider the space

$$X = H_{\text{per}}^2(0,1) := \{u \in H^2(0,1) : u(0) = u(1), \partial_x u(0) = \partial_x u(1)\}, \qquad (36.2)$$

i.e., we look for minimizers which are periodic. Note that one can also view (36.1)–(36.2) as the steady-state problem of a time-dependent PDE by viewing $\mathcal{E}_\varepsilon[u]$ as an energy and considering the associated gradient flow; see Chapter 23.

Proposition 36.1. *Critical points of $\mathcal{E}_\varepsilon$ (i.e., $D_u \mathcal{E}_\varepsilon[u^*] = 0$) satisfy the **Euler-Lagrange** equation associated to (36.1)–(36.2) given by*

$$\varepsilon^2 \frac{d^4 u^*}{dx^4} - \frac{1}{2} \frac{d}{dx} \left(\left(\frac{du^*}{dx} \right)^3 - \frac{du^*}{dx} \right) + u^* = 0 \qquad (36.3)$$

with periodic boundary conditions.

Proof. Let us suppose for ease of presentation here that $u \in C^2$ with periodic boundary conditions; the following argument can easily be carried out for H^2 as well. Let $X_2 = \{w \in C^2 : w(0) = w(1), \partial_x w(0) = \partial_x w(1)\}$ and consider the **first variation**

$$w = u^* + \delta v, \qquad v \in X_2, \ \delta \in \mathbb{R}.$$

Then we easily see that the real-valued function

$$\mathcal{E}_*(\delta) := \mathcal{E}_\varepsilon[u^* + \delta v]$$

has a minimum at $\delta = 0$ so that $\mathcal{E}_*'(0) = 0$. We can rewrite $\mathcal{E}_*$ as

$$\mathcal{E}_*(\delta) = \int_0^1 L((u^*)'' + \delta v'', (u^*)' + \delta v', u^* + \delta v) \, dx, \qquad ' = \frac{d}{dx},$$

where the so-called **Lagrangian** is given by

$$L(p, q, r) = \varepsilon^2 p^2 + \frac{1}{4}(q^2 - 1)^2 + r^2, \qquad p = \frac{\mathrm{d}^2 u}{\mathrm{d}x^2}, \ q = \frac{\mathrm{d}u}{\mathrm{d}x}, \ r = u.$$

Computing the derivative $\mathcal{E}'_*$ and evaluating at $\delta = 0$, we find

$$0 = \mathcal{E}'_*(0) = \int_0^1 v'' \partial_p L((u^*)'', (u^*)', u^*)$$
$$+ v' \partial_q L((u^*)'', (u^*)', u^*) + v L_r((u^*)'', (u^*)', u^*) \, \mathrm{d}x.$$

Integrating the last expression by parts, twice for the first term and once for the second term, and then using the periodicity of v to eliminate the boundary terms yields

$$0 = \int_0^1 v \left(\frac{\mathrm{d}^2}{\mathrm{d}x^2} \partial_p L((u^*)'', (u^*)', u^*) - \frac{\mathrm{d}}{\mathrm{d}x} \partial_q L((u^*)'', (u^*)', u^*) \right)$$
$$+ v L_r((u^*)'', (u^*)', u^*) \, \mathrm{d}x.$$

The last equality has to hold for any $v \in X_2$ so that we may conclude

$$\underbrace{\frac{\mathrm{d}^2}{\mathrm{d}x^2} \partial_p L((u^*)'', (u^*)', u^*)}_{=2\varepsilon^2 \frac{\mathrm{d}^4 u^*}{\mathrm{d}x^4}} - \underbrace{\frac{\mathrm{d}}{\mathrm{d}x} \partial_q L((u^*)'', (u^*)', u^*)}_{=\frac{\mathrm{d}}{\mathrm{d}x}\left(\left(\frac{\mathrm{d}u^*}{\mathrm{d}x}\right)^3 - \frac{\mathrm{d}u^*}{\mathrm{d}x}\right)} + \underbrace{L_r((u^*)'', (u^*)', u^*)}_{=2u} = 0,$$

which is precisely (36.3) upon dividing by 2. $\qquad \square$

Of course, the last derivation is substantially more general and links the class of Euler-Lagrange equations to minimizers. In many cases, it is useful to use dynamical systems methods to understand solutions of the Euler-Lagrange equations and, for our case, we end up with a fourth-order ODE with periodic boundary conditions, which can be rewritten in a more pleasant format.

Lemma 36.2. *The fourth-order ODE (36.3) can be rewritten in standard form as a (2,2)-fast-slow system, i.e., as a first-order ODE system with two fast and two slow variables, given by*

$$\begin{aligned}
\frac{\mathrm{d}x_1}{\mathrm{d}t} &= x_1' = \tfrac{1}{2}(x_2^3 - x_2) - y_1, \\
\frac{\mathrm{d}x_2}{\mathrm{d}t} &= x_2' = x_1, \\
\frac{\mathrm{d}y_1}{\mathrm{d}t} &= y_1' = \varepsilon y_2, \\
\frac{\mathrm{d}y_2}{\mathrm{d}t} &= y_2' = \varepsilon x_2,
\end{aligned} \tag{36.4}$$

where boundary conditions can be derived from the periodic conditions on u^*; see Exercise 36.5(b).

The proof of Lemma 36.2 is Exercise 36.5. As already indicated briefly in Example 17.1, systems in standard (m, n)-**fast-slow** form

$$\begin{aligned}
\frac{\mathrm{d}x}{\mathrm{d}t} &= f(x, y), \\
\frac{\mathrm{d}y}{\mathrm{d}t} &= \varepsilon g(x, y)
\end{aligned} \tag{36.5}$$

for $(x, y) \in \mathbb{R}^m \times \mathbb{R}^n$ can be treated using the multitude of approaches available for singularly perturbed systems when we are interested in the case $0 < \varepsilon \ll 1$, i.e., proving results for sufficiently small time scale separation.

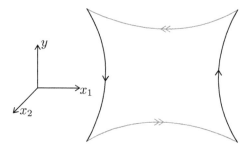

Figure 36.1. *Illustration of the singular periodic orbit for a simplified system arising from (36.4) by using a first integral; see the proof below.*

Theorem 36.3. *(fast-slow periodic orbits) There exists $\varepsilon_0 > 0$ such that, for all $\varepsilon \in (0, \varepsilon_0]$, the ODE (36.4) has a family of periodic orbits, each being $\mathcal{O}(\varepsilon)$-close to a singular orbit, where the singular orbit for $\varepsilon = 0$ consists of two fast orbit segments and two slow orbit segments; see also Figure 36.1.*

Proof. (Sketch [292]) A full proof is far beyond our scope here, but it is possible to illustrate the basic geometric ideas. First, note that (36.5) is actually a fast-slow Hamiltonian system of the form

$$
\begin{aligned}
x_1' &= \partial_{x_2} H, \\
x_2' &= -\partial_{x_1} H, \\
y_1' &= \varepsilon \partial_{y_2} H, \\
y_2' &= -\varepsilon \partial_{y_1} H
\end{aligned}
\tag{36.6}
$$

with Hamiltonian $H(x, y) = \frac{1}{8}(4y_2^2 - 8y_1 x_2 - 2x_2^2 + x_2^4 - 4x_1^2)$. Therefore, one can effectively reduce the system (36.4) to a $(2, 1)$-fast-slow system by considering each level set $\{H(x, y) =: \mu\}$ separately if we set

$$
y_1 = \frac{4y_2^2 - 8\mu - 2x_2^2 + x_2^4 - 4x_1^2}{8x_2}.
$$

The resulting fast-slow system

$$
\begin{aligned}
\frac{dx}{dt} &= f(x, y_2), \\
\frac{dy_2}{dt} &= \varepsilon g(x, y_2)
\end{aligned}
\tag{36.7}
$$

for $x = (x_1, x_2)^\top \in \mathbb{R}^2$ has the geometry shown in Figure 36.1. The critical manifold defined by $\mathcal{C}_0 = \{f = 0\}$ has two parts $\mathcal{C}_0^{l,r}$, which are of **saddle type**, i.e., the linearization $D_x f(p) \in \mathbb{R}^{2 \times 2}$ for $p \in \mathcal{C}_0^{l,r}$ has one negative and one positive eigenvalue. Points $p \in \mathcal{C}_0^{l,r}$ are equilibria of the **fast subsystem**

$$
\begin{aligned}
\frac{dx}{dt} &= f(x, y_2) = \left(x_2, \tfrac{1}{2}(x_2^3 - x_2) - y_1\right)^\top, \\
\frac{dy_2}{dt} &= 0.
\end{aligned}
\tag{36.8}
$$

One can then check, upon fixing some $\mu \in (-1/8, 1/24)$, that this system has a heteroclinic orbit from $\mathcal{C}_0^r$ to $\mathcal{C}_0^l$ at the fixed slow variable value

$$
y_{2,j} = \sqrt{2\mu + \frac{1}{4}}
$$

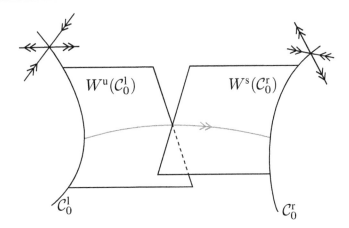

Figure 36.2. *Heteroclinic orbit of the fast subsystem lying in the transversal intersection of the stable and unstable manifolds of two parts of the saddle-type critical manifold.*

jumping at the steady state $(x_1, x_2) = (0, 1)$, and another jump at $-y_{2,j}$ starting at $(x_1, x_2) = (0, -1)$ and jumping from $\mathcal{C}_0^{\mathrm{l}}$ to $\mathcal{C}_0^{\mathrm{r}}$. The situation is illustrated in Figure 36.2.

The slow dynamics on $\mathcal{C}_0^{\mathrm{l,r}}$ follows quite easily by rescaling time $s = \varepsilon t$ in (36.7), then taking $\varepsilon \to 0$, which yields the **slow subsystem**

$$\frac{\mathrm{d}y_2}{\mathrm{d}s} = g(x, y_2) = x_2,$$

and x_2 is easily seen to be negative on $\mathcal{C}_0^{\mathrm{l}}$ and positive on $\mathcal{C}_0^{\mathrm{r}}$. Therefore, we have constructed a **singular** (or **candidate**) **orbit** for $\varepsilon = 0$ consisting of two fast and two slow segments as shown in Figure 36.1. Note carefully that this construction is valid for an entire range of level set values μ, which explains why we eventually obtain a whole family of orbits. However, we are now faced with the even more challenging task of showing persistence of the singular orbit for $0 < \varepsilon \ll 1$. This persistence can be proved using three main ingredients:

- The critical manifolds $\mathcal{C}_0^{\mathrm{l,r}}$ are **normally hyperbolic** in the considered regimes, i.e., $\mathrm{D}_x f(p) \in \mathbb{R}^{2 \times 2}$ for $p \in \mathcal{C}_0^{\mathrm{l,r}}$. So we may apply **Fenichel's Theorem** [195, 302, 346] to obtain perturbed **slow manifolds** $\mathcal{C}_\varepsilon^{\mathrm{l,r}}$; see Figures 36.1–36.2.

- The stable and unstable manifolds $W^{\mathrm{s,u}}(\mathcal{C}_0^{\mathrm{l,r}})$ also perturb by Fenichel's Theorem and can be shown to intersect transversally, i.e.,

$$W^{\mathrm{s}}(\mathcal{C}_0^{\mathrm{l}}) \pitchfork W^{\mathrm{u}}(\mathcal{C}_0^{\mathrm{r}}) \qquad \text{and} \qquad W^{\mathrm{s}}(\mathcal{C}_0^{\mathrm{r}}) \pitchfork W^{\mathrm{u}}(\mathcal{C}_0^{\mathrm{l}}),$$

where the intersections contain two heteroclinic singular connections of the fast subsystem; see Figure 36.2.

- Finally, one has to prove that the singular connection is robust to perturbations. One key element is that transversal intersections are robust to small perturbations and the second key element is the **Exchange Lemma** [304, 346], allowing one to track invariant manifolds and their tangent spaces also near slow dynamics.

For many more tools for proving persistence, also in singular cases, we refer to [346].

$\square$

A second example where a PDE gives rise to a fast-slow system is the **FitzHugh-Nagumo equation**

$$\begin{cases} \partial_t u = \partial_x^2 u + u(1-u)(u-p) - v, \\ \partial_t v = \varepsilon(u-v), \end{cases} \tag{36.9}$$

where $p \in (0, 1/2)$ is a parameter and $x \in \mathbb{R}$, so we work on an unbounded domain. An argument very similar to the one sketched in the proof of Theorem 36.3 yields the following.

Theorem 36.4. *(FHN (fast) pulse existence [84, 305]) For $\varepsilon > 0$ sufficiently small, there exists a speed s such that (36.9) has a travelling pulse $(u(x - st), v(x - st))$ corresponding to a homoclinic orbit with two fast and two slow segments as shown in Figure 36.3.*

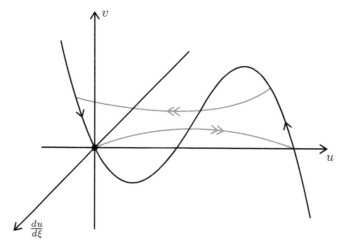

Figure 36.3. *Homoclinic orbit for the FitzHugh-Nagumo equation in the travelling wave frame coordinates $\xi = x - st$ corresponding to a travelling pulse solution of (36.9).*

Exercise 36.6 discusses the basic geometry of the pulse solution for the FitzHugh-Nagumo equation. Here we briefly want to highlight that homoclinic orbits, and hence also travelling pulses, can have a direct connection to **chaotic dynamics**. Figure 36.3 shows a sketch of a homoclinic orbit in a three-dimensional ODE, very similar to the case of the fast-slow travelling wave frame ODEs of the FitzHugh-Nagumo equation. Now consider the following general ODE:

$$z' = F(z), \qquad z \in \mathbb{R}^3. \tag{36.10}$$

Assume (36.10) has a homoclinic orbit $\gamma(t)$ with

$$\lim_{t \to \pm\infty} \gamma(t) = z^*$$

for a steady state z^*. Suppose the steady state $z^* \in \mathbb{R}^3$ is a **saddle focus**, i.e., $D_z F(z^*)$ has a complex conjugate pair of eigenvalues $\lambda_\pm = a \pm ib$, say with negative real part $a < 0$, and a positive real eigenvalue $\zeta > 0$. The manifolds $W^s(z^*)$ and $W^u(z^*)$ intersect along γ. Consider a small rectangle transverse to $W^u(z^*)$ and located near z^* as shown in Figure 36.4(a).

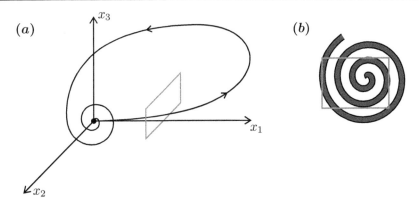

Figure 36.4. *Sketch of the situation arising for a Shilnikov homoclinic orbit. (a) Phase space of (36.10) with a homoclinic orbit and a cross-section. (b) Possible return map to the cross-section under suitable conditions, i.e., the spiral is the image of the original rectangle after it has been flown once around near the homoclinic.*

Note that this rectangle could be stretched and folded back to the same initial rectangle under the flow as shown in Figure 36.4(b). It is a highly nontrivial theorem to show that if the **saddle quantity**

$$\zeta + a$$

is positive, then the system has infinitely many periodic orbits near the homoclinic orbit. In fact, it is topologically conjugate to chaotic **Smale horseshoe** dynamics; see Exercise 36.7 and references for further details. This shows that PDE dynamics may inherit the complications arising from ODEs in a hidden form, e.g., it may not be apparent that (36.9) may generate infinitely many different periodic wave trains.

Exercise 36.5. (a) Prove Lemma 36.2. Hint: Define a new variable of the form

$$\alpha_1 \frac{\mathrm{d}^2 u^*}{\mathrm{d}x^2} + \alpha_2 \left(\left(\frac{\mathrm{d}u^*}{\mathrm{d}x} \right)^3 - \frac{\mathrm{d}^2 u^*}{\mathrm{d}x} \right),$$

where you have to determine the correct constants $\alpha_{1,2} \in \mathbb{R}$. (b) Find the boundary conditions for (36.4). ◊

Exercise 36.6. (a) Rewrite (36.9) in the travelling wave frame (cf. Chapter 7) and thereby verify that resulting ODEs are a $(2,1)$-fast-slow system. (b) Prove that there is a unique global steady state for this ODE system. (c) Find the critical manifold $\mathcal{C}_0 = \{f = 0\}$ and compute the linearization $\mathrm{D}_x f(p)$ for $p \in \mathcal{C}_0$. Prove that there are two branches of saddle type. (d) Try to sketch possible singular orbits combining the fast and slow subsystems. ◊

Exercise 36.7. This exercise discusses the **Bernoulli shift**, which is related to the return map dynamics generated by the homoclinic Smale horseshoe setup considered above. Consider the set Σ^2 of all bi-infinite sequences of two symbols

$$\omega = \cdots \omega_{-k} \cdots \omega_{-1} \omega_0 \omega_1 \cdots \omega_k \cdots, \qquad \omega_j \in \{0, 1\}.$$

The map $\beta(\omega) = \omega'$ defined via $\omega'_k = \omega_{k+1}$ for all $k \in \mathbb{Z}$ is called the Bernoulli shift and we view it as an iterated map. (a) Prove that $\|\omega\| = \sum_{k \in \mathbb{Z}} |\omega_k|/2^{|k|}$ is a norm on Σ^2 (and hence induces a metric). (b) Prove that β has an infinite number of periodic orbits, an infinite number of homoclinic orbits, and an infinite number of heteroclinic orbits. (c) Can you give a geometric sketch of why the return map indicated in Figure 36.4 could be topologically conjugate to the Bernoulli shift? ◊

Background and Further Reading: The first part of this chapter follows [292], which is motivated by the study of the energy functional in [418]. The derivation of the Euler-Lagrange equations is quite standard and can be found in many books on the calculus of variations [141, 186, 225, 309, 560]. The FitzHugh-Nagumo equation [202, 423] has been among the most prominent PDEs to motivate the development of fast-slow systems techniques [84, 185, 260, 302, 301, 304, 346, 470, 475, 520] and indeed supports chaotic waves [260]. Chaotic dynamics, homoclinic orbits, and Smale horseshoes [505, 504] are standard topics for advanced courses focusing solely on finite-dimensional dynamical systems; see, e.g., [15, 73, 247, 287, 320, 351, 563] for more details.

Appendix A

Finite Differences and Simulation

This appendix and the next cover some *very basic* numerical methods for PDEs and associated dynamics problems. Although the theme and focus of this book are analytical methods, it is often very helpful to be familiar with basic numerical schemes just to (I) visualize a result and/or (II) determine a new conjecture. Therefore, this appendix outlines some numerical methods which are already sufficient in many cases to accomplish (I)–(II). It is tremendously useful to be familiar with this background *before* using pre-packaged scientific computing software or programming suitable extensions.

As the first main motivating example in this section, suppose we want to simulate a one-dimensional reaction-diffusion equation

$$\partial_t u = \partial_x^2 u + f(u), \qquad u = u(x,t), \ (x,t) \in [0,1] \times [0,T], \qquad (A.1)$$

with an initial condition $u(x,0) = u_0(x)$, for a final time $T > 0$, and assuming homogeneous zero Neumann boundary conditions

$$\partial_x u(0,t) = 0 = \partial_x u(1,t) \qquad \forall t \in [0,T]. \qquad (A.2)$$

To apply a **finite-difference** numerical method we introduce a **mesh** of $H + 1$ points in the spatial domain ("**method of lines**")

$$x_i = ih, \qquad h = 1/H, \ i \in \{0, 1, 2, \ldots, H\}.$$

To approximate the second partial derivative in (A.1), a natural approximation is

$$\partial_x^2 u(x_i, t) \approx \frac{u(x_{i+1}, t) - 2u(x_i, t) + u(x_{i-1}, t)}{h^2} \qquad (A.3)$$

for an interior point $x_i \neq 0, 1$, and one easily checks that (A.3) does converge to the correct second derivative as $h \to 0$. For the boundary points, suppose we have "ghost points" x_{-1} and x_{H+1}, and let us focus on x_{-1}. One may approximate the derivative at $x_0 = 0$ by a **centered difference**

$$0 = \partial_x u(x_0, t) \approx \frac{u(x_{-1}, t) - u(x_1, t)}{2h}$$

and then use this equation to eliminate the ghost point in the formula (A.3) for the boundary point x_0, i.e., we have

$$\partial_x^2 u(x_0, t) \approx \frac{2u(x_1, t) - 2u(x_0, t)}{h^2} \qquad (A.4)$$

and similarly for the other boundary point x_H. In addition, we may simply evaluate $f(u)$ at (x_i, t) so we end up with the **semidiscrete** problem

$$\partial_t \begin{pmatrix} u_0 \\ u_1 \\ u_2 \\ \vdots \\ u_{H-1} \\ u_H \end{pmatrix} = \frac{1}{h^2} \underbrace{\begin{pmatrix} -2 & 2 & & & & \\ 1 & -2 & 1 & & & \\ & 1 & -2 & 1 & & \\ & & \ddots & \ddots & \ddots & \\ & & & 1 & -2 & 1 \\ & & & & 2 & -2 \end{pmatrix}}_{=:A_h} \begin{pmatrix} u_0 \\ u_1 \\ u_2 \\ \vdots \\ u_{H-1} \\ u_H \end{pmatrix} + \underbrace{\begin{pmatrix} f(u_0) \\ f(u_1) \\ f(u_2) \\ \vdots \\ f(u_{H-1}) \\ f(u_H) \end{pmatrix}}_{=:f_h(u.)},$$

where $u_i = u(x_i, t)$. We have obtained a set of ODEs which actually provides *another natural link* between PDEs and dynamical systems methods developed in the context of ODEs. The simplest time discretization is to use **Euler's method** and consider a time grid

$$t_j = jk, \qquad k = \frac{T}{K}, \; j \in \{0, 1, 2, \ldots, K\}.$$

Set $u_{i,j} := u(x_i, t_j)$; then the standard **forward difference** approximation in time gives

$$\partial_t u(x_i, t_j) \approx \frac{u(x_i, t_{j+1}) - u(x_i, t_j)}{k} =: \frac{u_{i,j+1} - u_{i,j}}{k}. \qquad (A.5)$$

In summary, we now have a *discrete-time map* on the space $\mathbb{R}^{H+1}$ given by

$$u_{.,j+1} = u_{.,j} + k\left(A_h u_{.,j} + f_h(u_{.,j})\right), \qquad (A.6)$$

where $u_{.,j}$ is simply the column vector containing all the values of u at the spatial mesh points at time t_j. Given a discretized version of u_0, i.e., $u_0(x_i)$ for $i \in \{0, 1, \ldots, H\}$, one can easily iterate (A.6) and thereby simulate an approximation of the original PDE.

Remark: Of course, more efficient implementations of finite-difference schemes exist. For example, one natural improvement would be to use an ODE algorithm, which is a lot more stable than Euler's method. Euler's method is known to suffer from very stringent time step restrictions for many ODEs/PDEs; we refer to the background references and Exercise A.3 for more details.

Instead of going into the error analysis of the numerical method, which is covered in full detail in many references mentioned in the background section below, we shall try the method on some examples.

Example A.1. Consider the case of a classical FKPP polynomial nonlinearity $f(u) = u(1 - u)$ (see Chapter 8), i.e., we consider

$$\partial_t u = \partial_x^2 u + u(1 - u), \qquad (x, t) \in [0, 1] \times [0, T], \qquad (A.7)$$

for the boundary conditions (A.2) and a simple initial condition

$$u_0(x) = \sin(2\pi x) + 0.8. \qquad (A.8)$$

Then it is natural to investigate what happens in the long-time dynamics just by numerical simulation of trajectories.

The results in Figure A.1 lead us to the conjecture that our chosen u_0 is within the basin of attraction of the constant $u \equiv 1$. In fact, numerical methods are frequently the only option to really describe the entire basin of attraction of locally stable patterns. ◆

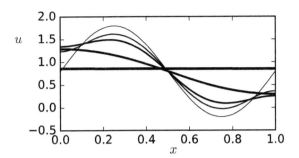

Figure A.1. *Time snapshots of the solution for (A.7) using the finite difference scheme (A.6) with initial condition (A.8). Thicker lines indicate later times, i.e., there are five times $t \in \{0, 0.005, 0.01, 0.05, 0.5\}$ and the final time is at $T = 0.5$ corresponding to the thickest, which is almost constant and seems to tend towards $u \equiv 1$. Algorithmic parameters are $H = 50$ and $K = 3 \cdot 10^4$, which satisfies the—very stringent and known—numerical stability criterion $k/h^2 \approx 0.04 \leq \frac{1}{2}$ for the forward Euler scheme we derived.*

Example A.2. Consider the case of a classical Nagumo polynomial nonlinearity $f(u) = u(1 - u)(u - p)$ for $p \in (0, \frac{1}{2})$ as discussed in many parts of this book (see, e.g., Chapter 7), i.e., consider

$$\partial_t u = \delta \partial_x^2 u + u(1 - u)(u - p), \qquad (x, t) \in [0, 10] \times [0, T], \qquad (A.9)$$

for $\delta = 0.001$. The same numerical finite-difference scheme can be used with relatively minor modifications. Note that the small diffusion coefficient is expected to make the interface of the (locally) stable front very thin, thereby avoiding boundary effects; cf. Chapters 20–21.

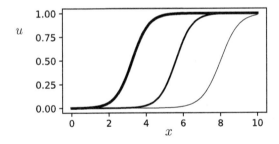

Figure A.2. *Time snapshots of the solution for (A.9) using the finite difference scheme (A.6) with initial condition (A.10). Thicker lines indicate later times, i.e., there are three times $t \in \{0, 15, 30\}$ and the final time is at $T = 30$ corresponding to the thickest curve. We clearly observe a left-propagating wave. Algorithmic parameters are $H = 200$ and $K = 10^5$.*

As an initial condition, we select

$$u_0(x) = \frac{1}{2} \left(\tanh(x - 8) + 1 \right). \qquad (A.10)$$

Figure A.2 shows that we indeed obtain a propagating wave front for the bistable Nagumo equation; see also Chapter 9. ◆

As another important class of problems deeply linking numerical methods to the analysis we have considered, let us consider scalar hyperbolic conservation laws (cf. Chapter 11)

$$\partial_t u + \partial_x(f(u)) = 0, \qquad u(x,0) = u_0(x),\ x \in [-L_1, L_2],\ t \in [0,T], \qquad \text{(A.11)}$$

and we select suitable boundary conditions as well as suitable $L_1, L_2 > 0$. As before, we consider a spatio-temporal mesh and set

$$u_{i,j} = u(x_i, t_j), \qquad x_i = ih,\ t_j = jk.$$

We have to discretize the derivatives with the added *dynamical* constraint to respect the propagation of information along characteristics. Indeed, suppose we consider Burgers' equation using $f(u) = \frac{1}{2}u^2$ and the classical initial condition

$$u_0(x) = \begin{cases} 1 & \text{if } x \le 0, \\ 1-x & \text{if } x \in (0,1), \\ 0 & \text{if } x \ge 1 \end{cases} \qquad \text{(A.12)}$$

already used in Example 11.4. Then characteristics travel from left to right, or are vertical. Hence, it is not advisable to approximate $u_{i,j+1}$ using $u_{i+1,j}$. One natural choice is to consider an **upwind scheme**

$$0 = \partial_t u + \partial_x(f(u)) \approx \frac{u_{i,j+1} - u_{i,j}}{k} + \frac{f(u_{i,j}) - f(u_{i-1,j})}{h}$$
$$\underset{\text{Burgers}}{=} \frac{u_{i,j+1} - u_{i,j}}{k} + \frac{u_{i,j}^2 - u_{i-1,j}^2}{2h}, \qquad \text{(A.13)}$$

under the assumption that the solution travels to the right.

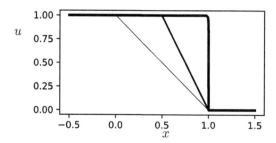

Figure A.3. *Time snapshots of the solution for Burgers' equation using the upwind scheme (A.13) with initial condition (A.12) on the spatial domain $[-0.5, 1.5]$. Thicker lines indicate later times, i.e., there are three times $t \in \{0, 0.5, 1.0\}$ and the final time is at $T = 1$ corresponding to the thickest. We clearly observe the formation of a shock. Algorithmic parameters are $H = 5 \cdot 10^3$ and $K = 1.6 \cdot 10^4$.*

Figure A.3 shows the numerical solution of Burgers' equation with initial condition (A.12). The numerical scheme represents the shock formation time at $t = 1$ and the shock shape very well. However, one should be careful since using poorly chosen numerical parameters may lead to numerical instability of the scheme; see also Exercise A.5 and the background references.

Exercise A.3. Consider the spatial semidiscretization of the reaction-diffusion problem (A.1) using (A.3). Instead of using the forward Euler scheme, use the backward Euler scheme, which uses

$$\partial_t u(x_i, t_j) \approx \frac{u(x_i, t_j) - u(x_i, t_{j-1})}{k} =: \frac{u_{i,j} - u_{i,j-1}}{k}$$

instead of (A.5). This yields and **implicit** method and one has to solve an algebraic equation for $u(\cdot, t_j)$ at each time step. Implement this method and compare its results to the forward Euler results on the Example A.1 for different time step sizes and a fixed spatial mesh. ◊

Exercise A.4. Extend the finite-difference method to two dimensions using an approximation of the Laplacian at a lattice point using this point and its four direct neighbors. Then use this method to simulate the FitzHugh-Nagumo equation (see also Chapter 36)

$$\begin{cases} \partial_t u = \Delta u + u(1-u)(u-p) - v, \\ \partial_t v = \varepsilon(u-v) \end{cases} \tag{A.14}$$

on a rectangle $\Omega = [0, L_1] \times [0, L_2]$ and starting from a random initial condition. You can select different parameters and aim to classify different long-term dynamical regimes. ◊

Exercise A.5. Consider the upwind scheme (A.13). Rewrite it as a time-discrete dynamical system. Prove that this dynamical system can produce growing oscillations if the time step is taken sufficiently large (or the spatial step is made sufficiently small). Hint: This question is related to the famous **Courant-Friedrichs-Lewy (CFL) condition**, which states that we obtain numerical stability if we have

$$\left| s\frac{k}{h} \right| \leq 1,$$

where s is the (characteristic) propagation speed. ◊

Background and Further Reading: The part on finite-difference methods as well as upwind schemes can be found in almost any textbook on the topics. A basic source is [355]. For more on finite-difference methods we refer to [208, 368, 404, 506, 531] and for entry points to the literature on numerical methods for conservation laws to [238, 271, 367, 498, 532]. Many software packages already offer a multitude of time-simulation algorithms for PDEs, but even for these tested methods it is often helpful to run some basic examples first, where the dynamical properties of the solution are well understood. For example, it is well known that the properties of a spatially discretized PDE may not correspond to the continuous PDE as in the case of propagation failure for waves in the Nagumo equation [321].

Appendix B

Finite Elements and Continuation

In addition to forward-time simulation discussed in Appendix A, another basic tool to understand PDE dynamics is to investigate steady states. As a motivating example, let us consider

$$0 = \Delta u + f(u; p), \qquad u = u(x), \ x \in \Omega = [-L_1, L_1] \times [-L_2, L_2] \subset \mathbb{R}^2, \quad \text{(B.1)}$$

with Dirichlet boundary conditions $u(x) = 0$ on $\partial\Omega$ and a given nonlinearity f with parameter $p \in \mathbb{R}$; see also Chapters 4–6. One natural approach to discretize (B.1) is to use finite differences as in Appendix A. However, we take the opportunity here to introduce **finite element methods (FEM)** as they are currently probably the standard approach to large classes of PDE problems. Let us write (B.1) in its weak formulation

$$\int_\Omega \nabla u \cdot \nabla v \, \mathrm{d}x - \int_{\partial\Omega} v \, \vec{n} \cdot \nabla u \, \mathrm{d}x = \int_\Omega v f(u; p) \, \mathrm{d}x, \qquad \text{(B.2)}$$

where v is a test function to be chosen. Consider a triangulation

$$\mathcal{T} := \{T_j\}_{j=1}^J, \qquad \Omega = \bigcup_{j=1}^J T_j$$

of Ω, i.e., each T_j is a triangle; frequently one assumes that the structure of the triangles is "nice," more precisely **conforming** and **shape regular**. A triangulation is conforming if intersections between triangles are either empty or consist of precisely one edge. The triangulation is shape regular if all the angles in the triangles are uniformly bounded from below; see also Figure B.1(a).

Furthermore, we introduce the (interior) nodes/vertices of the triangles

$$\{x_i : x_i \text{ is a node of some } T_j, \ x_i \notin \partial\Omega\} = \{x_i\}_{i=1}^N. \qquad \text{(B.3)}$$

To approximate (B.2) it is natural to consider test functions in

$$H_0^1(\Omega) = \{v \in L^2(\Omega) : \nabla v \in L^2(\Omega), v|_{\partial\Omega} = 0\}$$

so that the boundary term vanishes automatically. To obtain a numerical problem, the finite element method is based upon selecting a suitable subspace within the space of test functions, say

$$V \subset H_0^1, \qquad V := \{v \in C(\Omega) : v|_{T_j} \text{ is affine linear for all } T_j \in \mathcal{T}\},$$

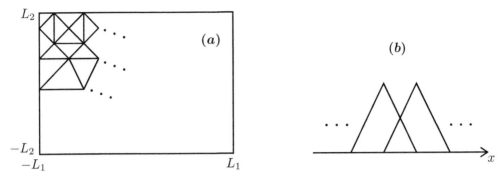

Figure B.1. *(a) Mesh of triangles to set up a finite element solution of a PDE in 2D. (b) Hat functions as basis elements for 1D.*

which is the space of **hat functions**; see Figure B.1(b). In particular, V just consists of piecewise-linear functions, but there are many other choices for possible subspaces V. We are left with the problem

$$\int_\Omega \nabla u \cdot \nabla v \, \mathrm{d}x = \int_\Omega v f(u; p) \, \mathrm{d}x. \tag{B.4}$$

To compute the solution of (B.4), let $\{\phi_k\}_{k=1}^N \subset V$ be the **nodal basis** defined by $\phi_k(x_i) = \delta_{ki}$, where δ_{ik} is the usual **Kronecker delta** ($\delta_{kk} = 1$ and $\delta_{ki} = 0$ if $k \neq i$). Using the basis, we make the ansatz

$$u(x) \approx \sum_{i=1}^N U_i \phi_i(x), \qquad U := (U_1, \ldots, U_N)^\top \tag{B.5}$$

and insert it into (B.4). Using as test functions $v = \phi_k$, we get

$$(AU)_k := \sum_{i=1}^N \left(\int_\Omega \nabla \phi_i \cdot \nabla \phi_k \, \mathrm{d}x \right) U_i = \int_\Omega \phi_k f \left(\sum_{i=1}^N U_i \phi_i; p \right) \mathrm{d}x =: \tilde{f}_k(U; p)$$

for each $k \in \{1, 2, \ldots, N\}$, where $A \in \mathbb{R}^{N \times N}$ is a matrix. The resulting system is just a nonlinear system of algebraic equations

$$AU - \tilde{f}(U; p) =: F(U; p) = 0, \qquad F : \mathbb{R}^N \times \mathbb{R} \to \mathbb{R}^N.$$

Remark: Of course, **assembling** the matrix A and evaluating the nonlinear map $\tilde{f}$ involves many practical tasks: generating a mesh/triangulation, labeling algorithms for the nodes, computing integrals, and so on. Here we shall focus on the mathematical basics, but see the background references for more details.

There is no completely simple "catch-all" strategy to compute all the solutions of a general nonlinear root-finding problem $F(U; p) = 0$. However, there is a wide variety of methods available. Among the most classical is **Newton's method**, which is an iterative method starting from an initial guess $U^{(0)}$:

$$U^{(k+1)} = U^{(k)} - \left(\mathrm{D}_u F(U^{(k)}; p) \right)^{-1} F(U^{(k)}; p).$$

A key ingredient for the success of Newton's method is to have a starting point $U^{(0)}$ which is sufficiently close to the desired solution. One option to obtain these starting points efficiently is to view the problem *parametrically*. If we also vary $p \in \mathbb{R}$, the goal is to compute an entire **branch** (or **curve**) of solutions

$$(U, p) = (U(s), p(s)) =: \gamma(s) \in \mathbb{R}^N, \qquad s \in \mathbb{R},$$

where s parametrizes the curve; see Figure B.2. Once a single solution $U = U(s_0)$ at the parameter $p = p(s_0)$ is known, it is attractive to consider **numerical continuation** via a **predictor-corrector method**, i.e., we look for a solution at $s_1 = s_0 + (\delta s)$ for some (sufficiently) small (δs) using a **predictor**. For example, the classical **tangent predictor** is

$$U(s_1) \approx U(s_0) + (\delta s)\dot{\gamma}(s_0), \qquad \dot{\gamma}(s_0) = \left.\frac{\mathrm{d}\gamma}{\mathrm{d}s}\right|_{s=s_0},$$

where $\dot{\gamma}(s_0)$ is the tangent vector to the curve γ; see Figure B.2.

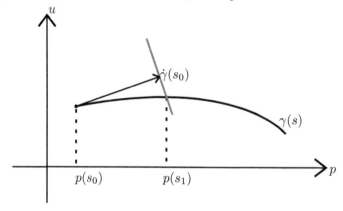

Figure B.2. *Predictor-corrector algorithm finding a one-dimensional curve γ of solutions to $F(U; p) = 0$, where p is a scalar parameter.*

Note that $\dot{\gamma}(s_0)$ is simply computable by solving a linear system since

$$0 = \left.\frac{\mathrm{d}}{\mathrm{d}s}F(u(s); p(s))\right|_{s=s_0} = \mathrm{D}_u F(\gamma(s_0))\dot{\gamma}(s_0).$$

It turns out to be beneficial in many situations to constrain the type of parametrization of the curve $\gamma(s)$. A typical choice is **arclength parametrization**, enforced by introducing another algebraic equation, defined locally near s_0 by setting

$$\mu(U; p, s) := \xi\langle\dot{U}(s_0), U(s) - U(s_0)\rangle + (1 - \xi)(p(s) - p(s_0))\dot{p}(s_0) - (s - s_0),$$

where $\xi \in (0, 1)$ is a weight factor to tune the continuation and $\langle\cdot, \cdot\rangle$ is the usual inner product, and then requiring

$$\mu(U; p, s) = 0; \tag{B.6}$$

see also Figure B.2. The interpretation of the formula (B.6) is that it defines a hyperplane perpendicular to $\dot{\gamma}(s_0)$ in the norm

$$\|\cdot\|_\xi^2 := \sqrt{\langle\cdot, \cdot\rangle_\xi}, \qquad \left\langle\begin{pmatrix}u \\ p\end{pmatrix}, \begin{pmatrix}\tilde{u} \\ \tilde{p}\end{pmatrix}\right\rangle_\xi := \xi\langle u, \tilde{u}\rangle + (1 - \xi)p\tilde{p}$$

and located at a distance $(s - s_0)$ from the known point $(U(s_0), p(s_0))$. Now one may consider the extended system

$$G(U; p) = \begin{pmatrix} F(U; p) \\ \mu(U; p, s) \end{pmatrix} = 0 \qquad (B.7)$$

and apply the same predictor-corrector idea as above to (B.7), i.e., compute the tangent vector, construct a predictor, and then use Newton's method. This numerical continuation approach computes the steady-state solution branches of (B.1) with high numerical efficiency. A second advantage is that it naturally lends itself to **numerical bifurcation detection**. Indeed, we may use FEM to discretize the spectral problem. The idea is then to use a **test function** and investigate its zeros

$$\Psi : \mathbb{R}^N \times \mathbb{R} \to \mathbb{R}, \qquad \Psi(U; p) \overset{!}{=} 0$$

to detect bifurcations. For example, if a single eigenvalue changes sign (see Chapters 4–6), then a natural test function to consider would be the determinant of the linearization around the steady state. However, finding good test functions to detect all possible bifurcation points reliably is essentially a branch of numerical mathematics in itself, so we shall not discuss it here any further; see the references for more details.

Example B.1. We consider the nonlinear elliptic PDE (B.1) for a classical example, the **cubic-quintic Allen-Cahn equation**

$$0 = \Delta u + 4(pu + u^3 - u^5), \qquad x \in \Omega = [-L_1, L_1] \times [-L_2, L_2] \qquad (B.8)$$

for $L_1 = 0.9$, $L_2 = 1.0$, and zero Dirichlet boundary conditions. Figure B.3 shows the results of combining FEM, with numerical continuation, bifurcation detection, and switching to different branches at each bifurcation point; see also Exercise B.4 for ideas regarding branch switching. For our example, we have computed the first three bifurcation points along the trivial branch $u \equiv 0$. All three bifurcations can be analytically treated using the Crandall-Rabinowitz Theorem 5.1. However, continuing bifurcation branches globally away from local bifurcation points is in many cases only possible numerically. Each of the three global branches is found to show a simple fold bifurcation. ◆

Exercise B.2. Consider the one-dimensional problem boundary value problem (BVP)

$$u'' := \frac{\mathrm{d}^2 u}{\mathrm{d} x^2} = g, \qquad u(0) = 0 = u(1) \qquad (B.9)$$

for some smooth function $g : [0, 1] \to \mathbb{R}$. (a) Write out the linear system arising from a FEM discretization using hat functions on an equally spaced mesh with spacing h. (b) Let u_h denote the FEM solution. Prove that

$$\|u' - u_h'\|_{L^2(0,1)} \lesssim h\|u\|_{H^2(0,1)}.$$

(c) As a bit more advanced exercise, try to prove that we have for the error

$$\|u - u_h\|_{L^2(0,1)} \lesssim h^2 \|u\|_{H^2(0,1)} \qquad (B.10)$$

using the previous result from (b). ◊

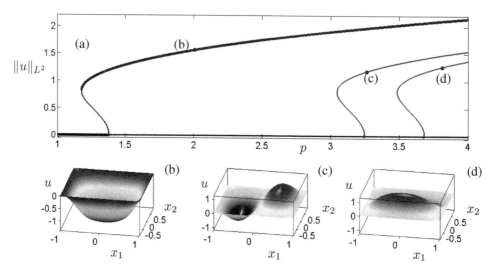

Figure B.3. *(a) Bifurcation diagram of the PDE* (B.8) *obtained via numerical continuation; the vertical axis shows the $L^2(\Omega)$-norm of the solution. The thick curves are locally stable, i.e., for the linearized problem the spectrum is contained in the left half of the complex plane. There are three simple branch/bifurcation points marked with circles. There are also three fold bifurcations on the nontrivial branches ($u \not\equiv 0$). The dots on the nontrivial branches indicate to which points the solutions in diagrams (b)–(d) are associated. (b)–(d) Solution visualized as level sets for parameter values $p = 2.0075, 3.2579, 3.8093$ on the upper parts of the nontrivial branches. Figure adapted from [347] using [541].*

Exercise B.3. (a) Implement the numerical method you have derived in Exercise B.2. Then design numerical tests to compare your implementation to the theoretical error bound (B.10). More advanced exercise: (b) Modify your method to compute solutions to the two-point BVP

$$-s\frac{d^2u}{dx^2} = \frac{du}{dx} + u(1 - u)(u - p) \tag{B.11}$$

for some $p \in (0, 1/2)$ and boundary conditions

$$\lim_{x \to -\infty} u(x) = 0, \qquad \lim_{x \to +\infty} u(x) = 1, \tag{B.12}$$

and determine s for which a solution exists. Hint: The BVP (B.11)–(B.12) arises from a travelling wave problem as discussed in Chapter 7, where also Lin's method has played a role. This method can be implemented using projection boundary conditions onto suitable stable and unstable eigenspaces near $(u, u') = (0, 0)$ and near $(u, u') = (1, 0)$, in combination with a root-finding algorithm (e.g., consider Newton's method). ◇

Exercise B.4. (a) Consider the ODE normal form of the transcritical bifurcation

$$\frac{du}{dt} = u(u - p), \qquad p \in \mathbb{R}, \ u = u(t), \tag{B.13}$$

where we already know everything about the bifurcation diagram as discussed in Example 5.5. Use numerical continuation to compute the trivial zero branch $(u, p) = (0, p)$ and implement an algorithm to branch switch at the bifurcation point $(0, 0)$ onto the nontrivial branch. (b) Can you generalize your algorithm *in theory* using FEM to the PDE case

$$\partial_t u = \Delta u + u(u - p), \qquad p \in \mathbb{R}, \ u = u(x, t), \tag{B.14}$$

on some smooth domain with zero Neumann boundary conditions? Hint: Use the Crandall-Rabinowitz Theorem 5.1 and Corollary 5.2. $\Diamond$

Background and Further Reading: There are many excellent texts developing the theoretical justification via error estimates and/or the practical implementation of FEM. Here we can just provide a few references [67, 117, 183, 284, 298, 512, 533]. Numerical continuation [14, 140, 243, 324, 499] is a standard tool in dynamical systems; see also [154, 157, 344, 351] for ODE-oriented directions and [41, 541] for PDE-focused approaches.

Remark: For this appendix, as well as the previous one, there are many software packages available providing sources codes and/or full software environments to numerically solve PDEs using FEM, finite differences, as well as many other methods. Furthermore, also for continuation, many codes exist. Just searching online directly for the relevant keywords is going to produce potential software tools. However, it frequently happens that the discovery of new PDE dynamics necessitates rethinking and/or extending current packages. Therefore, tools keep changing, and so we refrain from attempting to provide a list of PDE discretization software tools.

Bibliography

[1] J. Aaronson. *An Introduction to Infinite Ergodic Theory*. AMS, 1997. (Cited on p. 181)

[2] H. Abels, H. Garcke, and G. Grün. Thermodynamically consistent, frame indifferent diffuse interface models for incompressible two-phase flows with different densities. *Math. Mod. Meth. Appl. Sci.*, 22(3):1150013, 2012. (Cited on p. 120)

[3] M.J. Ablowitz and P.A. Clarkson. *Solitons, Nonlinear Evolution Equations and Inverse Scattering*. Cambridge University Press, 1991. (Cited on p. 37)

[4] M.J. Ablowitz and H. Segur. *Solitons and the Inverse Scattering Transform*. SIAM, 1981. (Cited on p. 37)

[5] R.H. Abraham and J.E. Marsden. *Foundations of Mechanics*. Benjamin-Cummings, 1978. (Cited on pp. 147, 148)

[6] R.A. Adams and J.J.F. Fournier. *Sobolev Spaces*. Elsevier, 2003. (Cited on p. 81)

[7] J.C. Alexander, R.A. Gardner, and C.K.R.T. Jones. A topological invariant arising in the stability analysis of travelling waves. *J. Reine Angew. Math.*, 410:167–212, 1990. (Cited on p. 55)

[8] J.C. Alexander, J.A. Yorke, Z. You, and I. Kan. Riddled basins. *Int. J. Bif. Chaos*, 2(4):795–813, 1992. (Cited on p. 99)

[9] N.D. Alikakos, P.W. Bates, and X. Chen. Convergence of the Cahn-Hilliard equation to the Hele-Shaw model. *Arch. Rat. Mech. Anal.*, 128(2):165–205, 1994. (Cited on p. 111)

[10] N.D. Alikakos, P.W. Bates, and G. Fusco. Slow motion for the Cahn-Hilliard equation in one space dimension. *J. Differential Equations*, 90(1):81–135, 1991. (Cited on pp. 111, 120)

[11] S. Alinhac. *Blowup for Nonlinear Hyperbolic Equations*. Springer, 2013. (Cited on p. 141)

[12] G. Allaire. Homogenization and two-scale convergence. *SIAM J. Math. Anal.*, 23(6):1482–1518, 1992. (Cited on pp. 182, 185)

[13] S.M. Allen and J.W. Cahn. A microscopic theory for antiphase boundary motion and its application to antiphase domain coarsening. *Acta Metallurgica*, 27(6):1085–1905, 1979. (Cited on p. 110)

[14] E.L. Allgower and K. Georg. *Introduction to Numerical Continuation Methods*. SIAM, 2003. (Cited on p. 209)

[15] K.T. Alligood, T.D. Sauer, and J.A. Yorke. *Chaos: An Introduction to Dynamical Systems*. Springer, 1996. (Cited on pp. 9, 198)

[16] A. Ambrosetti and P.H. Rabinowitz. Dual variational methods in critical point theory and applications. *J. Funct. Anal.*, 14(4):349–381, 1973. (Cited on pp. 136, 141)

[17] L. Ambrosio, N. Gigli, and G. Savaré. *Gradient Flows: In Metric Spaces and in the Space of Probability Measures.* Birkhäuser, 2006. (Cited on pp. 122, 125)

[18] S.N. Antontsev and S.I. Shmarev. A model porous medium equation with variable exponent of nonlinearity: Existence, uniqueness and localization properties of solutions. *Nonlinear Anal. Theory Methods Appl.*, 60(3):515–545, 2005. (Cited on p. 169)

[19] I.S. Aranson and L. Kramer. The world of the complex Ginzburg-Landau equation. *Rev. Mod. Phys.*, 74:99–143, 2002. (Cited on p. 67)

[20] D. Armbruster, J. Guckenheimer, and P. Holmes. Kuramoto-Sivashinsky dynamics on the center-unstable manifold. *SIAM J. Appl. Math.*, 49(3):676–691, 1989. (Cited on p. 94)

[21] A. Arnold, P. Markowich, G. Toscani, and A. Unterreiter. On convex Sobolev inequalities and the rate of convergence to equilibrium for Fokker-Planck type equations. *Comm. Partial Differential Equations*, 26(1):43–100, 2001. (Cited on pp. 125, 127, 129, 130, 159)

[22] L. Arnold. *Stochastic Differential Equations: Theory and Applications.* Wiley, 1974. (Cited on p. 130)

[23] L. Arnold. *Random Dynamical Systems.* Springer, 2003. (Cited on p. 105)

[24] V.I. Arnold. *Ordinary Differential Equations.* MIT Press, 1973. (Cited on pp. vii, 9, 176)

[25] V.I. Arnold. *Geometrical Methods in the Theory of Ordinary Differential Equations.* Springer, 1983. (Cited on p. 9)

[26] V.I. Arnold. *Mathematical Methods of Classical Mechanics.* Springer, 1997. (Cited on p. 148)

[27] V.I. Arnold, V.V. Kozlov, and A.I. Neishstadt. *Mathematical Aspects of Classical and Celestial Mechanics.* Springer, 3rd edition, 2006. (Cited on p. 148)

[28] D.G. Aronson. The porous medium equation. In *Nonlinear Diffusion Problems*, pages 1–46. Springer, 1986. (Cited on p. 169)

[29] D.G. Aronson and L.A. Peletier. Large time behaviour of solutions of the porous medium equation in bounded domains. *J. Differential Equations*, 39(3):378–412, 1981. (Cited on p. 169)

[30] D.G. Aronson and H.F. Weinberger. Nonlinear diffusion in population genetics, combustion, and nerve pulse propagation. In *Partial Differential Equations and Related Topics*, volume 446 of *Lecture Notes in Mathematics*, pages 5–49. Springer, 1974. (Cited on pp. 37, 47)

[31] D.G. Aronson and H.F. Weinberger. Multidimensional nonlinear diffusion arising in population genetics. *Adv. Math.*, 30(1):33–76, 1978. (Cited on p. 43)

[32] U.M. Ascher, R.M.M. Mattheij, and R.D. Russell. *Numerical Solution of Boundary Value Problems for Ordinary Differential Equations.* SIAM, 1987. (Cited on pp. 111, 189)

[33] H. Attouch, G. Buttazzo, and G. Michaille. *Variational Analysis in Sobolev and BV Spaces: Applications to PDEs and Optimization.* SIAM, 2014. (Cited on p. 185)

[34] A.V. Babin and M.I. Vishik. Attractors of partial differential evolution equations in an unbounded domain. *Proc. R. Soc. Edinburgh A*, 116(3):221–243, 1990. (Cited on p. 99)

[35] A.V. Babin and M.I. Vishik. *Attractors of Evolution Equations*. North-Holland, 1992. (Cited on pp. 83, 89, 99)

[36] N.S. Bakhvalov and G.P. Panasenko. *Homogenisation: Averaging Processes in Periodic Media: Mathematical Problems in the Mechanics of Composite Materials*. Kluwer, 1989. (Cited on p. 185)

[37] D. Bakry and M. Émery. Diffusions hypercontractives. In *Séminaire de Probabilités XIX 1983/84*, pages 177–206. Springer, 1985. (Cited on p. 130)

[38] D. Bakry, I. Gentil, and M. Ledoux. *Analysis and Geometry of Markov Diffusion Operators*. Springer, 2014. (Cited on p. 159)

[39] J.M. Ball. A version of the fundamental theorem for young measures. In M. Rascle, D. Serre, and M. Slemrod, editors, *PDEs and Continuum Models of Phase Transitions*, pages 207–215. Springer, 1989. (Cited on p. 185)

[40] J.M. Ball. Global attractors for damped semilinear wave equations. *Discrete Contin. Dyn. Syst.*, 10(1):31–52, 2004. (Cited on p. 99)

[41] R.E. Bank. *PLTMG: A Software Package for Solving Elliptic Partial Differential Equations. Users' Guide 10.0*. Department of Mathematics, University of California at San Diego, 2007. (Cited on p. 209)

[42] V. Barbu. *Nonlinear Semigroups and Differential Equations in Banach Spaces*. Noordhoff International, 1976. (Cited on p. 78)

[43] C. Bardos, F. Golse, and D. Levermore. Fluid dynamic limits of kinetic equations. I. Formal derivations. *J. Stat. Phys.*, 63(1):323–344, 1991. (Cited on p. 153)

[44] G.I. Barenblatt. *Scaling, Self-similarity, and Intermediate Asymptotics*. Cambridge University Press, 1996. (Cited on p. 169)

[45] D. Barkley. A model for fast computer simulation of waves in excitable media. *Physica D*, 49:61–70, 1991. (Cited on p. 48)

[46] A. Barone, F. Esposito, C.J. Magee, and A.C. Scott. Theory and applications of the sine-Gordon equation. *Riv. Nuovo Cimento (1971–1977)*, 1(2):227–267, 1971. (Cited on p. 72)

[47] P.W. Bates and C.K.R.T. Jones. Invariant manifolds for semilinear partial differential equations. In U. Kirchgraber and H.O. Walther, editors, *Dynamics Reported*, volume 2, pages 1–37. Wiley, 1989. (Cited on p. 89)

[48] P.W. Bates, K. Lu, and C. Zeng. Existence and persistence of invariant manifolds for semiflows in Banach spaces. *Mem. Amer. Math. Soc.*, 135:645, 1998. (Cited on p. 89)

[49] P.W. Bates, K. Lu, and C. Zeng. Approximately invariant manifolds and global dynamics of spike states. *Invent. Math.*, 174:355–433, 2008. (Cited on p. 141)

[50] C.M. Bender and S.A. Orszag. *Asymptotic Methods and Perturbation Theory*. Springer, 1999. (Cited on pp. 41, 43)

[51] S. Benzoni-Gavage and D. Serre. *Multi-dimensional Hyperbolic Partial Differential Equations*. Oxford University Press, 2007. (Cited on p. 61)

[52] H. Berestycki and F. Hamel. Front propagation in periodic excitable media. *Comm. Pure Appl. Math.*, 55(8):949–1032, 2002. (Cited on p. 43)

[53] M.S. Berger. *Nonlinearity and Functional Analysis*. Academic Press, 1977. (Cited on p. 18)

[54] N. Berglund and B. Gentz. *Noise-Induced Phenomena in Slow-Fast Dynamical Systems.* Springer, 2006. (Cited on p. 130)

[55] F. Bernis and A. Friedman. Higher order nonlinear degenerate parabolic equations. *J. Differential Equations*, 83(1):179–206, 1990. (Cited on p. 169)

[56] F. Bernis, L.A. Peletier, and S.M. Williams. Source type solutions of a fourth order nonlinear degenerate parabolic equation. *Nonlinear Anal. Theory Methods Appl.*, 18(3):217–234, 1992. (Cited on p. 169)

[57] A.L. Bertozzi. The mathematics of moving contact lines in thin liquid films. *Notices Amer. Math. Soc.*, 45(6):689–697, 1998. (Cited on p. 169)

[58] G.D. Birkhoff. Proof of the the ergodic theorem. *Proc. Natl. Acad. Sci. USA*, 17(12):656–660, 1931. (Cited on p. 181)

[59] M.S. Birman and M.Z. Solomjak. *Spectral Theory of Self-Adjoint Operators in Hilbert Space.* Springer, 2012. (Cited on p. 30)

[60] A. Blanchet, J. Dolbeault, and B. Perthame. Two-dimensional Keller-Segel model: Optimal critical mass and qualitative properties of the solutions. *Electron. J. Differential Equations*, 2006(44):1–33, 2006. (Cited on p. 164)

[61] D. Blömker. *Amplitude Equations for Stochastic Partial Differential Equations.* World Scientific, 2007. (Cited on p. 72)

[62] F. Bornemann. *Homogenization in Time of Singularly Perturbed Mechanical Systems.* Springer, 1998. (Cited on p. 185)

[63] R. Bowen. Hausdorff dimension of quasi-circles. *Publ. Math. IHES*, 50(1):11–25, 1979. (Cited on p. 105)

[64] A. Braides. *Γ-Convergence for Beginners.* Clarendon Press, 2002. (Cited on p. 185)

[65] M. Bramson. *Convergence of Solutions of the Kolmogorov Equation to Travelling Waves.* AMS, 1983. (Cited on p. 43)

[66] Y. Brenier. Convergence of the Vlasov-Poisson system to the incompressible Euler equations. *Comm. Partial Differential Equations*, 25(3):737–754, 2000. (Cited on p. 153)

[67] S.C. Brenner and L.R. Scott. *The Mathematical Theory of Finite Element Methods.* Springer, 3rd edition, 2008. (Cited on p. 209)

[68] A. Bressan. *Hyperbolic Systems of Conservation Laws: The One-dimensional Cauchy Problem.* Oxford University Press, 2000. (Cited on p. 61)

[69] A. Bressan and A. Constantin. Global conservative solutions of the Camassa-Holm equation. *Arch. Rat. Mech. Anal.*, 183(2):215–239, 2007. (Cited on p. 148)

[70] H. Brezis. *Functional Analysis, Sobolev Spaces and Partial Differential Equations.* Springer, 2011. (Cited on pp. 13, 81)

[71] M. Brin and G. Stuck. *Introduction to Dynamical Systems.* Cambridge University Press, 2002. (Cited on pp. 9, 89, 181)

[72] N.F. Britton. *Reaction-Diffusion Equations and Their Applications to Biology.* Academic Press, 1986. (Cited on p. 43)

[73] H. Broer and F. Takens. *Dynamical Systems and Chaos.* Springer, 2010. (Cited on p. 198)

[74] L. Bronsard and R.V. Kohn. On the slowness of phase boundary motion in one space dimension. *Comm. Pure Appl. Math.*, 43(8):983–997, 1990. (Cited on p. 110)

[75] L. Bronsard and R.V. Kohn. Motion by mean curvature as the singular limit of Ginzburg-Landau dynamics. *J. Differential Equations*, 90(2):211–237, 1991. (Cited on p. 110)

[76] J. Burke and E. Knobloch. Localized states in the generalized Swift-Hohenberg equation. *Phys. Rev. E*, 73:056211, 2006. (Cited on p. 67)

[77] L. Caffarelli, R. Kohn, and L. Nirenberg. Partial regularity of suitable weak solutions of the Navier-Stokes equations. *Comm. Pure Appl. Math.*, 35(6):771–831, 1982. (Cited on p. 83)

[78] G. Caginalp. An analysis of a phase field model of a free boundary. *Arch. Rat. Mech. Anal.*, 92:205–245, 1986. (Cited on p. 110)

[79] J.W. Cahn, C.M. Elliott, and A. Novick-Cohen. The Cahn-Hilliard equation with a concentration dependent mobility: Motion by minus the Laplacian of the mean curvature. *Eur. J. Appl. Math.*, 7(3):287–301, 1996. (Cited on p. 120)

[80] J.W. Cahn and J.E. Hilliard. Free energy of a nonuniform system. I. Interfacial free energy. *J. Chem. Phys.*, 28(2):258–267, 1958. (Cited on p. 111)

[81] R. Camassa and D.D. Holm. An integrable shallow water equation with peaked solitons. *Phys. Rev. Lett.*, 71(11):1661–1664, 1993. (Cited on p. 148)

[82] R.S. Cantrell and C. Cosner. *Spatial Ecology via Reaction-Diffusion Equations*. Wiley, 2004. (Cited on p. 43)

[83] E. Carlen. Superadditivity of Fisher's information and logarithmic Sobolev inequalities. *J. Funct. Anal.*, 101:194–211, 1991. (Cited on p. 125)

[84] G.A. Carpenter. A geometric approach to singular perturbation problems with applications to nerve impulse equations. *J. Differential Equations*, 23:335–367, 1977. (Cited on pp. 48, 196, 198)

[85] J. Carr. *Applications of Centre Manifold Theory*. Springer, 1981. (Cited on pp. 89, 176)

[86] J. Carr and R.L. Pego. Metastable patterns in solutions of $u_t = \epsilon^2 u_{xx} - f(u)$. *Comm. Pure Appl. Math.*, 42(5):523–576, 1989. (Cited on pp. 110, 115)

[87] J. Carr and R.L. Pego. Invariant manifolds for metastable patterns in $u_t = \epsilon^2 u_{xx} - f(u)$. *Proc. R. Soc. Edinburgh A*, 116(1):133–160, 1990. (Cited on p. 110)

[88] R. Carretero-Gonzalez, D.J. Frantzeskakis, and P.G. Kevrekidis. Nonlinear waves in Bose-Einstein condensates: Physical relevance and mathematical techniques. *Nonlinearity*, 21(7):R139, 2008. (Cited on p. 135)

[89] J.A. Carrillo, A. Jüngel, P.A. Markowich, G. Toscani, and A. Unterreiter. Entropy dissipation methods for degenerate parabolic problems and generalized Sobolev inequalities. *Monatsh. Math.*, 133(1):1–82, 2001. (Cited on p. 130)

[90] J.A. Carrillo, R.J. McCann, and C. Villani. Kinetic equilibration rates for granular media and related equations: Entropy dissipation and mass transportation estimates. *Rev. Mat. Iberoam.*, 19(3):971–1018, 2003. (Cited on p. 125)

[91] J.A. Carrillo, R.J. McCann, and C. Villani. Contractions in the 2-Wasserstein length space and thermalization of granular media. *Arch. Rat. Mech. Anal.*, 179(2):217–263, 2006. (Cited on p. 125)

[92] J.A. Carrillo and G. Toscani. Asymptotic L^1-decay of solutions of the porous medium equation to self-similarity. *Indiana Univ. Math. J.*, 49(1):113–142, 2000. (Cited on p. 169)

[93] J.A. Carrillo and G. Toscani. Long-time asymptotics for strong solutions of the thin film equation. *Comm. Math. Phys.*, 225(3):551–571, 2002. (Cited on p. 169)

[94] A. Carvalho, J.A. Langa, and J. Robinson. *Attractors for Infinite-Dimensional Non-Autonomous Dynamical Systems*. Springer, 2012. (Cited on p. 83)

[95] T. Cazenave. *Semilinear Schrödinger Equations*. AMS, 2003. (Cited on p. 72)

[96] T. Cazenave and A. Haraux. *An Introduction to Semilinear Evolution Equations*. Oxford University Press, 1998. (Cited on p. 78)

[97] C. Cercignani. *Mathematical Methods in Kinetic Theory*. Springer, 1969. (Cited on p. 153)

[98] C. Cercignani. *The Boltzmann Equation and Its Applications*. Springer, 1988. (Cited on p. 153)

[99] C. Cercignani, R. Illner, and M. Pulvirenti. *The Mathematical Theory of Dilute Gases*. Springer, 1994. (Cited on p. 153)

[100] J.M. Chadam and R.T. Glassey. Global existence of solutions to the Cauchy problem for time-dependent Hartree equations. *J. Math. Phys.*, 16(5):1122–1130, 1975. (Cited on p. 153)

[101] N. Chafee and E.F. Infante. A bifurcation problem for a nonlinear partial differential equation of parabolic type. *Appl. Anal.*, 4(1):17–37, 1974. (Cited on p. 25)

[102] F.A.C.C. Chalub, P.A. Markowich, B. Berthame, and C. Schmeiser. Kinetic models for chemotaxis and their drift-diffusion limits. *Monatsh. Math.*, 142(1):123–141, 2004. (Cited on p. 153)

[103] X. Chen. Generation and propagation of interfaces for reaction-diffusion equations. *J. Differential Equations*, 96(1):116–141, 1992. (Cited on p. 110)

[104] X. Chen. Spectrum for the Allen-Cahn, Cahn-Hilliard, and phase-field equations for generic interfaces. *Comm. Partial Differential Equations*, 19(7):1371–1392, 1994. (Cited on p. 110)

[105] X. Chen. Existence, uniqueness, and asymptotic stability of travelling waves in nonlocal evolution equations. *Adv. Differential Equations*, 2:125–160, 1997. (Cited on pp. 37, 47)

[106] C. Chicone. *Ordinary Differential Equations with Applications*. Texts in Applied Mathematics. Springer, 2nd edition, 2010. (Cited on pp. 9, 48)

[107] C. Chicone and Y. Latushkin. *Evolution Semigroups in Dynamical Systems and Differential Equations*. AMS, 1999. (Cited on p. 78)

[108] J.W. Cholewa and T. Dlotko. *Global Attractors in Abstract Parabolic Problems*. Cambridge University Press, 2000. (Cited on p. 99)

[109] A.J. Chorin and J.E. Marsden. *A Mathematical Introduction to Fluid Mechanics*. Springer, 1990. (Cited on p. 83)

[110] P. Chossat and R. Lauterbach. *Methods in Equivariant Bifurcations and Dynamical Systems*. World Scientific, 2000. (Cited on p. 176)

[111] B. Chow and V.I. Bogachev. *The Ricci Flow: Techniques and Applications*. AMS, 2010. (Cited on p. 130)

[112] B. Chow, P. Lu, and L. Ni. *Hamilton's Ricci Flow.* AMS, 2006. (Cited on p. 130)

[113] S.-N. Chow and J.K. Hale. *Methods of Bifurcation Theory.* Springer, 1982. (Cited on pp. 18, 136, 141)

[114] S.N. Chow and H. Leiva. Existence and roughness of the exponential dichotomy for skew-product semiflow in Banach spaces. *J. Differential Equations*, 120(2):429–477, 1995. (Cited on p. 55)

[115] S.N. Chow, K. Lu, and G.R. Sell. Smoothness of inertial manifolds. *J. Math. Anal. Appl.*, 169(1):283–312, 1992. (Cited on p. 94)

[116] I. Chueshov and I. Lasiecka. *Long-Time Behavior of Second Order Evolution Equations with Nonlinear Damping.* AMS, 2008. (Cited on p. 99)

[117] P.G. Ciarlet. *The Finite Element Method for Elliptic Problems.* SIAM, 2002. (Cited on p. 209)

[118] D. Cioranescu, A. Damlamian, and G. Griso. Periodic unfolding and homogenization. *Comptes Rend. Math.*, 335(1):99–104, 2002. (Cited on p. 185)

[119] P. Collet and J.P. Eckmann. *Instabilities and Fronts in Extended Systems.* Princeton University Press, 1990. (Cited on p. 66)

[120] P. Collet and J.P. Eckmann. The time dependent amplitude equation for the Swift-Hohenberg problem. *Comm. Math. Phys.*, 132(1):139–153, 1990. (Cited on pp. 68, 72)

[121] A. Constantin. On the scattering problem for the Camassa-Holm equation. *Proc. R. Soc. London A*, 457:953–970, 2001. (Cited on p. 148)

[122] A. Constantin and J. Escher. Global existence and blow-up for a shallow water equation. *Ann. Scuola Norm. Sci.*, 26(2):303–328, 1998. (Cited on p. 164)

[123] P. Constantin and C. Foias. Global Lyapunov exponents, Kaplan-Yorke formulas and the dimension of the attractors for 2D Navier-Stokes equations. *Comm. Pure Appl. Math.*, 38(1):1–27, 1985. (Cited on p. 105)

[124] P. Constantin and C. Foias. *Navier-Stokes Equations.* University of Chicago Press, 1988. (Cited on p. 83)

[125] P. Constantin, C. Foias, B. Nicolaenko, and R. Temam. *Integral Manifolds and Inertial Manifolds for Dissipative Partial Differential Equations.* Springer, 1989. (Cited on p. 94)

[126] P. Constantin, C. Foias, B. Nicolaenko, and R. Temam. Spectral barriers and inertial manifolds for dissipative partial differential equations. *J. Dyn. Differential Equations*, 1(1):45–73, 1989. (Cited on p. 94)

[127] J. B. Conway. *A Course in Functional Analysis.* Springer, 1990. (Cited on pp. 16, 18)

[128] W.A. Coppel. *Dichotomies in Stability Theory.* Springer, 1978. (Cited on p. 51)

[129] I.P. Cornfeld, S.V. Fomin, and Y.G. Sinai. *Ergodic Theory.* Springer, 2012. (Cited on p. 181)

[130] R. Courant and K.O. Friedrichs. *Supersonic Flow and Shock Waves.* Springer, 1999. (Cited on p. 61)

[131] M.G. Crandall and T.M. Liggett. Generation of semi-groups of nonlinear transformations on general Banach spaces. *Amer. J. Math.*, 93(2):265–298, 1971. (Cited on p. 78)

[132] M.G. Crandall and P.H. Rabinowitz. Bifurcation from simple eigenvalues. *J. Funct. Anal.*, 8(2):321–340, 1971. (Cited on p. 25)

[133] M.G. Crandall and P.H. Rabinowitz. Bifurcation, perturbation of simple eigenvalues and linearized stability. *Arch. Rat. Mech. Anal.*, 52:161–180, 1973. (Cited on p. 25)

[134] M. Cross and H. Greenside. *Pattern Formation and Dynamics in Nonequilibrium Systems.* Cambridge University Press, 2009. (Cited on pp. 66, 159)

[135] M.C. Cross and P.C. Hohenberg. Pattern formation outside of equilibrium. *Rev. Mod. Phys.*, 65(3):851–1112, 1993. (Cited on pp. 66, 159)

[136] J. Cuevas-Maraver, P.G. Kevrekidis, and F. Williams, editors. *The Sine-Gordon Model and Its Applications: From Pendula and Josephson Junctions to Gravity and High-energy Physics.* Springer, 2014. (Cited on p. 72)

[137] A.C. da Silva. *Lectures on Symplectic Geometry.* Springer, 2001. (Cited on p. 148)

[138] C.M. Dafermos. *Hyperbolic Conservation Laws in Continuum Physics.* Springer, 2010. (Cited on p. 61)

[139] R. Dal Passo, H. Garcke, and G. Grün. On a fourth-order degenerate parabolic equation: Global entropy estimates, existence, and qualitative behavior of solutions. *SIAM J. Math. Anal.*, 29(2):321–342, 1998. (Cited on p. 125)

[140] H. Dankowicz and F. Schilder. *Recipes for Continuation.* SIAM, 2013. (Cited on p. 209)

[141] B. Darcorogna. *Direct Methods in the Calculus of Variations.* Springer, 2007. (Cited on pp. 135, 198)

[142] R. Dautray and J.-L. Lions. *Mathematical Analysis and Numerical Methods for Science and Technology: Volume 3 Spectral Theory and Applications.* Springer, 2000. (Cited on p. 30)

[143] L. Debnath. *Nonlinear Partial Differential Equations.* Birkhäuser, 2005. (Cited on pp. 37, 148)

[144] A. Degasperis, D.D. Holm, and A.N. Hone. A new integrable equation with peakon solutions. *Theor. Math. Phys.*, 133(2):1463–1474, 2002. (Cited on p. 148)

[145] K. Deimling. *Nonlinear Functional Analysis.* Dover, 2010. (Cited on pp. 14, 18)

[146] M. del Pino and P.L. Felmer. Local mountain passes for semilinear elliptic problems in unbounded domains. *Calc. Var. Partial Differential Equations*, 4(2):121–137, 1996. (Cited on p. 141)

[147] M. del Pino and P.L. Felmer. Semi-classical states for nonlinear Schrödinger equations. *J. Funct. Anal.*, 149(1):245–265, 1997. (Cited on p. 72)

[148] C. Dellacherie and P.A. Meyer. *Probabilities and Potential, C: Potential Theory for Discrete and Continuous Semigroups.* Elsevier, 2011. (Cited on p. 78)

[149] M. Dellnitz and O. Junge. On the approximation of complicated dynamical behavior. *SIAM J. Numer. Anal.*, 36(2):491–515, 1999. (Cited on p. 181)

[150] A. Delshams, V. Gelfreich, A. Jorba, and T.M. Seara. Exponentially small splitting of separatrices under fast quasiperiodic forcing. *Comm. Math. Phys.*, 189(1):35–71, 1997. (Cited on p. 115)

[151] F. den Hollander. *Large Deviations.* AMS, 2008. (Cited on p. 115)

[152] B. Derrida, J.L. Lebowitz, E.R. Speer, and H. Spohn. Fluctuations of a stationary nonequilibrium interface. *Phys. Rev. Lett.*, 67(2):165–168, 1991. (Cited on p. 125)

[153] L. Desvillettes and C. Villani. On the trend to global equilibrium in spatially inhomogeneous entropy-dissipating systems. Part I: The linear Fokker-Planck equation. *Comm. Pure Appl. Math.*, 54(1):1–42, 2001. (Cited on p. 130)

[154] A. Dhooge, W. Govaerts, and Yu.A. Kuznetsov. MATCONT: A MATLAB package for numerical bifurcation analysis of ODEs. *ACM Trans. Math. Softw.*, 29:141–164, 2003. (Cited on p. 209)

[155] E. DiBenedetto. *Degenerate Parabolic Equations*. Springer, 2012. (Cited on p. 169)

[156] R. Dobrushin. Vlasov equations. *Funct. Anal. Appl.*, 13:115–123, 1979. (Cited on p. 153)

[157] E.J. Doedel, A. Champneys, F. Dercole, T. Fairgrieve, Y. Kuznetsov, B. Oldeman, R. Paffenroth, B. Sandstede, X. Wang, and C. Zhang. AUTO 2007p: Continuation and bifurcation software for ordinary differential equations (with HOMCONT). *http://cmvl.cs.concordia.ca/auto*, 2007. (Cited on p. 209)

[158] A. Doelman, R.A. Gardner, and T.J. Kaper. Stability analysis of singular patterns in the 1D Gray-Scott model: A matched asymptotics approach. *Physica D*, 122(1):1–36, 1998. (Cited on p. 191)

[159] A. Doelman, T.J. Kaper, and K. Promislow. Nonlinear asymptotic stability of the semistrong pulse dynamics in a regularized Gierer-Meinhardt model. *SIAM J. Math. Anal.*, 38(6):1760–1787, 2007. (Cited on p. 191)

[160] A. Doelman, T.J. Kaper, and P.A. Zegeling. Pattern formation in the one-dimensional Gray-Scott model. *Nonlinearity*, 10(2):523–563, 1997. (Cited on p. 191)

[161] C.R. Doering and J.D. Gibbon. *Applied Analysis of the Navier-Stokes Equations*. Cambridge University Press, 1995. (Cited on p. 83)

[162] C.R. Doering, J.D. Gibbon, D.D. Holm, and B. Nicolaenko. Low-dimensional behaviour in the complex Ginzburg-Landau equation. *Nonlinearity*, 1(2):279, 1988. (Cited on p. 67)

[163] J. Dolbeault, C. Mouhot, and C. Schmeiser. Hypocoercivity for linear kinetic equations conserving mass. *Trans. Amer. Math. Soc.*, 367(6):3807–3828, 2015. (Cited on p. 159)

[164] P.G. Drazin and R.S. Johnson. *Solitons: An Introduction*. Cambridge University Press, 1989. (Cited on p. 37)

[165] F. Dumortier. Compactification and desingularization of spaces of polynomial Liénard equations. *J. Differential Equations*, 224:296–313, 2006. (Cited on p. 176)

[166] N. Dunford and J.T. Schwartz. *Linear Operators Part I: General Theory*. Wiley, 1988. (Cited on p. 30)

[167] N. Dunford and J.T. Schwartz. *Linear Operators Part II: Spectral Theory*. Wiley, 1988. (Cited on p. 30)

[168] B. Düring, D. Matthes, and J.P. Milišic. A gradient flow scheme for nonlinear fourth order equations. *Discrete Contin. Dyn. Syst. B*, 14(3):935–959, 2010. (Cited on p. 125)

[169] S. Dyachenko, A.C. Newell, A. Pushkarev, and V.E. Zakharov. Optical turbulence: Weak turbulence, condensates and collapsing filaments in the nonlinear Schrödinger equation. *Physica D*, 57(1):96–160, 1992. (Cited on p. 72)

[170] U. Ebert and W. van Saarloos. Front propagation into unstable states: Universal algebraic convergence towards uniformly translating pulled fronts. *Physica D*, 146:1–99, 2000. (Cited on p. 43)

[171] J.P. Eckmann and D. Ruelle. Fundamental limitations for estimating dimensions and Lyapunov exponents in dynamical systems. *Physica D*, 56(2):185–187, 1992. (Cited on p. 105)

[172] D.E. Edmunds and W.D. Evans. *Spectral Theory and Differential Operators*. Clarendon Press, 1987. (Cited on p. 30)

[173] M.A. Efendiev and S.V. Zelik. The attractor for a nonlinear reaction-diffusion system in an unbounded domain. *Comm. Pure Appl. Math.*, 54(6):625–688, 2001. (Cited on p. 83)

[174] M. Einsiedler and T. Ward. *Ergodic Theory: With a View Towards Number Theory*. Springer, 2011. (Cited on p. 181)

[175] T. Eisner, B. Farkas, M. Haase, and R. Nagel. *Operator Theoretic Aspects of Ergodic Theory*. Springer, 2015. (Cited on p. 181)

[176] C.M. Elliott. The Cahn-Hilliard model for the kinetics of phase separation. In *Mathematical Models for Phase Change Problems*, pages 35–73. Birkhäuser, 1989. (Cited on p. 120)

[177] C.M. Elliott and Z. Songmu. On the Cahn-Hilliard equation. *Arch. Rat. Mech. Anal.*, 96(4):339–357, 1986. (Cited on p. 120)

[178] L.D. Elsgolc. *Calculus of Variations*. Dover, 2012. (Cited on p. 135)

[179] K.-J. Engel and R. Nagel. *Semigroups for Linear Evolution Equations*. Springer, 2000. (Cited on pp. 78, 180)

[180] K.-J. Engel and R. Nagel. *A Short Course on Operator Semigroups*. Springer, 2006. (Cited on p. 78)

[181] L. Erdös, B. Schlein, and H.T. Yau. Derivation of the Gross-Pitaevskii hierarchy for the dynamics of Bose-Einstein condensate. *Comm. Pure Appl. Math.*, 59(12):1659–1741, 2006. (Cited on p. 135)

[182] L. Erdös, B. Schlein, and H.T. Yau. Derivation of the Gross-Pitaevskii equation for the dynamics of Bose-Einstein condensate. *Ann. Math.*, 172(1):291–370, 2010. (Cited on p. 135)

[183] A. Ern and J.-L. Guermond. *Theory and Practice of Finite Elements*. Springer, 2004. (Cited on p. 209)

[184] J. Evans. Nerve axon equations III: Stability of nerve impulses. *Indiana Univ. Math. J.*, 22:577–594, 1972. (Cited on p. 55)

[185] J.W. Evans, N. Fenichel, and J.A. Feroe. Double impulse solutions in nerve axon equations. *SIAM J. Appl. Math.*, 42(2):219–234, 1982. (Cited on pp. 48, 198)

[186] L.C. Evans. *Partial Differential Equations*. AMS, 2002. (Cited on pp. vii, 11, 12, 13, 24, 28, 37, 61, 81, 136, 189, 198)

[187] L.C. Evans, H.M. Soner, and P.E. Souganidis. Phase transitions and generalized motion by mean curvature. *Comm. Pure Appl. Math.*, 45(9):1097–1123, 1992. (Cited on p. 110)

[188] D.J. Eyre. Systems of Cahn-Hilliard equations. *SIAM J. Appl. Math.*, 53(6):1686–1712, 1993. (Cited on p. 120)

[189] K.J. Falconer. The Hausdorff dimension of self-affine fractals. *Math. Proc. Cambridge Philos. Soc.*, 103(2):339–350, 1988. (Cited on p. 105)

[190] A. Favini and A. Yagi. *Degenerate Differential Equations in Banach Spaces*. CRC Press, 1998. (Cited on p. 78)

[191] C.L. Fefferman. Existence and smoothness of the Navier-Stokes equations. In *The Millennium Prize Problems*, pages 57–67. AMS, 2006. (Cited on p. 83)

[192] E. Feireisl. *Dynamics of Viscous Compressible Fluids*. Oxford University Press, 2004. (Cited on p. 83)

[193] W. Feller. *An Introduction to Probability Theory and its Applications*, volume 2. Wiley, 2nd edition, 1972. (Cited on p. 78)

[194] N. Fenichel. Persistence and smoothness of invariant manifolds for flows. *Indiana U. Math. J.*, 21:193–225, 1971. (Cited on p. 89)

[195] N. Fenichel. Geometric singular perturbation theory for ordinary differential equations. *J. Differential Equations*, 31:53–98, 1979. (Cited on pp. 94, 195)

[196] N. Fenichel. Vortex dynamics in three-dimensional continuous myocardium with fiber rotation: Filament instability and fibrillation. *Chaos*, 8(1):20–47, 1998. (Cited on p. 48)

[197] P. Fife and J.B. McLeod. The approach of solutions nonlinear diffusion equations to travelling front solutions. *Arch. Rat. Mech. Anal.*, 65:335–361, 1977. (Cited on p. 37)

[198] P.C. Fife. *Dynamics of Internal Layers and Diffusive Interfaces*. SIAM, 1988. (Cited on pp. 111, 191)

[199] P.C. Fife. Models for phase separation and their mathematics. *Electron. J. Differential Equations*, 48:1–26, 2000. (Cited on p. 120)

[200] P.C. Fife. *Mathematical Aspects of Reacting and Diffusing Systems*. Springer, 2013. (Cited on p. 83)

[201] R.A. Fisher. The wave of advance of advantageous genes. *Ann. Eugenics*, 7:353–369, 1937. (Cited on p. 43)

[202] R. FitzHugh. Mathematical models of threshold phenomena in the nerve membrane. *Bull. Math. Biophysics*, 17:257–269, 1955. (Cited on pp. 48, 198)

[203] A. Floer and A. Weinstein. Nonspreading wave packets for the cubic Schrödinger equation with a bounded potential. *J. Function. Anal.*, 69(3):397–408, 1986. (Cited on p. 72)

[204] C. Foias, O. Manley, R. Rosa, and R. Temam. *Navier-Stokes Equations and Turbulence*. Cambridge University Press, 2001. (Cited on p. 83)

[205] C. Foias, G.R. Sell, and R. Temam. Inertial manifolds for nonlinear evolutionary equations. *J. Differential Equations*, 73(2):309–353, 1988. (Cited on p. 94)

[206] C. Foias and E.S. Titi. Determining nodes, finite difference schemes and inertial manifolds. *Nonlinearity*, 4(1):135, 1991. (Cited on p. 94)

[207] G. Folland. *Real Analysis - Modern Techniques and Their Applications*. Wiley, 1999. (Cited on p. 122)

[208] G.E. Forsythe and W.R. Wasow. *Finite-difference Methods for Partial Differential Equations*. Literary Licensing, 2013. (Cited on p. 203)

[209] T.D. Frank. *Nonlinear Fokker-Planck Equations: Fundamentals and Applications.* Springer, 2005. (Cited on p. 130)

[210] H. Freistühler and D. Serre. L^1 stability of shock waves in scalar viscous conservation laws. *Comm. Pure Appl. Math.*, 51(3):291–301, 1998. (Cited on p. 61)

[211] A. Friedman. *Partial Differential Equations of Parabolic Type.* Dover, 1992. (Cited on p. 13)

[212] A. Friedman and B. McLeod. Blow-up of positive solutions of semilinear heat equations. *Indiana Univ. Math. J.*, 34(2):425–447, 1985. (Cited on p. 164)

[213] G. Friesecke, R.D. James, and S. Müller. A theorem on geometric rigidity and the derivation of nonlinear plate theory from three-dimensional elasticity. *Comm. Pure Appl. Math.*, 55(11):1461–1506, 2002. (Cited on p. 185)

[214] G. Friesecke, R.D. James, and S. Müller. A hierarchy of plate models derived from nonlinear elasticity by gamma-convergence. *Arch. Rat. Mech. Anal.*, 180(2):183–236, 2006. (Cited on p. 185)

[215] U. Frisch. *Turbulence: The Legacy of A.N. Kolmogorov.* Cambridge University Press, 1995. (Cited on p. 83)

[216] G. Froyland and K. Padberg. Almost-invariant sets and invariant manifolds-connecting probabilistic and geometric descriptions of coherent structures in flows. *Physica D*, 238(16):1507–1523, 2009. (Cited on p. 181)

[217] B. Fuchssteiner and A.S. Fokas. Symplectic structures, their Bäcklund transformations and hereditary symmetries. *Physica D*, 4(1):47–66, 1981. (Cited on p. 148)

[218] G. Fusco and J.K. Hale. Slow-motion manifolds, dormant instability, and singular perturbations. *J. Dyn. Differential Equations*, 1:75–94, 1989. (Cited on p. 110)

[219] V.A. Galaktionov and J.L. Vázquez. *A stability technique for evolution partial differential equations: a dynamical systems approach.* Springer, 2012. (Cited on p. 169)

[220] G.P. Galdi. *An Introduction to the Mathematical Theory of the Navier-Stokes Equations: Steady-State Problems.* Springer, 2011. (Cited on p. 83)

[221] T.W. Gamelin. *Complex Analysis.* Springer, 2001. (Cited on p. 41)

[222] C. Gardiner. *Stochastic Methods.* Springer, 4th edition, 2009. (Cited on p. 130)

[223] R. Gardner. The Brunn-Minkowski inequality. *Bull. Amer. Math. Soc.*, 39(3):355–405, 2002. (Cited on p. 125)

[224] R. Gardner and K. Zumbrun. The gap lemma and geometric criteria for instability of viscous shock profiles. *Comm. Pure Appl. Math.*, 51(7):797–855, 1998. (Cited on p. 55)

[225] I.M. Gel'fand and R.A. Silverman. *Calculus of Variations.* Courier Corporation, 2000. (Cited on pp. 135, 198)

[226] V. Gelfreich. Melnikov method and exponentially small splitting of separatrices. *Physica D*, 101(3):227–248, 1997. (Cited on p. 191)

[227] J.M. Ghidaglia and R. Temam. Attractors for damped nonlinear hyperbolic equations. *J. Math. Pure Appl.*, 66(3):273–319, 1987. (Cited on p. 105)

[228] U. Gianazza, G. Savaré, and G. Toscani. The Wasserstein gradient flow of the Fisher information and the quantum drift-diffusion equation. *Arch. Rat. Mech. Anal.*, 194(1):133–220, 2009. (Cited on p. 125)

[229] A. Gierer and H. Meinhardt. A theory of biological pattern formation. *Kybernetic*, 12:30–39, 1972. (Cited on p. 191)

[230] M.H. Giga, Y. Giga, and J. Saal. *Nonlinear Partial Differential Equations: Asymptotic Behavior of Solutions and Self-similar Solutions*. Springer, 2010. (Cited on p. 169)

[231] Y. Giga and R.V. Kohn. Asymptotically self-similar blow-up of semilinear heat equations. *Comm. Pure Appl. Math.*, 38(3):297–319, 1985. (Cited on p. 169)

[232] D. Gilbarg and N.S. Trudinger. *Elliptic Partial Differential Equations of Second Order*. Springer, 2001. (Cited on pp. 13, 138, 189)

[233] V.L. Ginzburg and L.D. Landau. On the theory of superconductivity. In *On Superconductivity and Superfluidity*, pages 113–137. Springer, 2009. (Cited on p. 67)

[234] E. Giusti. *Direct Methods in the Calculus of Variations*. World Scientific, 2003. (Cited on p. 135)

[235] R.T. Glassey. On the blowing up of solutions to the Cauchy problem for nonlinear Schrödinger equations. *J. Math. Phys.*, 18:1794–1797, 1977. (Cited on p. 72)

[236] P. Glendinning. *Stability, Instability and Chaos*. Cambridge University Press, 1994. (Cited on p. 9)

[237] T. Gneiting and M. Schlather. Stochastic models that separate fractal dimension and the Hurst effect. *SIAM Rev.*, 46(2):269–282, 2004. (Cited on p. 105)

[238] E. Godlewski and P.A. Raviart. *Numerical Approximation of Hyperbolic Systems of Conservation Laws*. Springer, 2013. (Cited on pp. 61, 203)

[239] F. Golse. On the dynamics of large particle systems in the mean field limit. arXiv:1301.5494v1, 2013. (Cited on pp. 150, 153)

[240] M. Golubitsky, V. LeBlanc, and I. Melbourne. Meandering of the spiral tip: An alternative approach. *J. Nonlinear Sci.*, 7(6):557–586, 1997. (Cited on p. 176)

[241] M. Golubitsky and D. Schaeffer. *Singularities and Groups in Bifurcation Theory*, volume 1. Springer, 1985. (Cited on p. 176)

[242] M. Golubitsky, D. Schaeffer, and I. Stewart. *Singularities and Groups in Bifurcation Theory*, volume 2. Springer, 1985. (Cited on p. 176)

[243] W.F. Govaerts. *Numerical Methods for Bifurcations of Dynamical Equilibria*. SIAM, 1987. (Cited on p. 209)

[244] H. Grad. Asymptotic theory of the Boltzmann equation. *Phys. Fluids*, 6(2):147–181, 1963. (Cited on p. 153)

[245] P. Grindrod. *Patterns and Waves: The Theory and Applications of Reaction-Diffusion Equations*. Clarendon Press, 1991. (Cited on p. 83)

[246] L. Gross. Logarithmic Sobolev inequalities. *Amer. J. Math.*, 97(4):1061–1083, 1975. (Cited on p. 125)

[247] J. Guckenheimer and P. Holmes. *Nonlinear Oscillations, Dynamical Systems, and Bifurcations of Vector Fields*. Springer, 1983. (Cited on pp. 9, 89, 176, 185, 191, 198)

[248] J. Guckenheimer and C. Kuehn. Homoclinic orbits of the FitzHugh-Nagumo equation: The singular limit. *Discrete Contin. Dyn. Syst. S*, 2(4):851–872, 2009. (Cited on p. 37)

[249] C. Gui and J. Wei. Multiple interior peak solutions for some singularly perturbed Neumann problems. *J. Differential Equations*, 158(1):1–27, 1999. (Cited on p. 141)

[250] M.E. Gurtin. Generalized Ginzburg-Landau and Cahn-Hilliard equations based on a microforce balance. *Physica D*, 92:178–192, 1996. (Cited on p. 120)

[251] M. Haase. *The Functional Calculus for Sectorial Operators*. Springer, 2006. (Cited on p. 78)

[252] K.P. Hadeler and F. Rothe. Travelling fronts in nonlinear diffusion equations. *J. Math. Biol.*, 2(3):251–263, 1975. (Cited on p. 43)

[253] J.K. Hale. *Ordinary Differential Equations*. Dover, 2009. (Cited on p. 9)

[254] J.K. Hale. *Asymptotic Behavior of Dissipative Systems*. AMS, 2010. (Cited on p. 83)

[255] J.K. Hale and H. Kocak. *Dynamics and Bifurcations*. Springer, 1991. (Cited on p. 9)

[256] J.K. Hale, L.T. Magelhaes, and W.M. Oliva. *Dynamics in Infinite Dimensions*. Springer, 2002. (Cited on p. 99)

[257] G. Haller. *Chaos Near Resonance*. Springer, 2012. (Cited on pp. 115, 148)

[258] M. Haragus and G. Iooss. *Local Bifurcations, Center Manifolds, and Normal Forms in Infinite-Dimensional Dynamical Systems*. Springer, 2010. (Cited on pp. 25, 176)

[259] P. Hartman. *Ordinary Differential Equations*. SIAM, 2nd edition, 2002. (Cited on p. 45)

[260] S.P. Hastings. Single and multiple pulse waves for the FitzHugh-Nagumo equations. *SIAM J. Appl. Math.*, 42(2):247–260, 1982. (Cited on p. 198)

[261] S.P. Hastings and J.B. McLeod. An elementary approach to a model problem of Lagerstrom. *SIAM J. Math. Anal.*, 40(6):2421–2436, 2009. (Cited on p. 115)

[262] B.T. Hayes and P. LeFloch. Non-classical shocks and kinetic relations: scalar conservation laws. *Arch. Rat. Mech. Anal.*, 139:1–56, 1997. (Cited on pp. 59, 61)

[263] D. Henry. *Geometric Theory of Semilinear Parabolic Equations*. Springer, 1981. (Cited on pp. 25, 48, 74, 75, 76, 78)

[264] F. Hérau and F. Nier. Isotropic hypoellipticity and trend to equilibrium for the Fokker-Planck equation with a high-degree potential. *Arch. Rat. Mech. Anal.*, 171(2):151–218, 2004. (Cited on p. 130)

[265] M.A. Herrero and J.J. Velázquez. Singularity patterns in a chemotaxis model. *Math. Ann.*, 306(1):583–623, 1996. (Cited on p. 164)

[266] E. Hille and R.S. Phillips. *Functional Analysis and Semi-Groups*. AMS, 1996. (Cited on p. 78)

[267] T. Hillen and K. Painter. A user's guide to PDE models for chemotaxis. *J. Math. Biol*, 58(1):183–217, 2009. (Cited on p. 164)

[268] E.J. Hinch. *Perturbation Methods*. Cambridge University Press, 1991. (Cited on p. 191)

[269] M.W. Hirsch, C.C. Pugh, and M. Shub. *Invariant Manifolds*. Springer, 1977. (Cited on p. 89)

[270] M.W. Hirsch, S. Smale, and R. Devaney. *Differential Equations, Dynamical Systems, and an Introduction to Chaos*. Academic Press, 2nd edition, 2003. (Cited on pp. vii, 4, 9)

[271] H. Holden and N.H. Risebro. *Front Tracking for Hyperbolic Conservation Laws.* Springer, 2015. (Cited on pp. 61, 203)

[272] P.J. Holmes and J.E. Marsden. Melnikov's method and Arnold diffusion for perturbations of integrable Hamiltonian systems. *J. Math. Phys.*, 23(4):669–675, 1982. (Cited on p. 191)

[273] P.J. Holmes, J.E. Marsden, and J. Scheurle. Exponentially small splittings of separatrices with applications to KAM theory and degenerate bifurcations. *Contemp. Math.*, pages 213–243, 1988. (Cited on p. 115)

[274] E. Hopf. The partial differential equation $u_t + uu_x = \mu_{xx}$. *Comm. Pure Appl. Math.*, 3(3):201–230, 1950. (Cited on p. 61)

[275] L. Hörmander. *The Analysis of Linear Partial Differential Operators II.* Springer, 1983. (Cited on p. 13)

[276] L. Hörmander. *The Analysis of Linear Partial Differential Operators III.* Springer, 1985. (Cited on p. 13)

[277] L. Hörmander. *The Analysis of Linear Partial Differential Operators I.* Springer, 1990. (Cited on p. 13)

[278] L. Hörmander. *The Analysis of Linear Partial Differential Operators IV.* Springer, 1994. (Cited on p. 13)

[279] U. Hornung, editor. *Homogenization and Porous Media.* Springer, 2012. (Cited on p. 185)

[280] D. Horstmann. From 1970 until present: The Keller-Segel model in chemotaxis and its consequences I. *Jahresbericht DMV*, 105(3):103–165, 2003. (Cited on p. 164)

[281] D. Horstmann. From 1970 until present: The Keller-Segel model in chemotaxis and its consequences II. *Jahresbericht DMV*, 106(2):51–69, 2004. (Cited on p. 164)

[282] D. Horstmann and M. Winkler. Boundedness vs. blow-up in a chemotaxis system. *J. Differential Equations*, 215:52–107, 2005. (Cited on p. 164)

[283] R. Hoyle. *Pattern Formation: An Introduction to Methods.* Cambridge University Press, 2006. (Cited on pp. 66, 159, 176)

[284] T.J. Hughes. *The Finite Element Method: Linear Static and Dynamic Finite Element Analysis.* Courier Corporation, 2012. (Cited on p. 209)

[285] J.M. Hyman and B. Nicolaenko. The Kuramoto-Sivashinsky equation: A bridge between PDE's and dynamical systems. *Physica D*, 18(1):113–126, 1986. (Cited on p. 94)

[286] T. Ilmanen. Convergence of the Allen-Cahn equation to Brakke's motion by mean curvature. *J. Differential Geom.*, 38(2):417–461, 1993. (Cited on p. 110)

[287] Yu. Ilyashenko and W. Li. *Nonlocal Bifurcations.* AMS, 1999. (Cited on p. 198)

[288] Yu. Ilyashenko and S. Yakovenko. *Lectures on Analytic Differential Equations.* AMS, 2008. (Cited on p. 176)

[289] G. Iooss and D.D. Joseph. *Elementary Stability and Bifurcation Theory.* Springer, 1997. (Cited on pp. 18, 25)

[290] D. Iron, M.J. Ward, and J. Wei. The stability of spike solutions to the one-dimensional Gierer-Meinhardt model. *Physica D*, 150(1):25–62, 2001. (Cited on p. 191)

[291] K. Itô. *Diffusion Processes.* Wiley, 1974. (Cited on p. 130)

[292] A. Iuorio, C. Kuehn, and P. Szmolyan. Geometry and numerical continuation of multiscale orbits in a nonconvex variational problem. *Discrete Contin. Dyn. Syst. S, to appear.* (Cited on pp. 194, 198)

[293] Y. Jabri. *The Mountain Pass Theorem: Variants, Generalizations and Some Applications.* Cambridge University Press, 2003. (Cited on p. 141)

[294] W. Jäger and S. Luckhaus. On explosions of solutions to a system of partial differential equations modelling chemotaxis. *Trans. Amer. Math. Soc.*, 329(2):819–824, 1992. (Cited on p. 164)

[295] V.V.E. Jikov, S.M. Kozlov, and O.A.E. Oleinik. *Homogenization of Differential Operators and Integral Functionals.* Springer, 2012. (Cited on p. 185)

[296] F. John. Blow-up of solutions of nonlinear wave equations in three space dimensions. *Proc. Natl. Acad. Sci. USA*, 76(4):1559–1560, 1979. (Cited on p. 141)

[297] F. John. *Partial Differential Equations.* Springer, 1982. (Cited on p. 13)

[298] C. Johnson. *Numerical Solution of Partial Differential Equations by the Finite Element Method.* Dover, 2012. (Cited on p. 209)

[299] R.A. Johnson, K.J. Palmer, and G.R. Sell. Ergodic properties of linear dynamical systems. *SIAM J. Math. Anal.*, 18(1):1–33, 1987. (Cited on p. 55)

[300] M.S. Jolly, I.G. Kevrekidis, and E.S. Titi. Approximate inertial manifolds for the Kuramoto-Sivashinsky equation: Analysis and computations. *Physica D*, 44(1):38–60, 1990. (Cited on p. 94)

[301] C.K.R.T. Jones. Stability of the travelling wave solution of the FitzHugh-Nagumo system. *Trans. Amer. Math. Soc.*, 286(2):431–469, 1984. (Cited on pp. 48, 198)

[302] C.K.R.T. Jones. Geometric singular perturbation theory. In *Dynamical Systems (Montecatini Terme, 1994)*, volume 1609 of *Lecture Notes in Mathematics*, pages 44–118. Springer, 1995. (Cited on pp. 94, 195, 198)

[303] C.K.R.T. Jones, T.J. Kaper, and N. Kopell. Tracking invariant manifolds up to exponentially small errors. *SIAM J. Math. Anal.*, 27(2):558–577, 1996. (Cited on p. 115)

[304] C.K.R.T. Jones and N. Kopell. Tracking invariant manifolds with differential forms in singularly perturbed systems. *J. Differential Equations*, 108(1):64–88, 1994. (Cited on pp. 195, 198)

[305] C.K.R.T. Jones, N. Kopell, and R. Langer. Construction of the FitzHugh-Nagumo pulse using differential forms. In *Multiple-Time-Scale Dynamical Systems*, pages 101–113. Springer, 2001. (Cited on pp. 48, 196)

[306] R. Jordan, D. Kinderlehrer, and F. Otto. The variational formulation of the Fokker–Planck equation. *SIAM J. Math. Anal.*, 29(1):1–17, 1998. (Cited on p. 125)

[307] D.D. Joseph and D.H. Sattinger. Bifurcating time periodic solutions and their stability. *Arch. Rat. Mech. Anal.*, 45(2):79–109, 1972. (Cited on p. 141)

[308] J. Jost. *Partial Differential Equations.* Springer, 2006. (Cited on pp. 13, 81)

[309] J. Jost and X. Li-Jost. *Calculus of Variations.* Cambridge University Press, 1998. (Cited on pp. 135, 198)

[310] C.B. Morrey Jr. *Multiple Integrals in the Calculus of Variations.* Springer, 2009. (Cited on p. 135)

[311] A. Jüngel. The boundedness-by-entropy method for cross-diffusion systems. *Nonlinearity*, 28:1963–2001, 2015. (Cited on p. 164)

[312] A. Jüngel. *Entropy Methods for Diffusive Partial Differential Equations*. Springer, 2016. (Cited on p. 130)

[313] O. Kallenberg. *Foundations of Modern Probability*. Springer, 2nd edition, 2002. (Cited on p. 130)

[314] I. Kan. Open sets of diffeomorphisms having two attractors, each with an everywhere dense basin. *Bull. Amer. Math. Soc.*, 31(1):68–74, 1994. (Cited on p. 99)

[315] H. Kantz. A robust method to estimate the maximal Lyapunov exponent of a time series. *Phys. Lett. A*, 185(1):77–87, 1994. (Cited on p. 105)

[316] H. Kantz and T. Schreiber. *Nonlinear Time Series Analysis*. Cambridge University Press, 2004. (Cited on p. 181)

[317] T. Kapitula. Stability analysis of pulses via the Evans function: dissipative systems. In *Dissipative Solitons*, volume 661 of *Lecture Notes in Physics*, pages 407–427. Springer, 2005. (Cited on pp. 48, 55)

[318] T. Kapitula and K. Promislow. *Spectral and Dynamical Stability of Nonlinear Waves*. Springer, 2013. (Cited on p. 48)

[319] T. Kato. *Perturbation Theory for Linear Operators*. Springer, 1980. (Cited on pp. 30, 53)

[320] A. Katok and B. Hasselblatt. *Introduction to the Modern Theory of Dynamical Systems*. Cambridge University Press, 1995. (Cited on pp. 8, 9, 89, 181, 198)

[321] J.P. Keener. Propagation and its failure in coupled systems of discrete excitable cells. *SIAM J. Appl. Math.*, 47:556–572, 1987. (Cited on p. 203)

[322] E. Keller and S. Segel. Model for chemotaxis. *J. Theor. Biol.*, 30(2):225–234, 1971. (Cited on p. 164)

[323] G. Keller. *Equilibrium States in Ergodic Theory*. Cambridge University Press, 1998. (Cited on p. 181)

[324] H. Keller. *Lectures on Numerical Methods in Bifurcation Problems*. Springer, 1986. (Cited on p. 209)

[325] H.J. Kelley. Flight path optimization with multiple time scales. *J. Aircraft*, 8(4):238–240, 1971. (Cited on p. 89)

[326] C.E. Kenig and F. Merle. Global well-posedness, scattering and blow-up for the energy-critical, focusing, non-linear Schrödinger equation in the radial case. *Invent. Math.*, 166(3):645–675, 2006. (Cited on p. 72)

[327] C.E. Kenig and F. Merle. Global well-posedness, scattering and blow-up for the energy-critical focusing non-linear wave equation. *Acta Math.*, 201(2):147–212, 2008. (Cited on p. 141)

[328] J. Kevorkian and J.D. Cole. *Multiple Scale and Singular Perturbation Methods*. Springer, 1996. (Cited on p. 191)

[329] I.G. Kevrekidis, B. Nicolaenko, and J.C. Scovel. Back in the saddle again: A computer assisted study of the Kuramoto-Sivashinsky equation. *SIAM J. Appl. Math.*, 50(3):760–790, 1990. (Cited on p. 94)

[330] P.G. Kevrekidis. *The Discrete Nonlinear Schrödinger Equation: Mathematical Analysis, Numerical Computations and Physical Perspectives.* Springer, 2009. (Cited on p. 72)

[331] P.G. Kevrekidis and D.J. Frantzeskakis. Pattern forming dynamical instabilities of Bose-Einstein condensates. *Mod. Phys. Lett. B*, 18:173–202, 2004. (Cited on p. 135)

[332] P.G. Kevrekidis, D.J. Frantzeskakis, and R. Carretero-González, editors. *Emergent Nonlinear Phenomena in Bose-Einstein Condensates: Theory and Experiment.* Springer, 2007. (Cited on p. 135)

[333] H. Kielhoefer. *Bifurcation Theory: An Introduction with Applications to PDEs.* Springer, 2004. (Cited on pp. 18, 23, 24, 25, 30)

[334] Yu. Kifer. *Ergodic Theory of Random Transformations.* Springer, 2012. (Cited on p. 181)

[335] P. Kirrmann, G. Schneider, and A. Mielke. The validity of modulation equations for extended systems with cubic nonlinearities. *Proc. R. Soc. Edinburgh A*, 122(1):85–91, 1992. (Cited on pp. 70, 71, 72, 78)

[336] B. Kleiner and J. Lott. Notes on Perelman's papers. *Geom. Topol.*, 12(5):2587–2855, 2008. (Cited on p. 130)

[337] P.E. Kloeden and M. Rasmussen. *Nonautonomous Dynamical Systems.* AMS, 2011. (Cited on p. 55)

[338] H. Koch and D. Tataru. Well-posedness for the Navier-Stokes equations. *Adv. Math.*, 157(1):22–35, 2001. (Cited on p. 83)

[339] R.V. Kohn and F. Otto. Upper bounds on coarsening rates. *Comm. Math. Phys.*, 229(3):375–395, 2002. (Cited on pp. 117, 120)

[340] A. Kolmogorov, I. Petrovskii, and N. Piscounov. A study of the diffusion equation with increase in the amount of substance, and its application to a biological problem. In V.M. Tikhomirov, editor, *Selected Works of A. N. Kolmogorov I*, pages 248–270. Kluwer, 1991. Translated by V. M. Volosov from Bull. Moscow Univ. Math. Mech. 1:1–25, 1937. (Cited on p. 43)

[341] T. Kolokolnikov, M.J. Ward, and J. Wei. The existence and stability of spike equilibria in the one-dimensional Gray-Scott model: The low feed rate regime. *Stud. Appl. Math.*, 115(1):21–71, 2005. (Cited on p. 191)

[342] M. Kot. *Elements of Mathematical Ecology.* Cambridge University Press, 2003. (Cited on p. 43)

[343] G. Kovacic and S. Wiggins. Orbits homoclinic to resonances, with an application to chaos in a model of the forced and damped sine-Gordon equation. *Physica D*, 57(1):185–225, 1992. (Cited on p. 105)

[344] B. Krauskopf, H.M. Osinga, and J. Galán-Vique, editors. *Numerical Continuation Methods for Dynamical Systems: Path Following and Boundary Value Problems.* Springer, 2007. (Cited on p. 209)

[345] B. Krauskopf and T. Riess. A Lin's method approach to finding and continuing heteroclinic connections involving periodic orbits. *Nonlinearity*, 21(8):1655–1690, 2008. (Cited on p. 37)

[346] C. Kuehn. *Multiple Time Scale Dynamics.* Springer, 2015. 814 pp. (Cited on pp. 48, 64, 89, 94, 176, 185, 191, 195, 198)

[347] C. Kuehn. Numerical continuation and SPDE stability for the 2D cubic-quintic Allen-Cahn equation. *SIAM/ASA J. Uncertainty Quantification*, 3:762–789, 2015. (Cited on p. 208)

[348] A. Kufner. *Weighted Sobolev Spaces*. Teubner, 1980. (Cited on p. 81)

[349] S. Kuksin. *Nearly Integrable Infinite-Dimensional Hamiltonian Systems*. Springer, 1993. (Cited on p. 148)

[350] S. Kuksin and J. Pöschel. Invariant Cantor manifolds of quasi-periodic oscillations for a nonlinear Schrödinger equation. *Ann. Math.*, 142:149–179, 1995. (Cited on pp. 145, 148)

[351] Yu.A. Kuznetsov. *Elements of Applied Bifurcation Theory*. Springer, 3rd edition, 2004. (Cited on pp. 9, 176, 198, 209)

[352] O.A. Ladyzhenskaya. On the determination of minimal global attractors for the Navier-Stokes and other partial differential equations. *Russ. Math. Surv.*, 42(6):27, 1987. (Cited on p. 105)

[353] Y.C. Lai and C. Grebogi. Intermingled basins and two-state on-off intermittency. *Phys. Rev. E*, 52(4):R3313, 1995. (Cited on p. 99)

[354] D. Lannes. *The Water Waves Problem: Mathematical Analysis and Asymptotics*. AMS, 2013. (Cited on p. 72)

[355] S. Larsson and V. Thomée. *Partial Differential Equations with Numerical Methods*. Springer, 2008. (Cited on p. 203)

[356] A. Lasota and M.C. Mackey. *Chaos, Fractals and Noise*. Springer, 1994. (Cited on pp. 179, 180, 181)

[357] R.S. Laugesen. Spectral theory of partial differential equations. arXiv:1203.2344v1, 2012. (Cited on p. 30)

[358] P.D. Lax. Weak solutions of nonlinear hyperbolic equations and their numerical computation. *Comm. Pure Appl. Math.*, 7(1):159—193, 1954. (Cited on p. 61)

[359] P.D. Lax. Hyperbolic systems of conservation laws II. *Comm. Pure Appl. Math.*, 10(4):537–566, 1957. (Cited on p. 61)

[360] P.D. Lax. Development of singularities of solutions of nonlinear hyperbolic partial differential equations. *J. Math. Phys.*, 5(5):611–613, 1964. (Cited on p. 164)

[361] P.D. Lax. *Hyperbolic Systems of Conservation Laws and the Mathematical Theory of Shock Waves*. SIAM, 1973. (Cited on p. 61)

[362] P.D. Lax. Periodic solutions of the KdV equation. *Comm. Pure Appl. Math.*, 28(1):141–188, 1975. (Cited on p. 61)

[363] P.D. Lax. *Functional Analysis*. Wiley, 2002. (Cited on p. 18)

[364] M. Ledoux. *The Concentration of Measure Phenomenon*. AMS, 2005. (Cited on p. 125)

[365] J.M. Lee. *Introduction to Smooth Manifolds*. Springer, 2006. (Cited on pp. 8, 142, 148)

[366] P.G. LeFloch. *Hyperbolic Systems of Conservation Laws: The Theory of Classical and Nonclassical Shock Waves*. Springer, 2002. (Cited on p. 61)

[367] R.J. LeVeque. *Numerical Methods for Conservation Laws*. Birkhäuser, 1992. (Cited on pp. 59, 61, 203)

[368] R.J. LeVeque. *Finite Difference Methods for Ordinary and Partial Differential Equations: Steady-State and Time-Dependent Problems*. SIAM, 2007. (Cited on p. 203)

[369] C.D. Levermore. Moment closure hierarchies for kinetic theories. *J. Stat. Phys.*, 83(5):1021–1065, 1996. (Cited on p. 153)

[370] C.D. Levermore and M. Oliver. The complex Ginzburg-Landau equation as a model problem. In P. Deift, C.D. Levermore, and C.E. Wayne, editors, *Dynamical Systems and Probabilistic Methods in Partial Differential Equations (Berkeley, 1994)*, pages 141–190. AMS, 1998. (Cited on p. 67)

[371] H.A. Levine. Instability and nonexistence of global solutions to nonlinear wave equations of the form $Pu_{tt} = -Au + \mathfrak{F}(u)$. *Trans. Amer. Math. Soc.*, 192:1–21, 1974. (Cited on p. 141)

[372] H.A. Levine. The role of critical exponents in blowup theorems. *SIAM Rev.*, 32(2):262–288, 1990. (Cited on p. 141)

[373] A.J. Lichtenberg and M.A. Lieberman. *Regular and Chaotic Dynamics*. Springer, 1992. (Cited on p. 9)

[374] E.H. Lieb. Thomas-Fermi and related theories of atoms and molecules. *Rev. Mod. Phys.*, 53(4):603–641, 1981. (Cited on p. 135)

[375] E.H. Lieb and M. Loss. *Analysis*. AMS, 2nd edition, 2001. (Cited on pp. 13, 139)

[376] E.H. Lieb, R. Seiringer, J.P. Solovej, and J. Yngvason. *The Mathematics of the Bose Gas and its Condensation*. Springer, 2005. (Cited on p. 135)

[377] E.H. Lieb, R. Seiringer, and J. Yngvason. Bosons in a trap: A rigorous derivation of the Gross-Pitaevskii energy functional. In *The Stability of Matter: From Atoms to Stars*, pages 685–697. Springer, 2001. (Cited on p. 135)

[378] C.S. Lin, W.M. Ni, and I. Takagi. Large amplitude stationary solutions to a chemotaxis system. *J. Differential Equations*, 72(1):1–27, 1988. (Cited on p. 141)

[379] C.S. Lin, W.M. Ni, and I. Takagi. On the shape of least-energy solutions to a semilinear Neumann problem. *Comm. Pure Appl. Math.*, 44(7):819–851, 1991. (Cited on p. 141)

[380] X.-B. Lin. Using Melnikov's method to solve Shilnikov's problems. *Proc. R. Soc. Edinburgh A*, 116:295–325, 1990. (Cited on p. 37)

[381] P.-L. Lions and B. Perthame. Propagation of moments and regularity for the 3-dimensional Vlasov-Poisson system. *Invent. Math.*, 105(1):415–430, 1991. (Cited on p. 153)

[382] T.P. Liu. Nonlinear stability of shock waves for viscous conservation laws. *Bull. Amer. Math. Soc.*, 12(2):233–236, 1985. (Cited on p. 61)

[383] T.P. Liu. Hyperbolic conservation laws with relaxation. *Comm. Math. Phys.*, 108(1):153–175, 1987. (Cited on p. 61)

[384] D.J.B. Lloyd, B. Sandstede, D. Avitabile, and A.R. Champneys. Localized hexagon patterns of the planar Swift-Hohenberg equation. *SIAM J. Appl. Dyn. Syst.*, 7(3):1049–1100, 2008. (Cited on p. 67)

[385] E. Lombardi. *Oscillatory Integrals and Phenomena Beyond All Algebraic Orders: With Applications to Homoclinic Orbits in Reversible Systems*. Springer, 2007. (Cited on p. 115)

[386] J. Lott and C. Villani. Ricci curvature for metric-measure spaces via optimal transport. *Ann. Math.*, 169(3):903–991, 2009. (Cited on p. 125)

[387] A. Lunardi. *Analytic Semigroups and Optimal Regularity in Parabolic Problems.* Springer, 1995. (Cited on p. 78)

[388] A.M. Lyapunov. *Problème géneral de la stabilité du mouvement.* Princeton University Press, 1947. (Cited on p. 89)

[389] J. Maas. Gradient flows of the entropy for finite Markov chains. *J. Funct. Anal.,* 261(8):2250–2292, 2011. (Cited on p. 125)

[390] W. Magnus and S. Winkler. *Hill's Equation.* Dover, 1979. (Cited on p. 48)

[391] A.J. Majda. *Compressible Fluid Flow and Systems of Conservation Laws in Several Space Variables.* Springer, 2012. (Cited on p. 61)

[392] A.J. Majda and A.L. Bertozzi. *Vorticity and Incompressible Flow.* Cambridge University Press, 2002. (Cited on p. 153)

[393] J. Mallet-Paret and G.R. Sell. Inertial manifolds for reaction diffusion equations in higher space dimensions. *J. Amer. Math. Soc.,* 1(4):805–866, 1988. (Cited on p. 94)

[394] B.B. Mandelbrot. Self-affine fractals and fractal dimension. *Physica Scripta,* 32(4):257, 1985. (Cited on p. 105)

[395] R. Mañé. *Ergodic Theory and Differentiable Dynamics.* Springer, 2012. (Cited on p. 181)

[396] C. Marchioro and M. Pulvirenti. *Mathematical Theory of Incompressible Nonviscous Fluids.* Springer, 1994. (Cited on p. 153)

[397] M. Marion. Approximate inertial manifolds for reaction-diffusion equations in high space dimension. *J. Dyn. Differential Equations,* 1(3):245–267, 1989. (Cited on p. 94)

[398] J. Marsden and T. Ratiu. *Introduction to Mechanics and Symmetry: A Basic Exposition of Classical Mechanical Systems.* Springer, 2013. (Cited on p. 148)

[399] D. Matthes, R.J. McCann, and G. Savaré. A family of nonlinear fourth order equations of gradient flow type. *Comm. Partial Differential Equations,* 34(11):1352–1397, 2009. (Cited on p. 125)

[400] C. McMullen. The Hausdorff dimension of general Sierpiński carpets. *Nagoya Math. J.,* 96:1–9, 1984. (Cited on p. 105)

[401] J.D. Meiss. *Differential Dynamical Systems.* SIAM, 2007. (Cited on p. 9)

[402] I. Mezić. Spectral properties of dynamical systems, model reduction and decompositions. *Nonlinear Dynamics,* 41(1):309–325, 2005. (Cited on p. 181)

[403] D. Michelson. Steady solutions of the Kuramoto-Sivashinsky equation. *Physica D,* 19(1):89–111, 1986. (Cited on p. 94)

[404] R.E. Mickens. *Nonstandard Finite Difference Models of Differential Equations.* World Scientific, 1994. (Cited on p. 203)

[405] A. Mielke. The Ginzburg-Landau equation in its role as a modulation equation. In *Handbook of Dynamical Systems 2,* pages 759–834. Elsevier, 2002. (Cited on p. 67)

[406] A. Mielke and G. Schneider. Attractors for modulation equations on unbounded domains-existence and comparison. *Nonlinearity,* 8(5):743, 1995. (Cited on p. 67)

[407] A. Mielke and G. Schneider. Derivation and justification of the complex Ginzburg-Landau equation as a modulation equation. In P. Deift, C.D. Levermore, and C.E. Wayne, editors, *Dynamical Systems and Probabilistic methods in Partial Differential Equations (Berkeley, CA, 1994)*, pages 191–216. AMS, 1996. (Cited on p. 72)

[408] P.D. Miller. *Applied Asymptotic Analysis*. AMS, 2006. (Cited on p. 43)

[409] J. Milnor. On the concept of attractor. *Comm. Math. Phys.*, 99:177–195, 1985. (Cited on p. 99)

[410] N. Van Minh, F. Räbiger, and R. Schnaubelt. Exponential stability, exponential expansiveness, and exponential dichotomy of evolution equations on the half-line. *Integral Equations Operator Theory*, 32(3):332–353, 1998. (Cited on p. 55)

[411] A. Miranville and S. Zelik. Attractors for dissipative partial differential equations in bounded and unbounded domains. In *Handbook of Differential Equations: Evolutionary Equations*, pages 103–200. Elsevier, 2008. (Cited on p. 83)

[412] K. Mischaikow and M. Mrozek. Conley Index Theory. In B. Fiedler, editor, *Handbook of Dynamical Systems*, volume 2, pages 393–460. North-Holland, 2002. (Cited on p. 37)

[413] R.M. Miura. Korteweg-de Vries equation and generalizations. I. A remarkable explicit nonlinear transformation. *J. Math. Phys.*, 9(8):1202–1204, 1968. (Cited on p. 61)

[414] R.M. Miura, C.S. Gardner, and M.D. Kruskal. Korteweg-de Vries equation and generalizations. II. Existence of conservation laws and constants of motion. *J. Math. Phys.*, 9(8):1204–1209, 1968. (Cited on p. 61)

[415] J.W. Morgan and G. Tian. *Ricci flow and the Poincaré conjecture*. AMS, 2007. (Cited on p. 130)

[416] P. De Mottoni and M. Schatzman. Geometrical evolution of developed interfaces. *Trans. Amer. Math. Soc.*, 347(5):1533–1589, 1995. (Cited on p. 110)

[417] C. Mouhot and C. Villani. On Landau damping. *Acta Math.*, 207:29–201, 2011. (Cited on p. 159)

[418] S. Müller. Singular perturbations as a selection criterion for periodic minimizing sequences. *Calc. Var. Partial Differential Equations*, 1(2):169–204, 1993. (Cited on p. 198)

[419] S. Müller. Variational models for microstructure and phase transitions. In S. Hildebrandt and M. Struwe, editors, *Calculus of Variations and Geometric Evolution Problems*, pages 85–210. Springer, 1999. (Cited on p. 185)

[420] J.D. Murray. *Mathematical Biology I: An Introduction*. Springer, 3rd edition, 2002. (Cited on p. 43)

[421] J.D. Murray. *Mathematical Biology II: Spatial Models and Biomedical Applications*. Springer, 3rd edition, 2003. (Cited on p. 159)

[422] T. Nagai, T. Senba, and K. Yoshida. Application of the Trudinger-Moser inequality to a parabolic system of chemotaxis. *Funkcialaj Ekvacioj*, 40:411–433, 1997. (Cited on p. 164)

[423] J. Nagumo, S. Arimoto, and S. Yoshizawa. An active pulse transmission line simulating nerve axon. *Proc. IRE*, 50:2061–2070, 1962. (Cited on pp. 37, 48, 198)

[424] K. Nakanishi and W. Schlag. *Invariant Manifolds and Dispersive Hamiltonian Evolution Equations*. Eur. Math. Soc., 2011. (Cited on p. 72)

[425] V. Nanjundiah. Chemotaxis, signal relaying and aggregation morphology. *J. Theor. Biol.*, 42(1):63–105, 1973. (Cited on p. 164)

[426] A.H. Nayfeh. *Perturbation Methods*. Wiley, 2004. (Cited on p. 191)

[427] A.I. Neishtadt. Persistence of stability loss for dynamical bifurcations. I. *Differential Equations Translations*, 23:1385–1391, 1987. (Cited on p. 115)

[428] A.C. Newell. *Solitons in Mathematics and Physics*. SIAM, 1985. (Cited on p. 37)

[429] G. Nguetseng. A general convergence result for a functional related to the theory of homogenization. *SIAM J. Math. Anal.*, 20(3):608–623, 1989. (Cited on p. 185)

[430] W.-M. Ni. Diffusion, cross-diffusion and their spike-layer steady states. *Notices Amer. Math. Soc.*, 45(1):9–18, 1998. (Cited on pp. 141, 164)

[431] W.M. Ni and I. Takagi. Locating the peaks of least-energy solutions to a semilinear Neumann problem. *Duke Math. J.*, 70(2):247–281, 1993. (Cited on p. 141)

[432] W.M. Ni and J. Wei. On the location and profile of spike-layer solutions to singularly perturbed semilinear Dirichlet problems. *Comm. Pure Appl. Math.*, 48(7):731–768, 1995. (Cited on p. 141)

[433] W.M. Ni and J. Wei. On positive solutions concentrating on spheres for the Gierer-Meinhardt system. *J. Differential Equations*, 221(1):158–189, 2006. (Cited on p. 191)

[434] B. Nicolaenko, B. Scheurer, and R. Temam. Some global dynamical properties of the Kuramoto-Sivashinsky equations: Nonlinear stability and attractors. *Physica D*, 16(2):155–183, 1985. (Cited on p. 94)

[435] F. Nier and B. Helffer. *Hypoelliptic estimates and spectral theory for Fokker-Planck operators and Witten Laplacians*. Springer, 2005. (Cited on p. 130)

[436] Y. Nishiura and D. Ueyama. Spatio-temporal chaos for the Gray-Scott model. *Physica D*, 150(3):137–162, 2001. (Cited on p. 191)

[437] A. Novick-Cohen and L.A. Segel. Nonlinear aspects of the Cahn-Hilliard equation. *Physica D*, 10(3):277–298, 1984. (Cited on p. 120)

[438] T. Ogawa and Y. Tsutsumi. Blow-up of H^1 solution for the nonlinear Schrödinger equation. *J. Differential Equations*, 92(2):317–330, 1991. (Cited on p. 72)

[439] O.A. Oleinik, A.S. Shamaev, and G.A. Yosifian. *Mathematical Problems in Elasticity and Homogenization*. North-Holland, 1992. (Cited on p. 185)

[440] A.R. Osborne and A. Provenzale. Finite correlation dimension for stochastic systems with power-law spectra. *Physica D*, 35(3):357–381, 1989. (Cited on p. 105)

[441] F. Otto. The geometry of dissipative evolution equations: The porous medium equation. *Comm. Partial Differential Equations*, 26(1):101–174, 2001. (Cited on p. 125)

[442] F. Otto and C. Villani. Generalization of an inequality by Talagrand and links with the logarithmic Sobolev inequality. *J. Funct. Anal.*, 173(2):361–400, 2000. (Cited on p. 125)

[443] K.J. Palmer. Exponential separation, exponential dichotomy and spectral theory for linear systems of ordinary differential equations. *J. Differential Equations*, 46(3):324–345, 1982. (Cited on p. 55)

[444] K.J. Palmer. Exponential dichotomies and transversal homoclinic points. *J. Differential Equations*, 55:225–256, 1984. (Cited on pp. 51, 55)

[445] K.J. Palmer. Exponential dichotomies and Fredholm operators. *Proc. Amer. Math. Soc.*, 104:149–156, 1988. (Cited on pp. 51, 55)

[446] G.A. Pavliotis and A.M. Stuart. *Multiscale Methods: Averaging and Homogenization.* Springer, 2008. (Cited on pp. 179, 181, 185)

[447] L.E. Payne and D.H. Sattinger. Saddle points and instability of nonlinear hyperbolic equations. *Israel J. Math.*, 22(3):273–303, 1975. (Cited on p. 141)

[448] A. Pazy. *Semigroups of Linear Operators and Applications to Partial Differential Equations.* Springer, 1983. (Cited on pp. 78, 180)

[449] P. Pedregal. *Parametrized Measures and Variational Principles.* Birkhäuser, 2012. (Cited on p. 185)

[450] R.L. Pego. Front migration in the nonlinear Cahn-Hilliard equation. *Proc. R. Soc. London A*, 422(1863):261–278, 1989. (Cited on pp. 111, 120)

[451] R.L. Pego and M.I. Weinstein. Eigenvalues, and instabilities of solitary waves. *Proc. R. Soc. London A*, 340(1656):47–94, 1992. (Cited on p. 55)

[452] R.L. Pego and M.I. Weinstein. Asymptotic stability of solitary waves. *Comm. Math. Phys.*, 164(2):305–349, 1994. (Cited on p. 55)

[453] M.A. Peletier. Variational modelling: Energies, gradient flows, and large deviations. arXiv:1402.1990, 2014. (Cited on p. 125)

[454] G. Perelman. The entropy formula for the Ricci flow and its geometric applications. arXiv:math/0211159, 2002. (Cited on p. 130)

[455] L. Perko. *Differential Equations and Dynamical Systems.* Springer, 2001. (Cited on p. 9)

[456] O. Perron. Über Stabilität und asymptotisches Verhalten der Integrale von Differentialgleichungssystemen. *Math. Zeitschr.*, 29(1):129–160, 1929. (Cited on p. 89)

[457] Ya.B. Pesin. Characteristic Lyapunov exponents and smooth ergodic theory. *Russ. Math. Surv.*, 32(4):55–114, 1977. (Cited on p. 105)

[458] Ya.B. Pesin. On rigorous mathematical definitions of correlation dimension and generalized spectrum for dimensions. *J. Stat. Phys.*, 71(3):529–547, 1993. (Cited on p. 105)

[459] K. Petersen. *Ergodic Theory.* Cambridge University Press, 1989. (Cited on p. 181)

[460] K. Pfaffelmoser. Global classical solutions of the Vlasov-Poisson system in three dimensions for general initial data. *J. Differential Equations*, 95(2):281–303, 1992. (Cited on p. 153)

[461] M. Del Pino and J. Dolbeault. Best constants for Gagliardo-Nirenberg inequalities and applications to nonlinear diffusions. *J. Math. Pures Appl.*, 81(9):847–875, 2002. (Cited on p. 125)

[462] L.M. Pismen. *Patterns and Interfaces in Dissipative Dynamics.* Springer, 2006. (Cited on p. 66)

[463] J. Pöschel. A KAM-theorem for some nonlinear partial differential equations. *Ann. Scuola Norm. Sci.*, 23(1):119–148, 1996. (Cited on p. 148)

[464] G. Da Prato and J. Zabczyk. *Stochastic Equations in Infinite Dimensions.* Cambridge University Press, 1992. (Cited on pp. 78, 115)

[465] G. Da Prato and J. Zabczyk. *Second Order Partial Differential Equations in Hilbert Spaces*. Cambridge University Press, 2002. (Cited on p. 78)

[466] M.H. Protter and H.F. Weinberger. *Maximum Principles in Differential Equations*. Springer, 1999. (Cited on p. 13)

[467] P.H. Rabinowitz. Some global results for nonlinear eigenvalue problems. *J. Funct. Anal.*, 7(3):487–513, 1971. (Cited on p. 25)

[468] P.H. Rabinowitz. On a class of nonlinear Schrödinger equations. *Z. Angew. Math. Phys.*, 43(2):270–291, 1992. (Cited on p. 141)

[469] D. Rand. Dynamics and symmetry. Predictions for modulated waves in rotating fluids. *Arch. Rat. Mech. Anal.*, 79(1):1–37, 1982. (Cited on p. 176)

[470] J. Rauch and J. Smoller. Qualitative theory of the FitzHugh-Nagumo equations. *Adv. Math.*, 27(1):12–44, 1978. (Cited on p. 198)

[471] G. Raugel and G.R. Sell. Navier-Stokes equations on thin 3D domains. I. Global attractors and global regularity of solutions. *J. Amer. Math. Soc.*, 6(3):503–568, 1993. (Cited on p. 99)

[472] M. Renardy and R.C. Rogers. *An Introduction to Partial Differential Equations*. Springer, 2004. (Cited on pp. vii, 13, 90)

[473] L.G. Reyna and M.J. Ward. On the exponentially slow motion of a viscous shock. *Comm. Pure Appl. Math.*, 48(2):79–120, 1995. (Cited on p. 111)

[474] F. Riesz and B.Sz. Nagy. *Functional Analysis*. Ungar, 1955. (Cited on p. 78)

[475] J. Rinzel and D. Terman. Propagation phenomena in a bistable reaction-diffusion system. *SIAM J. Appl. Math.*, 42(5):1111–1137, 1982. (Cited on p. 198)

[476] H. Risken. *The Fokker-Planck Equation*. Springer, 1996. (Cited on p. 130)

[477] J.C. Robinson. *Infinite-Dimensional Dynamical Systems*. Cambridge University Press, 2001. (Cited on pp. 83, 94, 105)

[478] R.C. Robinson. *An Introduction to Dynamical Systems: Continuous and Discrete*. AMS, 2013. (Cited on pp. 9, 99)

[479] T. Roubicek. *Relaxation in Optimization Theory and Variational Calculus*. DeGruyter, 1997. (Cited on p. 185)

[480] J. Rubinstein, P. Sternberg, and J.B. Keller. Fast reaction, slow diffusion, and curve shortening. *SIAM J. Appl. Math.*, 49(1):116–133, 1989. (Cited on p. 111)

[481] W. Rudin. *Principles of Mathematical Analysis*. McGraw-Hill, 1976. (Cited on p. 15)

[482] W. Rudin. *Functional Analysis*. McGraw-Hill, 1991. (Cited on pp. 16, 18, 22)

[483] R.J. Sacker and G.R. Sell. Dichotomies for linear evolutionary equations in Banach spaces. *J. Differential Equations*, 113(1):17–67, 1994. (Cited on p. 55)

[484] J.A. Sanders, F. Verhulst, and J. Murdock. *Averaging Methods in Nonlinear Dynamical Systems*. Springer, 2007. (Cited on p. 185)

[485] E. Sandier and S. Serfaty. Gamma-convergence of gradient flows with applications to Ginzburg-Landau. *Comm. Pure Appl. Math.*, 57(12):1627–1672, 2004. (Cited on pp. 125, 185)

[486] B. Sandstede. Stability of travelling waves. In B. Fiedler, editor, *Handbook of Dynamical Systems*, volume 2, pages 983–1055. Elsevier, 2001. (Cited on pp. 48, 51, 55)

[487] B. Sandstede and A. Scheel. Absolute and convective instabilities of waves on unbounded and large bounded domains. *Physica D*, 145(3):233–277, 2000. (Cited on p. 55)

[488] B. Sandstede, A. Scheel, and C. Wulff. Dynamics of spiral waves on unbounded domains using center-manifold reductions. *J. Differential Equations*, 141:122–149, 1997. (Cited on p. 176)

[489] D.H. Sattinger. On the stability of waves of nonlinear parabolic systems. *Adv. Math.*, 22(3):312–355, 1976. (Cited on p. 48)

[490] D.H. Sattinger. *Group Theoretic Methods in Bifurcation Theory*. Springer, 2006. (Cited on p. 176)

[491] D.H. Sattinger. *Topics in Stability and Bifurcation Theory*. Springer, 2006. (Cited on p. 25)

[492] R. Schaaf. Stationary solutions of chemotaxis systems. *Trans. Amer. Math. Soc.*, 292(2):531–556, 1985. (Cited on p. 164)

[493] A. Scheel. Bifurcation to spiral waves in reaction-diffusion systems. *SIAM J. Math. Anal.*, 29(6):1399–1418, 1998. (Cited on pp. 174, 176)

[494] G. Schneider. Justification of modulation equations for hyperbolic systems via normal forms. *Nonlinear Differential Equations Appl. NoDEA*, 5(1):69–82, 1998. (Cited on p. 72)

[495] G. Schneider and H. Uecker. *Nonlinear PDEs: A Dynamical Systems Approach*. AMS, 2017. (Cited on p. 72)

[496] G.R. Sell. Global attractors for the three-dimensional Navier-Stokes equations. *J. Dyn. Differential Equations*, 8(1):1–33, 1996. (Cited on pp. 99, 105)

[497] G.R. Sell and Y. You. *Dynamics of Evolutionary Equations*. Springer, 2002. (Cited on pp. 99, 105)

[498] J.A. Sethian. *Level Set Methods and Fast Marching Methods*. Cambridge University Press, 1999. (Cited on p. 203)

[499] R. Seydel. *Practical Bifurcation and Stability Analysis*. Springer, 1994. (Cited on p. 209)

[500] N. Shigesada, K. Kawasaki, and E. Teramoto. Spatial segregation of interacting species. *J. Theor. Biol.*, 79:83–99, 1979. (Cited on p. 164)

[501] M.A. Shubin. *Pseudodifferential Operators and Spectral Theory*. Springer, 1987. (Cited on p. 30)

[502] N. Sidorov, B.Loginov, A.V. Sinitsyn, and M.V. Falaleev. *Lyapunov-Schmidt Methods in Nonlinear Analysis and Applications*. Springer, 2013. (Cited on p. 18)

[503] J. Sijbrand. Properties of center manifolds. *Trans. Amer. Math. Soc.*, 289:431–469, 1985. (Cited on p. 89)

[504] S. Smale. Diffeomorphisms with many periodic points. In S.S. Cairns, editor, *Differential and Combinatorial Topology*, pages 63–80. Princeton University Press, 1963. (Cited on p. 198)

[505] S. Smale. Differentiable dynamical systems. *Bull. Amer. Math. Soc.*, 289:747–817, 1967. (Cited on p. 198)

[506] G.D. Smith. *Numerical Solution of Partial Differential Equations: Finite Difference Methods*. Oxford University Press, 1985. (Cited on p. 203)

[507] J. Smoller. *Shock Waves and Reaction-Diffusion Equations*. Springer, 1994. (Cited on pp. 37, 59, 61, 83)

[508] H. Sohr. *The Navier-Stokes Equations: An Elementary Functional Analytic Approach*. Springer, 2012. (Cited on p. 83)

[509] Y. Sone. *Singular Perturbations in Manufacturing*. Springer, 2012. (Cited on p. 153)

[510] S. Sternberg. *Dynamical Systems*. Dover, 2010. (Cited on p. 9)

[511] A. Stevens and H.G. Othmer. Aggregation, blowup, and collapse: The ABC's of taxis in reinforced random walks. *SIAM J. Appl. Math.*, 57(4):1044–1081, 1997. (Cited on p. 164)

[512] G. Strang and G.J. Fix. *An Analysis of the Finite Element Method*. SIAM, 2nd edition, 2008. (Cited on p. 209)

[513] W.A. Strauss. *Nonlinear Wave Equations*. AMS, 1990. (Cited on p. 164)

[514] W.A. Strauss. *Partial Differential Equations: An Introduction*. Wiley, 2008. (Cited on p. 13)

[515] S.H. Strogatz. *Nonlinear Dynamics and Chaos*. Westview Press, 2000. (Cited on p. 9)

[516] M. Struwe. *Variational Methods*. Springer, 1990. (Cited on p. 135)

[517] K.T. Sturm. On the geometry of metric measure spaces. *Acta Math.*, 196(1):65–131, 2006. (Cited on p. 125)

[518] C. Sulem and P.L. Sulem. *The nonlinear Schrödinger equation: Self-focusing and wave collapse*. Springer, 2007. (Cited on p. 72)

[519] J. Swift and P.C. Hohenberg. Hydrodynamic fluctuations at the convective instability. *Phys. Rev. A*, 15(1):319–328, 1977. (Cited on p. 66)

[520] P. Szmolyan. Transversal heteroclinic and homoclinic orbits in singular perturbation problems. *J. Differential Equations*, 92:252–281, 1991. (Cited on p. 198)

[521] F. Takens. On the numerical determination of the dimension of an attractor. In *Dynamical Systems and Bifurcations*, pages 99–106. Springer, 1985. (Cited on pp. 99, 105)

[522] L. Tartar. *The General Theory of Homogenization - A Personal Introduction*. Springer, 2009. (Cited on p. 185)

[523] M. Taylor. *Partial Differential Equations I: Basic Theory*. Springer, 2010. (Cited on p. 13)

[524] M. Taylor. *Partial Differential Equations II: Qualitative Studies of Linear Equations*. Springer, 2013. (Cited on p. 13)

[525] M. Taylor. *Partial Differential Equations III: Nonlinear Equations*. Springer, 2013. (Cited on p. 13)

[526] R. Temam. *Navier-Stokes Equations*. North-Holland, 1984. (Cited on p. 105)

[527] R. Temam. *Navier-Stokes Equations and Nonlinear Functional Analysis*. SIAM, 1995. (Cited on p. 83)

[528] R. Temam. *Infinite-Dimensional Dynamical Systems in Mechanics and Physics*. Springer, 1997. (Cited on pp. 83, 94, 95, 96, 98, 99, 103, 105, 120)

[529] G. Teschl. *Ordinary Differential Equations and Dynamical Systems*. AMS, 2012. (Cited on pp. 8, 48, 89, 158)

[530] G. Theocharis, Z. Rapti, P.G. Kevrekidis, D.J. Frantzeskakis, and V.V. Konotop. Modulational instability of Gross-Pitaevskii-type equations in 1+1 dimensions. *Phys. Rev. A*, 67(6):063610, 2003. (Cited on p. 135)

[531] J.W. Thomas. *Numerical Partial Differential Equations: Finite Difference Methods*. Springer, 1995. (Cited on p. 203)

[532] J.W. Thomas. *Numerical Partial Differential Equations: Conservation Laws and Elliptic Equations*. Springer, 2013. (Cited on p. 203)

[533] V. Thomée. *Galerkin Finite Element Methods for Parabolic Problems*. Springer, 2006. (Cited on p. 209)

[534] A.N. Tikhonov. Systems of differential equations containing small small parameters in the derivatives. *Mat. Sbornik N. S.*, 31:575–586, 1952. (Cited on p. 89)

[535] E.S. Titi. On approximate inertial manifolds to the Navier-Stokes equations. *J. Math. Anal. Appl.*, 149(2):540–557, 1990. (Cited on p. 94)

[536] P. Topping. *Lectures on the Ricci Flow*. Cambridge University Press, 2006. (Cited on p. 130)

[537] H.C. Tuckwell. *Introduction to Theoretical Neurobiology*. Cambridge University Press, 1988. (Cited on p. 43)

[538] M. Tucsnak and G. Weiss. *Observation and Control for Operator Semigroups*. Springer, 2009. (Cited on p. 78)

[539] A.M. Turing. The chemical basis of morphogenesis. *Philos. Trans. R. Soc. B*, 237(641):37–72, 1952. (Cited on p. 159)

[540] J.J. Tyson and J.P. Keener. Singular perturbation theory of traveling waves in excitable media (a review). *Physica D*, 32(3):327–361, 1988. (Cited on p. 43)

[541] H. Uecker, D. Wetzel, and J.D.M. Rademacher. pde2path - A MATLAB package for continuation and bifurcation in 2D elliptic systems. *Numer. Math. Theor. Meth. Appl.*, 7:58–106, 2014. (Cited on pp. 208, 209)

[542] A. van Harten. On the validity of the Ginzburg-Landau equation. *J. Nonlinear Sci.*, 1(4):397–422, 1991. (Cited on p. 72)

[543] W. van Saarloos. Front propagation into unstable states. *Phys. Rep.*, 386:29–222, 2003. (Cited on p. 43)

[544] W. van Saarloos and P.C. Hohenberg. Fronts, pulses, sources and sinks in generalized complex Ginzburg-Landau equations. *Physica D*, 56(4):303–367, 1992. (Cited on p. 67)

[545] A. Vanderbauwhede and G. Iooss. Center manifold theory in infinite dimensions. In *Dynamics Reported*, pages 125–163. Springer, 1992. (Cited on pp. 89, 176)

[546] S.R.S. Varadhan. *Large Deviations and Applications*. SIAM, 1984. (Cited on p. 115)

[547] J.L. Vázquez. *Smoothing and Decay Estimates for Nonlinear Diffusion Equations: Equations of Porous Medium Type*. Oxford University Press, 2006. (Cited on p. 169)

[548] J.L. Vázquez. *The Porous Medium Equation: Mathematical Theory*. Oxford University Press, 2007. (Cited on pp. 168, 169)

[549] F. Verhulst. *Nonlinear Differential Equations and Dynamical Systems.* Springer, 2006. (Cited on p. 9)

[550] C. Villani. A review of mathematical topics in collisional kinetic theory. In S. Friedlander and D. Serre, editors, *Handbook of Mathematical Fluid Dynamics I*, pages 71–305. Elsevier, 2002. (Cited on p. 153)

[551] C. Villani. *Topics in Optimal Transportation.* AMS, 2003. (Cited on pp. 125, 153)

[552] C. Villani. *Optimal Transport: Old and New.* Springer, 2008. (Cited on pp. 125, 153)

[553] C. Villani. Hypocoercivity. *Mem. Amer. Math. Soc.*, 202:950, 2009. (Cited on p. 159)

[554] A.I. Volpert, V. Volpert, and V.A. Volpert. *Traveling Wave Solutions of Parabolic Systems.* AMS, 1994. (Cited on pp. 37, 43)

[555] P. Walters. *An Introduction to Ergodic Theory.* Springer, 2000. (Cited on p. 181)

[556] B. Wang. Attractors for reaction-diffusion equations in unbounded domains. *Physica D*, 128(1):41–52, 1999. (Cited on p. 83)

[557] C.E. Wayne. Periodic and quasi-periodic solutions of nonlinear wave equations via KAM theory. *Comm. Math. Phys.*, 127(3):479–528, 1990. (Cited on p. 148)

[558] J. Wei. On single interior spike solutions of the Gierer-Meinhardt system: Uniqueness and spectrum estimates. *Eur. J. Appl. Math.*, 10(4):353–378, 1999. (Cited on p. 191)

[559] J. Weidmann. *Spectral Theory of Ordinary Differential Equations.* Springer, 1987. (Cited on pp. 30, 48)

[560] R. Weinstock. *Calculus of Variations.* Dover, 1974. (Cited on p. 198)

[561] A.D. Wentzell and M.I. Freidlin. On small random perturbations of dynamical systems. *Russ. Math. Surv.*, 25:1–55, 1970. (Cited on pp. 115, 130)

[562] S. Wiggins. *Normally Hyperbolic Invariant Manifolds in Dynamical Systems.* Springer, 1994. (Cited on p. 89)

[563] S. Wiggins. *Introduction to Applied Nonlinear Dynamical Systems and Chaos.* Springer, 2nd edition, 2003. (Cited on pp. 9, 191, 198)

[564] M. Willem. *Minimax Theorems.* Springer, 1997. (Cited on p. 141)

[565] M. Winkler. Aggregation vs. global diffusive behavior in the higher-dimensional Keller-Segel model. *J. Differential Equations*, 248:2889–2905, 2010. (Cited on p. 164)

[566] E. Yanagida. Stability of travelling front solutions of the FitzHugh-Nagumo equations. *Math. Comput. Modelling*, 12(3):289–301, 1989. (Cited on p. 48)

[567] K. Yosida. *Functional Analysis.* Springer, 1995. (Cited on p. 78)

[568] L.-S. Young. Dimension, entropy and Lyapunov exponents. *Ergodic Theor. Dyn. Syst.*, 2(1):109–124, 1982. (Cited on p. 105)

[569] V.E. Zakharov and L.D. Faddeev. Korteweg-de Vries equation: A completely integrable Hamiltonian system. *Function. Anal. Appl.*, 5(4):280–287, 1971. (Cited on p. 61)

[570] E. Zeidler. *Nonlinear Functional Analysis and its Applications I: Fixed Point Theorems.* Springer, 1986. (Cited on p. 18)

[571] E. Zeidler. *Nonlinear Functional Analysis and its Applications II/A: Linear Monotone Operators*. Springer, 1989. (Cited on p. 18)

[572] E. Zeidler. *Applied Functional Analysis: Main Principles and Their Applications*. Springer, 1995. (Cited on p. 18)

[573] E. Zeidler. *Applied Functional Analysis: Applications of Mathematical Physics*. Springer, 1999. (Cited on p. 18)

[574] E. Zeidler. *Nonlinear Functional Analysis and its Applications II/B: Nonlinear Monotone Operators*. Springer, 2013. (Cited on p. 18)

[575] E. Zeidler. *Nonlinear Functional Analysis and its Applications III: Variational Methods and Optimization*. Springer, 2013. (Cited on p. 18)

[576] S. Zelik. Inertial manifolds and finite-dimensional reduction for dissipative PDEs. arXiv:1303.4457v1, 2013. (Cited on pp. 86, 89, 92, 94)

[577] K. Zumbrun. Stability and dynamics of viscous shock waves. In *Nonlinear Conservation Laws and Applications*, pages 123–167. Springer, 2011. (Cited on p. 61)

[578] K. Zumbrun and P. Howard. Pointwise semigroup methods and stability of viscous shock waves. *Indiana Univ. Math. J.*, 47(3):741–872, 1998. (Cited on p. 55)

Index